Topics in Applied Physics Volume 34

Topics in Applied Physics Founded by Helmut K. V. Lotsch

Nonlinear Methods of Spectral Analysis

Edited by S. Haykin

With Contributions by

J. Capon S. Haykin S. Kesler R. N. McDonough
M. Ooe E. A. Robinson T. J. Ulrych

Second Corrected and Updated Edition

With 45 Figures

Springer-Verlag Berlin Heidelberg GmbH 1983

Simon Haykin, Ph. D., D. Sc.
Communication Research Laboratory, McMaster University, 1280 Main Street West, Hamilton, Ontario L8S 4L7, Canada

ISBN 978-3-540-12386-6 ISBN 978-3-540-70752-3 (eBook)
DOI 10.1007/978-3-540-70752-3

Library of Congress Cataloging in Publication Data. Main entry under title: Nonlinear methods of spectral analysis. (Topics in applied physics; v. 34). Includes bibliographical references and index. 1. Spectral theory (Mathematics) 2. Entropy (Information theory) 3. Nonlinear theories. I. Haykin, S. S. II. Capon, J. (Jack) III. Series. QC20.7.S64N66 1983 515.7'222 83-4649

Preface to the Second Edition

In this second edition, corrections have been made in the various chapters of the book.

Also, a new chapter (Chap. 7) has been added, highlighting some recent advances in spectral estimation, which have been reported since the appearance of the first edition of the book.

Hamilton, Ontario
March 1983

Simon Haykin

Preface to the First Edition

This Topics volume, written by seven active research workers, is the first book to be completely devoted to a detailed treatment of nonlinear methods of spectral analysis. These methods have found increasing application in a variety of disciplines during the last 10 or 15 years. The topics covered in the volume are as follows:

1) Prediction-error filtering (Chap. 2)
2) Maximum-entropy spectral estimation (Chap. 2, 3, 6)
3) Modeling of discrete time series (Chaps. 3, 4)
4) Maximum-likelihood spectral estimation (Chaps. 5, 6)
5) Array signal processing (Chaps. 5, 6).

Chapter 1 is included as an introduction to the volume. The discussions presented on the maximum-entropy method in Chaps. 2, 3 are designed to be somewhat of a complementary nature so as to enhance the presentation; similarly, for the discussions on mixed autoregressive-moving average spectra presented in Chaps. 3 and 4, and for the discussions on the maximum-likelihood method presented in Chaps. 5 and 6.

Each chapter is essentially self-contained. Wherever it is appropriate, however, cross-references between chapters of the volume have been included in order to help the reader develop a more complete understanding of the pertinent topic.

Much of the material included in the volume has not appeared in book form before. It is hoped that the volume will be found useful by research workers in the field of spectral analysis and its applications, as well as newcomers to the field.

In conclusion, I would like to express my deep gratitude to my co-contributors of the volume, and to Dr. H. Lotsch, of Springer Verlag for making the publication of this volume possible.

Hamilton, Ontario
April 1979

Simon Haykin

Contents

Contributors

Capon, Jack
Lincoln Laboratory, MIT, Lexington, MA 02173, USA

Haykin, Simon
Communications Research Laboratory,
McMaster University, 1280 Main Street West,
Hamilton, Ontario L8S 4L7, Canada

Kesler, Stanislav
Department of Electrical and Computer Engineering,
Drexel University, Philadelphia, PA 19104, USA

McDonough, Robert N.
Applied Physics Laboratory, Johns Hopkins University,
Laurel, MD 20810, USA

Ooe, Masatsugu
International Latitude Observatory, Mizusawa-Shi, Iwate-Ken, Japan

Robinson, Enders A.
Mathematics Department, University of Tulsa,
Tulsa, OK 74104, USA

Ulrych, Tad J.
Department of Geophysics and Astronomy,
The University of British Columbia, 2075 Westbrook Place,
Vancouver, B.C. V6T 1W5, Canada

1. Introduction

S. Haykin

With 3 Figures

1.1 Time Series

A time series consists of a set of observations made sequentially in time. The time series is said to be continuous if the set of observations is continuous. On the other hand, if the set of observations is discrete, the time series is said to be discrete. In this book we will be largely concerned with the analysis of discrete time series.

The *discrete time series*

$$\{x_n\}=\{x_1, x_2, \ldots, x_N\} \tag{1.1}$$

denotes a set of observations made at equidistant time intervals Δt, $2\Delta t, \ldots, N\Delta t$. Such a set of numbers may be generated in practice by uniformly sampling a continuous-time signal $x(t)$ at the rate $1/\Delta t$ samples per second. Alternatively, the signal may be in discrete form to begin with.

The time series $\{x_n\}$ is said to be *deterministic* if future values of the series can be exactly described by some mathematical function. On the other hand, $\{x_n\}$ is said to be a *statistical time series* if future values of the series can be described only in terms of a probability distribution. A statistical time series represents one particular realization of a *stochastic process*, that is, a statistical phenomenon that evolves in time according to probabilistic laws.

A stochastic process is *strictly stationary* if its properties are unaffected by a change of time origin. Thus, for a discrete stochastic process to be strictly stationary, the *joint-probability distribution function* of any set of observations must be unaffected by shifting all the times of observation forward or backward by any integer amount [1.1]. An important consequence of strict stationarity is that the whole probability structure of the process depends only on time differences. However, a less restrictive requirement, called *weak stationarity* of order q, is that the *statistical moments* of the process, up to order q, depend only on time differences.

1.1.1 Autocorrelation and Autocovariance Functions

Consider the time series $\{x_n\}$ as one particular realization of a discrete stochastic process. The mean of such a process is defined by

$$\mu_x(n)=\mathrm{E}[x_n], \tag{1.2}$$

where E denotes the *expectation operator*. The *autocorrelation function* of the process, for lag m and time origin n, is defined by

$$R_x(m,n)=\mathrm{E}[x_{n+m}x_n^*], \tag{1.3}$$

where the asterisk denotes the complex conjugate operation. The corresponding *autocovariance function* of the process is defined by

$$C_x(m,n)=\mathrm{E}\{[x_{n+m}-\mu_x(n+m)]\,[x_n^*-\mu_x^*(n)]\}. \tag{1.4}$$

In the case of a weakly stationary process of order two, the mean $\mu_x(n)$ and autocorrelation function $R_x(m,n)$ are both independent of the time origin n, in which case we may write

$$\mu_x(n)=\mu_x=\text{constant}, \tag{1.5}$$

$$R_x(m,n)=R_x(m), \tag{1.6}$$

and

$$C_x(m,n)=C_x(m). \tag{1.7}$$

For the case of $m=0$, we find that $R_x(0)$ is equal to the mean-square value of the random variable x_n, whereas $C_x(0)$ is equal to the variance of x_n. Also, when the mean μ_x is zero, the autocorrelation and autcovariance functions become the same.

The autocorrelation function $R_x(m)$ of a weakly stationary process exhibits *conjugate symmetry*, that is,

$$R_x(-m)=R_x^*(m). \tag{1.8}$$

A function having this property is said to be Hermitian.

1.1.2 Spectral Density

The autocorrelation function $R_x(m)$ provides a time-domain description of the second-order statistics of a weakly stationary process. Equivalently, we may use the *spectral density* $S_x(f)$ to provide a frequency-domain description of the second-order statistics of the process. The spectral density $S_x(f)$ and autocorrelation function $R_x(m)$ form a Fourier transform pair [1.2]. That is, the spectral density of a time series $\{x_n\}$ of infinite length is defined in terms of its autocorrelation function by the formula

$$S_x(f)=\Delta t\sum_{m=-\infty}^{\infty}R_x(m)\exp(-\mathrm{j}2\pi fm\Delta t), \tag{1.9}$$

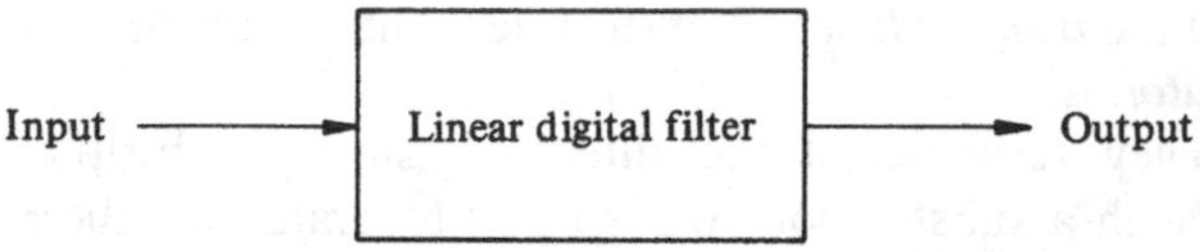

Fig. 1.1. Transmission of a time series through a linear filter

where Δt is the time between the equispaced samples of the series. To recover the autocorrelation function from the spectral density, we may use the inverse formula

$$R_x(m) = \int_{-1/2\Delta t}^{1/2\Delta t} S_x(f) \exp(j2\pi f m \Delta t) df . \tag{1.10}$$

1.1.3 Linear Filtering

The processing of a time series is often accomplished by means of a linear filter. It is customary to characterize such a device in the time domain by specifying its *impulse response*, which is defined as the response of the filter produced by a unit impulse. Consider then a time series $\{x_n\}$ of length N applied to a linear filter of impulse response $\{h_n\}$ of length M, as illustrated in Fig. 1.1. The resulting response of the filter, denoted by $\{y_n\}$, is given by the *convolution sum*

$$y_n = \sum_{k=1}^{M} h_k x_{n-k+1} , \qquad 1 \leqq n \leqq M+N-1 . \tag{1.11}$$

If $\{x_n\}$ or $\{h_n\}$ or both are complex valued, then $\{y_n\}$ will also be complex valued.

Another useful way of describing the input-output relation of a linear filter is by transforming the time-domain description of (1.11) into the z domain. To do this, we take the z transform of both sides of (1.11), obtaining the compact relation [1.3]

$$Y(z) = H(z)X(z) , \tag{1.12}$$

where

$$X(z) = \sum_{n=1}^{N} x_n z^{-n} , \tag{1.13}$$

$$H(z) = \sum_{n=1}^{M} h_n z^{-n} , \tag{1.14}$$

and

$$Y(z) = \sum_{n=1}^{M+N-1} y_n z^{-n} . \tag{1.15}$$

The function $H(z)$ is called the *transfer function* of the filter. The variable z^{-1} is called the *unit delay operator.*

To obtain the *frequency response* of the filter, we simply substitute $z=\exp(\mathrm{j}2\pi f \Delta t)$ in (1.12). Such a substitution corresponds to evaluating the z transforms $X(z)$, $H(z)$, and $Y(z)$ on the unit circle of the z plane.

1.2 Modeling of Time Series

A stochastic process may be modeled by using the idea that a time series, in which successive values are highly dependent, can be generated from a series of independent shocks or impulses. These impulses are random drawings from a fixed distribution, usually assumed Gaussian and having zero mean and variance σ^2. Such a set of random variables is called a *white noise* process. We may thus model a stochastic process as the output of a linear filter produced by a white noise process of zero mean. However, for such a model to be useful in practice, it is rather important to employ the smallest possible number of filter parameters for an adequate representation of the process. We now consider how to introduce this principle of *parsimony* and yet retain models which are representatively useful [1.1]. In particular, we may distinguish three types of a stochastic process:

1) The filter consists of a *recursive* structure with feedback paths only, as illustrated in Fig. 1.2. We refer to a process modeled as the output of such a filter in response to a white noise process as an *autoregressive* (AR) process of order p, where p refers to the number of unit-delay elements contained in the filter.

2) The filter consists of a *nonrecursive* structure, that is, one with feed-forward paths only, as illustrated in Fig. 1.3. We refer to a process modeled in this way as a *moving average* (MA) process of order q, where q refers to the number of unit-delay elements contained in the filter.

3) In order to have adequate flexibility in the modeling of a statistical time series, it is sometimes advantageous to include both autoregressive and moving average components in the model. That is, we may model the process as the output of a filter with both feedback and feedforward paths. Such a process is called a mixed *autoregressive-moving average* (ARMA) process of order (p, q), where p and q refer to the autoregressive and moving average components of the process, respectively.

A detailed treatment of the AR, MA, and ARMA models is given in Chap. 3.

1.3 Linear and Nonlinear Methods of Spectral Analysis

In the characterization of a second-order weakly stationary stochastic process, use of the spectral density is often preferred to the autocorrelation function, because a spectral representation may reveal such useful information as hidden

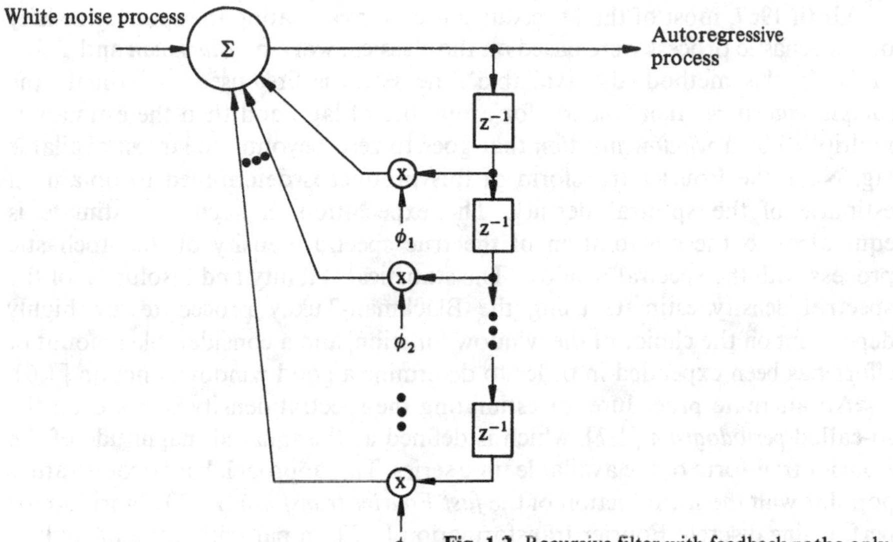

Fig. 1.2. Recursive filter with feedback paths only

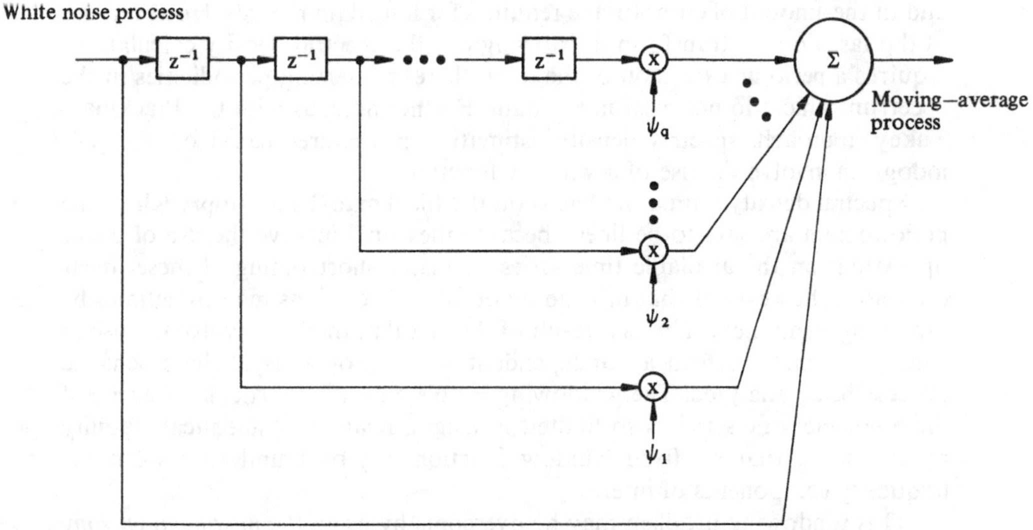

Fig. 1.3. Nonrecursive filter

periodicities or close spectral peaks. In using the spectral density to characterize the process, however, we have to design a reliable *estimator* based on a finite length of data. Indeed, the estimation of spectral density of a stationary time series of limited length has been a classical problem ever since *Wiener* [1.4] established the fact that the spectral density and autorcorrelation function of the process form a Fourier transform pair.

Until 1967, most of the procedures used for estimating the spectral density of a stochastic process were based on the classical work by *Blackman* and *Tukey* [1.5]. In this method, the available time series is first used to estimate the *sample autocorrelation function* for a number of lags, and then the estimate is multiplied by a *window function* that goes to zero beyond the largest available lag. Next, the Fourier transform of this product is determined to obtain an estimate of the spectral density. The expectation of such an estimate is equivalent to the convolution of the true spectral density of the stochastic process with the spectral window. The statistical stability and resolution of the spectral density estimate using the Blackman-Tukey procedure are highly dependent on the choice of the window function, and a considerable amount of effort has been expended in order to determine a good window function [1.6].

An alternate procedure for estimating the spectral density is based on the so-called *periodogram* [1.7], which is defined as the squared magnitude of the Fourier transform of the available time series. This approach has become rather popular with the introduction of the *fast Fourier transform* (FFT) algorithm for performing discrete Fourier transformation [1.3]. In particular, the procedure described by *Welch* [1.8], based on time averaging over several short modified periodograms, results in a significant reduction in the number of computations and in the amount of core storage required for long data records. However, use of the fast Fourier transform in carrying out the spectral density calculations requires a periodic extension of the data, thereby inserting periodicities in the spectrum which do not exist in the data. Furthermore, as with the Blackman-Tukey approach, spectral density estimation procedures based on the periodogram involve the use of a window function.

Spectral density estimators based on the Blackman-Tukey approach or the periodogram are said to be linear because they only involve the use of linear operations on the available time series. A major shortcoming of these linear estimators, however, is that misleading or false conclusions may sometimes be drawn by using them. This is a result of the fact that they all involve the use of window functions which are independent of the properties of the stochastic process being analyzed. The windowing problem may be particularly accute if the available time series is so limited in length, relative to statistical stability requirements, that no fixed window function can be found to resolve the frequency components of interest.

This windowing problem may be overcome by using the *maximum-entropy method* (MEM) or the *maximum-likelihood method* (MLM)[1]. Spectral density estimators based on these two methods are said to be nonlinear because their design is data-dependent.

The basic idea of the MEM is to choose the spectrum which corresponds to the most random or the most unpredictable time series whose autocorrelation function agrees with a set of known values [1.9]. This condition is equivalent to

1 The IEEE Press has published a volume, a large portion of which consists of selected reprints of papers published during 1967–1978, on the maximum-entropy and maximum-likelihood methods of spectral analysis and their applications. For details of this volume see [1.13].

an extrapolation of the autocorrelation function of the available time series by maximizing the *entropy* (which is a measure of the average information content) of the process. Thus, the MEM differs from the conventional linear methods of spectral analysis in that it avoids such assumptions as periodic extension of the data or that the data outside of the available record length is zero. The MEM is also referred to as an autoregressive spectral estimator [1.10].

The MLM was originally devised by *Capon* [1.11] for the purpose of frequency wave number analysis with arrays. The method may be represented in terms of a minimum-variance unbiased estimator of the spectral components of a stationary time series. Indeed, at a particular frequency, it turns out that the maximum-likelihood spectral estimate is the power which would come through a filter designed to pass that frequency undistorted and to reject all other frequency components of the input in an optimal manner.

Both the MEM and MLM show considerable promise for estimating spectra, particularly when the length of the available time series is limited [1.12]. Such a limitation may arise in practice because the pertinent physical phenomena may be justifiably considered stationary only over a limited time interval, or because data may be collected only over a limited time interval, or both.

1.4 Organization of the Volume

The present volume is devoted to a detailed treatment of nonlinear methods of spectral analysis.

Chapter 2 presents a tutorial development of the maximum-entropy method and its relationship to *prediction-error filtering*. It also includes some experimental results obtained by applying the MEM to classify *clutter* data generated in a radar environment.

Chapter 3 presents an extensive survey of the modeling and spectral analysis aspects of both autoregressive (AR) and mixed autoregressive–moving average (ARMA) processes, giving detailed attention to the pertinent literature. It concludes with a numerical example, which illustrates the application of autoregressive and mixed autoregressive-moving average models to the polar motion of the earth.

Chapter 4 is devoted to the analysis of mixed autoregressive–moving average (ARMA) processes. An algorithm is developed to compute the moving average (MA) and autoregressive (AR) components of a non invertible ARMA system from a description of its impulse response. This algorithm is applied to a reflection seismogram, which may be regarded as the impulse response of a non invertible ARMA system.

Chapter 5 is concerned with a theoretical treatment of some signal processing methods in large arrays. In particular, it is shown that the structure of the optimum detector for a known signal in additive Gaussian noise includes a maximum-likelihood filter. When the noise is stationary in both time and space, the performance of the maximum-likelihood filter may be explained in terms of a frequency wave number spectrum which characterizes the noise.

Finally, Chap. 6 gives an account of conventional linear methods for estimating the frequency wave number spectrum in array processing systems, and their basic limitations. This dicussion prepares the way for a detailed treatment of the application of the maximum-likelihood and maximum-entropy methods to array processing system. A critical assessment of the pertinent literature is also included.

References

1.1 G.E.P.Box, G.M.Jenkins: *Time Series Analysis: Forecasting and Control*, revised ed. (Holden-Day, San Francisco 1976)

1.2 S.Haykin: *Communication Systems* (Wiley and Sons, New York 1978)

1.3 A.V.Oppenheim, R.W.Schafer: *Digital Signal Processing* (Prentice-Hall, Englewood Cliffs, N.J. 1975)

1.4 N.Wiener: *Extrapolation, Interpolation, and Smoothing of Stationary Time Series With Engineering Applications* (Wiley and Sons, New York 1949)

1.5 R.B.Blackman, J.W.Tukey: *The Measurement of Power Spectra From the Point of View of Communications Engineering* (Dover, New York 1959)

1.6 G.M.Jenkins, D.G.Watts: *Spectral Analysis and Its Applications* (Holden-Day, San Francisco 1969)

1.7 R.H.Jones: Technometrics **7**, 531–542 (1965)

1.8 P.D.Welch: IEEE Trans. AU-**15** (3), 70–73 (1967)

1.9 J.P.Burg: "Maximum Entropy Spectral Analysis", 37th Ann. Intern. Meeting, Soc. of Explor. Geophys., Oklahoma City, Oklahoma (1967)

1.10 E.Parzen: "Multiple Time Series Modeling", in *Multivariate Analysis-II*, ed. by P.R.Krishnaiah (Academic Press, New York 1969)

1.11 J.Capon: Proc. IEEE **57** (8), 1408–1418 (1969)

1.12 R.T.Lacoss: Geophysics **36** (4), 661–675 (1971)

1.13 D.G.Childers (ed.): *Modern Spectrum Analysis* (The Institute of Electrical and Electronics Engineers, Inc., IEEE Press, New York 1978)

2. Prediction-Error Filtering and Maximum-Entropy Spectral Estimation

S. Haykin and S. Kesler

With 16 Figures

In this chapter we develop the basic theory of a nonlinear method of spectral analysis known as the maximum-entropy method (MEM), which was originally developed by *Burg* [2.1]. The idea of the method is to choose the spectrum (in the form of a nonnegative function of frequency) which corresponds to the most random or the most unpredictable time series whose autocorrelation function agrees with a set of known values. The method derives its name from the fact that this condition corresponds to the concept of maximum entropy as used in information theory.

An important attribute of the maximum-entropy spectral estimator is that it has a greater (and quite often much greater) resolution than conventional linear spectral estimators. Another attribute of the method is that it is usually well-suited for the spectral analysis of relatively short data records. Furthermore, the computation time required by the method is of the same order of magnitude as conventional linear spectral estimators. We accordingly find that the maximum-entropy spectral analysis is a highly practical theory with numerous applications.

We begin the chapter, however, by reviewing the theory of Wiener filters, with particular emphasis on prediction. The concept of prediction is rather basic to the development of the maximum-entropy method.

2.1 Wiener Filters

The subject of linear least-squares estimation is rather vast [2.2, 3]. A least-squares criterion for estimation was first used by *Gauss* [2.4] in his efforts to determine the orbital parameters of the asteroid Ceres. In the 1940s, *Kolmogoroff* [2.5] and *Wiener* [2.6] reintroduced and reformulated the least-squares problem for the estimation and prediction of stochastic processes. In this section, we will develop the discrete-time domain solution to the Wiener filtering problem where the requirement is to filter a desired signal from a weakly stationary complex-valued time series for which certain statistical measures are known.

Consider a linear digital filter whose impulse response is denoted by the sequence $\{h_n\}$ of length M. Suppose that a weakly stationary stochastic process, represented by the complex-valued series $\{x_n\}$ of length N, is applied as input to

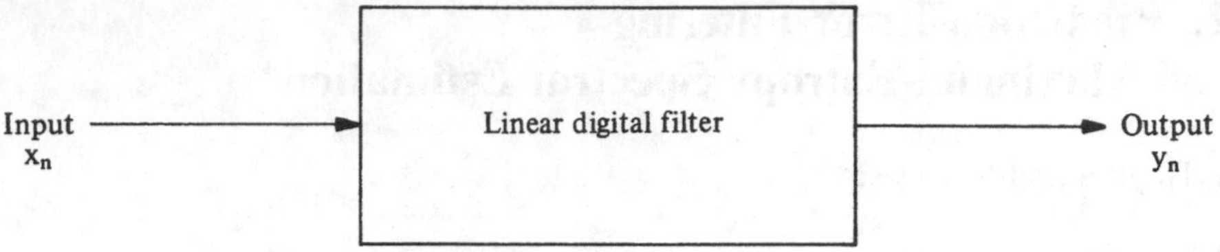

Fig. 2.1. Linear filtering of a time series

this filter (see Fig. 2.1). We assume that $\{x_n\}$ has zero mean. Let the time series $\{y_n\}$ represent the resulting output of the filter and $\{d_n\}$ represent a desired form of the output. Both $\{y_n\}$ and $\{d_n\}$ are of length $M+N-1$. The resulting error between the filter output and the desired output is defined by

$$e_n = d_n - y_n . \tag{2.1}$$

The requirement is to determine the particular form of impulse response for the filter which will make the mean-square value of the error e_n as small as possible. We refer to this condition as the least-mean-square (LMS) error criterion.

2.1.1 Wiener-Hopf Equation

The mean-square value of the error e_n is a real and positive scalar quantity, as shown by

$$P = \mathrm{E}[|e_n|^2] , \tag{2.2}$$

where E denotes the expectation operator. We note that

$$|e_n|^2 = e_n e_n^* , \tag{2.3}$$

where the asterisk denotes the complex conjugate operation. Therefore, substituting (2.1) in (2.2), we may write

$$\begin{aligned} P &= \mathrm{E}[(d_n - y_n)(d_n^* - y_n^*)] \\ &= \mathrm{E}[d_n d_n^*] - \mathrm{E}[d_n y_n^*] - \mathrm{E}[y_n d_n^*] + \mathrm{E}[y_n y_n^*] . \end{aligned} \tag{2.4}$$

However, the time-domain form of the input-output relation of the filter is defined by the convolution sum

$$y_n = \sum_{k=1}^{M} h_k x_{n-k+1} . \tag{2.5}$$

Therefore, substituting (2.5) in (2.4), we get

$$P=\mathrm{E}[d_n d_n^*]-\sum_{k=1}^{M} h_k^* \mathrm{E}[d_n x_{n-k+1}^*]-\sum_{k=1}^{M} h_k \mathrm{E}[x_{n-k+1} d_n^*]$$

$$+\sum_{k=1}^{M}\sum_{l=1}^{M} h_k h_l^* \mathrm{E}[x_{n-k+1} x_{n-l+1}^*]\,. \tag{2.6}$$

The four terms on the right-hand side of (2.6) may be interpreted as follows:

1) The expectation $\mathrm{E}[d_n d_n^*]$ is equal to the power of the desired time series, as shown by

$$\mathrm{E}[d_n d_n^*]=R_d(0)\,, \tag{2.7}$$

where $R_d(0)$ is the autocorrelation function of d_n for zero lag.

2) The expectation $\mathrm{E}[d_n x_{n-k+1}^*]$ is equal to the cross-correlation function of the desired output and the filter input for lag $k-1$, as shown by

$$\mathrm{E}[d_n x_{n-k+1}^*]=R_{dx}(k-1)\,. \tag{2.8}$$

Let

$$\boldsymbol{h}=\begin{bmatrix} h_1 \\ h_2 \\ \vdots \\ h_M \end{bmatrix} \tag{2.9}$$

and

$$\boldsymbol{R}_{dx}=\begin{bmatrix} R_{dx}(0) \\ R_{dx}(1) \\ \vdots \\ R_{dx}(M-1) \end{bmatrix}. \tag{2.10}$$

Then, in terms of the vectors $\boldsymbol{h}$ and $\boldsymbol{R}_{dx}$, we may express the second term on the right-hand side of (2.6) in matrix form as follows:

$$\sum_{k=1}^{M} h_k^* \mathrm{E}[d_n x_{n-k+1}^*]=\sum_{k=1}^{M} h_k^* R_{dx}(k-1)$$

$$=\boldsymbol{h}^{\mathrm{H}}\boldsymbol{R}_{dx}\,, \tag{2.11}$$

where the superscript H denotes the operation of Hermitian transposition.

3) The third term on the right-hand side of (2.6) may be similarly expressed in matrix form as follows:

$$\sum_{k=1}^{M} h_k \mathrm{E}[x_{n-k+1} d_n^*] = \boldsymbol{R}_{dx}^{\mathrm{H}} \boldsymbol{h} . \tag{2.12}$$

4) The expectation $\mathrm{E}[x_{n-k+1} x_{n-l+1}^*]$ is equal to the autocorrelation function of the filter input for the lag $l-k$, as shown by

$$\mathrm{E}[x_{n-k+1} x_{n-l+1}^*] = R_x(l-k) . \tag{2.13}$$

Accordingly, we may express the fourth term on the right-hand side of (2.6) in the following matrix form

$$\sum_{k=1}^{M} \sum_{l=1}^{M} h_k h_l^* \mathrm{E}[x_{n-k+1} x_{n-l+1}^*] = \boldsymbol{h}^{\mathrm{H}} \boldsymbol{R}_x \boldsymbol{h} , \tag{2.14}$$

where $\boldsymbol{R}_x$ is the autocorrelation matrix of the filter input, defined by

$$\boldsymbol{R}_x = \begin{bmatrix} R_x(0) & R_x(-1) & \dots & R_x(1-M) \\ R_x(1) & R_x(0) & \dots & R_x(2-M) \\ \vdots & \vdots & \ddots & \vdots \\ R_x(M-1) & R_x(M-2) & \dots & R_x(0) \end{bmatrix} . \tag{2.15}$$

The matrix $\boldsymbol{R}_x$ is an equidiagonal matrix. Such a matrix is called a Toeplitz matrix. In our case, the matrix $\boldsymbol{R}_x$ is also Hermitian. The properties of this matrix are discussed in Appendix A.

We may thus rewrite (2.6) in matrix form simply as follows:

$$P = R_d(0) - \boldsymbol{h}^{\mathrm{H}} \boldsymbol{R}_{dx} - \boldsymbol{R}_{dx}^{\mathrm{H}} \boldsymbol{h} + \boldsymbol{h}^{\mathrm{H}} \boldsymbol{R}_x \boldsymbol{h} . \tag{2.16}$$

The error power P will be at a minimum when the derivative of P with respect to the vector $\boldsymbol{h}$ is zero. Let $\boldsymbol{h}_{\mathrm{o}}$ denote the particular value of $\boldsymbol{h}$ for which this condition is satisfied. Then, using the rule for differentiation with respect to a vector [2.7], we find, by differentiating both sides of (2.16) with respect to the vector $\boldsymbol{h}$, that

$$-\boldsymbol{R}_{dx}^{\mathrm{H}} + \boldsymbol{h}_{\mathrm{o}}^{\mathrm{H}} \boldsymbol{R}_x = 0 ,$$

so that

$$\boldsymbol{h}_{\mathrm{o}}^{\mathrm{H}} \boldsymbol{R}_x = \boldsymbol{R}_{dx}^{\mathrm{H}} . \tag{2.17}$$

As mentioned above, the autocorrelation matrix $\boldsymbol{R}_x$ is Hermitian, that is, $\boldsymbol{R}_x^{\mathrm{H}} = \boldsymbol{R}_x$. Therefore, taking the Hermitian transpose of both sides of (2.17), we obtain

$$(\boldsymbol{h}_{\mathrm{o}}^{\mathrm{H}} \boldsymbol{R}_x)^{\mathrm{H}} = (\boldsymbol{R}_{dx}^{\mathrm{H}})^{\mathrm{H}}.$$

Hence

$$\boldsymbol{R}_x \boldsymbol{h}_{\mathrm{o}} = \boldsymbol{R}_{dx}. \tag{2.18}$$

Equation (2.18) is the matrix form of the well-known Wiener-Hopf equation [2.8–10]. It represents a system of linear complex-valued simultaneous equations which can be solved for $\boldsymbol{h}_{\mathrm{o}}$ to yield the formal solution

$$\boldsymbol{h}_{\mathrm{o}} = \boldsymbol{R}_x^{-1} \boldsymbol{R}_{dx}, \tag{2.19}$$

where $\boldsymbol{R}_x^{-1}$ is the inverse of the autocorrelation matrix $\boldsymbol{R}_x$. We call the resulting filter defined by the impulse response vector $\boldsymbol{h}_{\mathrm{o}}$ the Wiener filter. This filter is optimum in the sense that its output vector $\boldsymbol{y}$ is the best approximation (in the LMS error sense) to a desired vector $\boldsymbol{d}$.

2.1.2 Minimum Error Power

The mean-square value of the error is given by (2.16). Therefore, substituting (2.18) in (2.16), we find that the minimum value of the error power, produced by the Wiener filter, is given by

$$P_{\min} = R_d(0) - \boldsymbol{R}_{dx}^{\mathrm{H}} \boldsymbol{h}_{\mathrm{o}} \tag{2.20}$$

or, equivalently,

$$P_{\min} = R_d(0) - \boldsymbol{R}_{dx}^{\mathrm{H}} \boldsymbol{R}_x^{-1} \boldsymbol{R}_{dx}. \tag{2.21}$$

2.2 Prediction-Error Filter

Consider the special case of a Wiener filter that is designed to predict the sample value of a weakly stationary time series $\{x_n\}$, one time unit ahead, by using the present and past values of the time series $\{x_n\}$ [2.11]. That is

$$d_n = x_{n+1}. \tag{2.22}$$

Then we may express the cross-correlation function of the desired time series $\{d_n\}$ and the input time series $\{x_n\}$ as follows:

$$\begin{aligned} R_{dx}(m) &= \mathrm{E}[x_{n+1+m}x_n^*] \\ &= R_x(m+1), \qquad m=0,1,\ldots,M-1. \end{aligned} \tag{2.23}$$

Accordingly, we may rewrite (2.18) in expanded form as follows:

$$\begin{bmatrix} R_x(0) & R_x(-1) & \ldots & R_x(1-M) \\ R_x(1) & R_x(0) & \ldots & R_x(2-M) \\ \vdots & \vdots & \ddots & \vdots \\ R_x(M-1) & R_x(M-2) & \ldots & R_x(0) \end{bmatrix} \begin{bmatrix} h_{o,1} \\ h_{o,2} \\ \vdots \\ h_{o,M} \end{bmatrix} = \begin{bmatrix} R_x(1) \\ R_x(2) \\ \vdots \\ R_x(M) \end{bmatrix}. \tag{2.24}$$

That is:

$$\sum_{k=1}^{M} h_{o,k} R_x(m-k) = R_x(m), \qquad m=1,2,\ldots,M. \tag{2.25}$$

The corresponding value of the error e_n is therefore

$$\begin{aligned} e_n &= d_n - y_n \\ &= x_n - \sum_{m=1}^{M} h_{o,m} x_{n-m} \\ &= \sum_{m=0}^{M} a_m x_{n-m}, \end{aligned} \tag{2.26}$$

where the coefficient a_m is defined by:

$$a_m = \begin{cases} 1, & m=0 \\ -h_{o,m} & m=1,2,\ldots,M. \end{cases} \tag{2.27}$$

The digital filter defined by the impulse response vector

$$\boldsymbol{a} = \begin{bmatrix} a_0 \\ a_1 \\ \vdots \\ a_M \end{bmatrix} \tag{2.28}$$

is called the prediction-error filter. According to (2.26), the error e_n may be viewed as the response produced by passing the input x_n through this filter. The functional relationship between the linear predictive filter, characterized by the impulse response vector $\boldsymbol{h}_o$, and the prediction-error filter, characterized by the

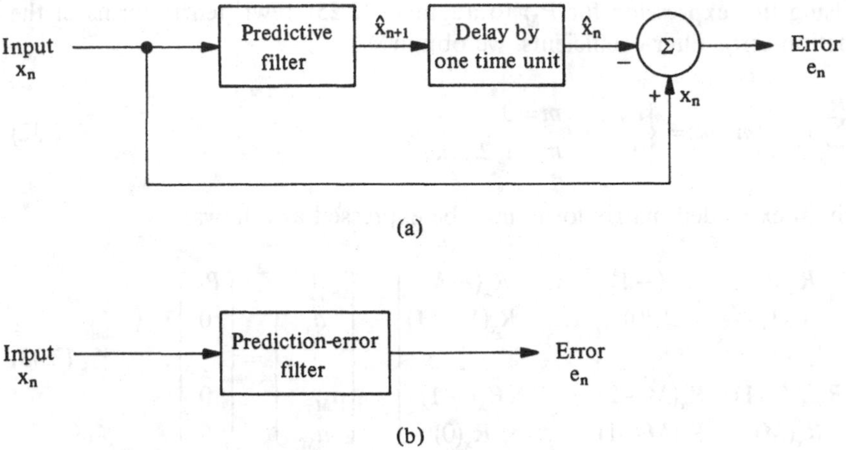

Fig. 2.2. Illustrating the relationship between a predictive filter and prediction-error filter

impulse response vector $\boldsymbol{a}$, is illustrated in Fig. 2.2. Note that the length of the impulse response vector $\boldsymbol{a}$ is greater than that of the impulse response vector $\boldsymbol{h}_o$ by one. Nevertheless, we will describe both filters as having the same order, namely, M.

2.2.1 Prediction-Error Filter Equations

Let P_M denote the output power of a prediction-error filter of order M. As mentioned above, the output of this filter is equal to the error e_n. It follows therefore that P_M is equal to the minimum error power $P_{\min}$ given by (2.20). For the special case of a Wiener filter designed to predict the sample value of the input x_n, one time unit ahead, as in (2.22), we find that

$$R_d(0) = R_x(0) \tag{2.29}$$

and

$$\begin{aligned} \boldsymbol{R}_{dx}^{\mathrm{H}} &= [R_x^*(1) \quad R_x^*(2) \ldots R_x^*(M)] \\ &= [R_x(-1) \quad R_x(-2) \ldots R_x(-M)] \,. \end{aligned} \tag{2.30}$$

Hence, substituting (2.29, 30) in (2.20), using the relationship between the filter coefficients $h_{o,k}$ and a_k as defined by (2.27), and noting that $P_{\min} = P_M$, we may write

$$P_M = \sum_{k=0}^{M} a_k R_x(-k) \,. \tag{2.31}$$

Using this expression for P_M to augment (2.25), rewritten in terms of the prediction-error filter coefficients, we obtain

$$\sum_{k=0}^{M} a_k R_x(m-k) = \begin{cases} P_M, & m=0 \\ 0, & m=1,2,\ldots,M \end{cases} \tag{2.32}$$

which, in expanded matrix form, may be expressed as follows:

$$\begin{bmatrix} R_x(0) & R_x(-1) & \ldots & R_x(-M) \\ R_x(1) & R_x(0) & \ldots & R_x(1-M) \\ \vdots & \vdots & \ddots & \vdots \\ R_x(M-1) & R_x(M-2) & \ldots & R_x(-1) \\ R_x(M) & R_x(M-1) & \ldots & R_x(0) \end{bmatrix} \begin{bmatrix} 1 \\ a_1 \\ \vdots \\ a_{M-1} \\ a_M \end{bmatrix} = \begin{bmatrix} P_M \\ 0 \\ \vdots \\ 0 \\ 0 \end{bmatrix}. \tag{2.33}$$

We refer to this system of $(M+1)$ linear simultaneous equations as the prediction-error filter equations.

2.3 Maximum-Entropy Spectral Estimator

Suppose that we are given $2M+1$ values of the autocorrelation function of a weakly stationary time series $\{x_n\}$ of zero mean. The problem which we wish to solve is to obtain a spectral density estimate which corresponds to the most random or most unpredictable time series whose autocorrelation function is consistent with a set of known values. In terms of information theory, this statement corresponds to the principle of maximum entropy [2.1, 12].

In Appendix B, it is shown that in the case of a set of Gaussian-distributed random variables of zero mean, the entropy (which provides a measure of the information content of the process) is given by, except for a scaling factor,

$$H = \tfrac{1}{2}\ln[\det(\boldsymbol{R}_x)], \tag{2.34}$$

where $\boldsymbol{R}_x$ is the autocorrelation matrix of the process. When the process is of infinite duration, however, we find that the entropy H diverges, and so we cannot use it as a measure of information content. In such a case, we may use the entropy rate defined by [2.13]

$$\begin{aligned} h &= \lim_{M\to\infty} \frac{H}{M+1} \\ &= \lim_{M\to\infty} \tfrac{1}{2}\ln[\det(\boldsymbol{R}_x)]^{1/M+1}. \end{aligned} \tag{2.35}$$

The limiting form of the determinant of the autocorrelation matrix $\boldsymbol{R}_x$ is related to the spectral density $S_x(f)$ of the input time series $\{x_n\}$ to the prediction-error filter as follows (see Appendix A):

$$\lim_{M\to\infty} [\det(\boldsymbol{R}_x)]^{1/M+1} = 2B \exp\left\{\frac{1}{2B}\int_{-B}^{B} \ln[S_x(f)]df\right\}, \tag{2.36}$$

where it is assumed that the time series $\{x_n\}$ is limited to the frequency band $-B \leqq f \leqq B$. Hence, substituting (2.35) in (2.36) yields the entropy rate of the process as

$$h = \tfrac{1}{2}\ln(2B) + \frac{1}{4B}\int_{-B}^{B} \ln[S_x(f)]df. \tag{2.37}$$

Although this relation depends on the assumption that the time series $\{x_n\}$ is Gaussian distributed, nevertheless, the form of the relation is valid for any stationary time series.

The problem is to determine a real positive-valued spectral density estimate $\hat{S}_x(f)$ characterized by an entropy rate h that is both stationary with respect to the unknown autocorrelation values and consistent with respect to the known autocorrelation values. One approach to solving this constrained optimization problem is to use Lagrange multipliers. This procedure is described in Chaps. 3, 6. However, we will use a different approach in this section [2.12, 14].

The spectral density $S_x(f)$ is related to the autocorrelation function $R_x(m)$ by

$$S_x(f) = \Delta t \sum_{m=-\infty}^{\infty} R_x(m) \exp(-\mathrm{j}\,2\pi m f \Delta t), \tag{2.38}$$

where Δt is the sampling period. When the stochastic process is sampled at the Nyquist rate in order to generate the time series $\{x_n\}$, we have $\Delta t = 1/2B$ [2.15]. Substituting (2.38) in (2.37), and assuming the use of sampling at the Nyquist rate, we get

$$h = \frac{1}{4B}\int_{-B}^{B} \ln\left[\sum_{m=-\infty}^{+\infty} R_x(m) \exp(-\mathrm{j}\,2\pi m f \Delta t)\right] df. \tag{2.39}$$

As previously stated, suppose that we know the first $2M+1$ values of the autocorrelation function $R_x(m)$. Then, the most reasonable choice for the unknown values of the autocorrelation function is one that adds no information or entropy to the process. That is,

$$\frac{\partial h}{\partial R_x(m)} = 0, \quad |m| \geqq M+1. \tag{2.40}$$

Hence, carrying out this differentiation, we find that the conditions for an extremum are as follows:

$$\int_{-B}^{B} \frac{\exp(-\mathrm{j}\,2\pi m f \Delta t)}{\hat{S}_x(f)} df = 0, \quad |m| \geqq M+1, \tag{2.41}$$

where $\hat{S}_x(f)$ is the spectral density estimate constrained by (2.40). Equation (2.41) implies that the reciprocal of $\hat{S}_x(f)$ is expressible in the form of a truncated Fourier series as follows:

$$\frac{1}{\hat{S}_x(f)} = \sum_{n=-M}^{M} c_n \exp(-j2\pi n f \Delta t), \tag{2.42}$$

where $c_{-n} = c_n^*$ so as to ensure that $\hat{S}_x(f)$ is a real quantity. The c_n are the Fourier coefficients of the expansion.

The next step in the derivation is to determine values for the set of coefficients $\{c_n\}$ such that the spectral density estimate $\hat{S}_x(f)$ is consistent with the known values of the autocorrelation function, that is,

$$R_x(m) = \int_{-B}^{B} \hat{S}_x(f) \exp(j2\pi m f \Delta t) df, \quad 0 \leq m \leq M. \tag{2.43}$$

Therefore, substituting (2.42) in (2.43), we get

$$R_x(m) = \int_{-B}^{B} \frac{\exp(j2\pi m f \Delta t)}{\sum_{n=-M}^{M} c_n \exp(-j2\pi n f \Delta t)} df, \quad 0 \leq m \leq M. \tag{2.44}$$

At this point, it is convenient to use z-transform notation. Thus, with the variable z defined by

$$z = \exp(j2\pi f \Delta t) \tag{2.45}$$

and therefore

$$df = \frac{1}{j2\pi\Delta t}\left(\frac{dz}{z}\right) = \frac{B}{j\pi}\left(\frac{dz}{z}\right) \tag{2.46}$$

we may rewrite (2.44) in the form

$$R_x(m) = \frac{B}{j\pi} \oint \frac{z^{m-1}}{\sum_{n=-M}^{M} c_n z^{-n}} dz, \quad 0 \leq m \leq M, \tag{2.47}$$

where the contour integration is carried out around the unit circle in the z plane in a counterclockwise direction, as shown in Fig. 2.3. Because of the Hermitian property of the set of coefficients $\{c_n\}$, we may express the summation in the denominator of the integrand in (2.47) as the product of two polynomials as follows:

$$\sum_{n=-M}^{M} c_n z^{-n} = G_M(z) G_M^*(1/z^*), \tag{2.48}$$

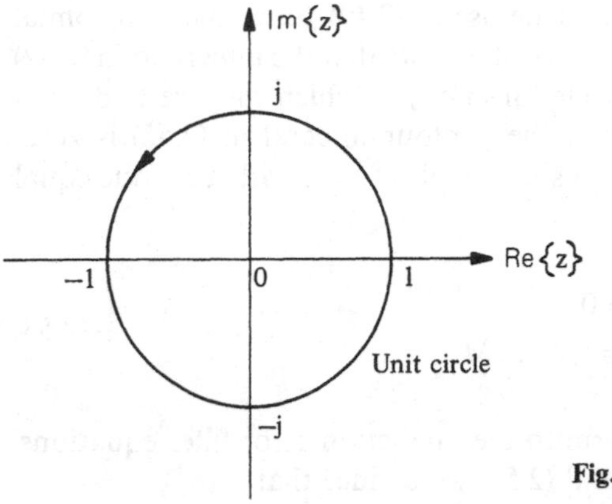

Fig. 2.3. Z plane

where

$$G_M(z) = \sum_{n=0}^{M} g_n z^{-n} \tag{2.49}$$

and

$$G_M^*(1/z^*) = \sum_{n=0}^{M} g_n^* z^n . \tag{2.50}$$

The first polynomial $G_M(z)$ is chosen to be minimum phase[1] (i.e., with its zeros all located inside the unit circle), whereas the second polynomial $G_M^*(1/z^*)$ is chosen to be maximum phase (i.e., with its zeros all located outside the unit circle). Furthermore, the zeros of the two polynomials are the inverse of each other with respect to the unit circle. Thus, substituting (2.48) in (2.47), we get

$$R_x(m) = \frac{B}{\mathrm{j}\pi} \oint \frac{z^{m-1}}{G_M(z) G_M^*(1/z^*)} dz, \quad 0 \leqq m \leqq M . \tag{2.51}$$

Using this relation, we next form the summation

$$\begin{aligned} \sum_{k=0}^{M} g_k R_x(m-k) &= \frac{B}{\mathrm{j}\pi} \oint \frac{z^{m-1} \sum_{k=0}^{M} g_k z^{-k}}{G_M(z) G_M^*(1/z^*)} dz \\ &= \frac{B}{\mathrm{j}\pi} \oint \frac{z^{m-1}}{G_M^*(1/z^*)} dz, \quad 0 \leqq m \leqq M . \end{aligned} \tag{2.52}$$

1 A function of the complex frequency variable is said to be minimum phase if, for a specified amplitude response, it has the minimum phase shift possible, so that any change in the amplitude response produces a corresponding change in the phase response, and vice versa [2.16].

In the second line of (2.52), we have made use of (2.49). Since the polynomial $G_M^*(1/z^*)$ has no zeros inside the unit circle, it follows that the integrand in (2.52) is analytic on and inside the unit circle for $m \geqq 1$, in which case we find from Cauchy's residue theorem [2.17] that the contour integral in (2.52) is zero. However, for $m=0$ the integrand has a simple pole at $z=0$ with a residue equal to $1/g_0^*$. We thus conclude that

$$\sum_{k=0}^{M} g_k R_x(m-k) = \begin{cases} 2B/g_0^*, & m=0 \\ 0 & m=1,2,\ldots,M. \end{cases} \tag{2.53}$$

This set of equations is similar in form to the prediction-error filter equations. In particular, by comparing (2.32) and (2.53) we deduce that

$$g_k = \frac{2B}{g_0^* P_M} a_k \quad 0 \leqq k \leqq M. \tag{2.54}$$

Thus, noting that $a_0 = 1$, we find that for $k=0$,

$$|g_0|^2 = \frac{2B}{P_M}. \tag{2.55}$$

Hence, using (2.54, 55), we may rewrite (2.48) in the form

$$\sum_{n=-M}^{M} c_n z^{-n} = \frac{2B}{P_M} A_M(z) A_M^*(1/z^*), \tag{2.56}$$

where $A_M(z)$ is the transfer function of the prediction-error filter:

$$A_M(z) = \sum_{k=0}^{M} a_k z^{-k} \tag{2.57}$$

and

$$A_M^*(1/z^*) = \sum_{k=0}^{M} a_k^* z^k. \tag{2.58}$$

Finally, substituting (2.56), with $z=\exp(\mathrm{j}2\pi f \Delta t)$, in (2.42), we get the desired expression for the spectral density estimate $\hat{S}_x(f)$, namely,

$$\hat{S}_x(f) = \frac{P_M}{2B\left|1 + \sum_{m=1}^{M} a_m \exp(-\mathrm{j}2\pi m f \Delta t)\right|^2}, \tag{2.59}$$

where P_M is the output power of a prediction-error filter of order M, and a_m, $m=0, 1, 2, \ldots, M$, are the corresponding filter coefficients, B is the bandwidth of the stochastic process, and Δt is the sampling period equal to $1/2B$. We refer to $\hat{S}_x(f)$ as the maximum-entropy spectral density estimate of the process.

In Chap. 3, it is shown that by least-squares fitting an autoregressive model to the given time series $\{x_n\}$, the spectral density estimate so obtained takes on an identical form to the maximum-entropy spectral density estimate of (2.59); this is basically the procedure suggested by *Parzen* [2.18]. In other words, the principle of maximum entropy and the representation of the process by an autoregressive model are equivalent.

2.4 Recursive Formulas for Calculating the Prediction-Error Filter Coefficients and Minimum Error Power

In order to use (2.59) to calculate the maximum-entropy spectral density estimate $\hat{S}_x(f)$, we obviously need to know the output power P_M of the prediction-error filter and also the corresponding values of filter coefficients a_m, $m=0, 1, \ldots, M$. In this section, we will derive recursive formulas for the efficient computation of these quantities.

Suppose that we know the solution to the set of M equations pertaining to a prediction-error filter of order $M-1$, as shown by:

$$\sum_{k=0}^{M-1} a_k^{(M-1)} R_x(m-k) = \begin{cases} P_{M-1}, & m=0 \\ 0, & m=1, \ldots, M-1, \end{cases} \tag{2.60}$$

where the superscript $(M-1)$ has been assigned to the filter coefficients in order to display the filter order. We may rewrite (2.60) in matrix form as follows:

$$\begin{bmatrix} R_x(0) & R_x(-1) & \ldots & R_x(1-M) \\ R_x(1) & R_x(0) & \ldots & R_x(2-M) \\ \vdots & \vdots & \ddots & \vdots \\ R_x(M-1) & R_x(M-2) & \ldots & R_x(0) \end{bmatrix} \begin{bmatrix} 1 \\ a_1^{(M-1)} \\ \vdots \\ a_{M-1}^{(M-1)} \end{bmatrix} = \begin{bmatrix} P_{M-1} \\ 0 \\ \vdots \\ 0 \end{bmatrix}. \tag{2.61}$$

We next modify (2.60) by performing three operations: 1) We take the complex conjugate of both sides of (2.60) and recognize that the prediction-error power P_{M-1} is a real quantity and that $R_x^*(m) = R_x(-m)$. 2) We substitute $M-1-k$ for k. 3) We substitute m for $M-1-m$. We thus obtain

$$\sum_{k=0}^{M-1} a_{M-1-k}^{(M-1)*} R_x(m-k) = \begin{cases} 0, & m=0, 1, \ldots, M-2 \\ P_{M-1}, & m=M-1 \end{cases} \tag{2.62}$$

which may be written in matrix form as follows:

$$\begin{bmatrix} R_x(0) & R_x(-1) & \dots & R_x(1-M) \\ R_x(1) & R_x(0) & \dots & R_x(2-M) \\ \vdots & \vdots & \ddots & \vdots \\ R_x(M-1) & R_x(M-2) & \dots & R_x(0) \end{bmatrix} \begin{bmatrix} a_{M-1}^{(M-1)*} \\ a_{M-2}^{(M-1)*} \\ \vdots \\ 1 \end{bmatrix} = \begin{bmatrix} 0 \\ 0 \\ \vdots \\ P_{M-1} \end{bmatrix}. \quad (2.63)$$

The autocorrelation matrix in (2.61) and (2.63) is now of exactly the same form. We may distinguish between these 2 sets of relations by noting that the first set of relations pertains to a prediction-error filter of $M-1$, operated in the forward direction. On the other hand, the second set of relations defined by (2.63) pertains to a prediction-error filter of the same order, except that it is operated in the backward direction.

We may combine (2.61, 63) to expand the number of prediction-error filter equations by one as follows:

$$\begin{bmatrix} R_x(0) & R_x(-1) & \dots & R_x(1-M) & R_x(-M) \\ R_x(1) & R_x(0) & \dots & R_x(2-M) & R_x(1-M) \\ \vdots & \vdots & \ddots & \vdots & \vdots \\ R_x(M-1) & R_x(M-2) & \dots & R_x(0) & R_x(-1) \\ R_x(M) & R_x(M-1) & \dots & R_x(1) & R_x(0) \end{bmatrix}$$

$$\cdot \left\{ \begin{bmatrix} 1 \\ a_1^{(M-1)} \\ \vdots \\ a_{M-1}^{(M-1)} \\ 0 \end{bmatrix} + \varrho_M \begin{bmatrix} 0 \\ a_{M-1}^{(M-1)*} \\ \vdots \\ a_1^{(M-1)*} \\ 1 \end{bmatrix} \right\} = \left[\begin{bmatrix} P_{M-1} \\ 0 \\ \vdots \\ 0 \\ \Delta_M \end{bmatrix} + \varrho_M \begin{bmatrix} \Delta_M^* \\ 0 \\ \vdots \\ 0 \\ P_{M-1} \end{bmatrix} \right]. \quad (2.64)$$

However, for the corresponding prediction-error filter of order M we have:

$$\begin{bmatrix} R_x(0) & R_x(-1) & \dots & R_x(1-M) & R_x(-M) \\ R_x(1) & R_x(0) & \dots & R_x(2-M) & R_x(1-M) \\ \vdots & \vdots & \ddots & \vdots & \vdots \\ R_x(M-1) & R_x(M-2) & \dots & R_x(0) & R_x(-1) \\ R_x(M) & R_x(M-1) & \dots & R_x(1) & R_x(0) \end{bmatrix} \begin{bmatrix} 1 \\ a_1^{(M)} \\ \vdots \\ a_{M-1}^{(M)} \\ a_M^{(M)} \end{bmatrix} = \begin{bmatrix} P_M \\ 0 \\ \vdots \\ 0 \\ 0 \end{bmatrix}. \quad (2.65)$$

Therefore, comparing (2.64) and (2.65), we deduce that

$$a_m^{(M)} = a_m^{(M-1)} + \varrho_M a_{M-m}^{(M-1)*}, \qquad m = 0, 1, \dots, M, \quad (2.66)$$

$$P_M = P_{M-1} + \varrho_M \Delta_M^*, \quad (2.67)$$

and

$$0=\Delta_M+\varrho_M P_{M-1}. \tag{2.68}$$

In the recursive formula of (2.66), note that for all values of M

$$a_m^{(M)}=\begin{cases}1, & \text{for} \quad m=0\\ \varrho_M, & \text{for} \quad m=M.\\ 0, & \text{for} \quad m>M\end{cases} \tag{2.69}$$

The recursive procedure based on (2.66) is called Levinson's recursion [2.19, 20].

Equation (2.68) will always have a solution provided that $P_M>0$. Thus, using (2.68) to eliminate Δ_M^* from (2.67), we get the following recursive formula for the prediction-error power [2.21]

$$P_M=P_{M-1}(1-|\varrho_M|^2). \tag{2.70}$$

The prediction-error power decreases (or remains the same) as the order of prediction-error filter increases [2.19]. That is, we have $P_M \leqq P_{M-1}$. Hence, from (2.70) we deduce that $|\varrho_M| \leqq 1$. By analogy with the transmission of power through a terminated two-port network, we may therefore view ϱ_M as a reflection coefficient. The parameter ϱ_M is also sometimes referred to as a partial correlation (PARCOR) coefficient [2.77].

Note that (2.66) is reversible provided that $|\varrho_M| \neq 1$. That is, given the coefficients of a prediction-error filter of order M, we can obtain the corresponding coefficients of the prediction-error filter of order $M-1$.

To initiate the recursive algorithms of (2.66, 70), we start with $M=0$, which corresponds to the elementary case of a prediction-error filter of zero order, and then increase the value of M by one at each recursion step up to a specified value of M. We shall have more to say on this issue in Sect. 2.5, 10.

The recursive formulas of (2.66, 70) assume that the available time series $\{n_n\}$ is complex valued, which means that the corresponding values of the prediction-error filter coefficients will also be complex valued. If, however, $\{x_n\}$ is real valued, the filter coefficients will, in a corresponding way, assume real values too.

2.5 Optimum Procedure for Calculating the Reflection Coefficients

An important feature of Levinson's recursion is that if we know the values of the set of reflection coefficients $\{\varrho_m\}$, $m=1, 2, \ldots, M$, then we may recursively compute the corresponding values of the complete set of prediction-error filter

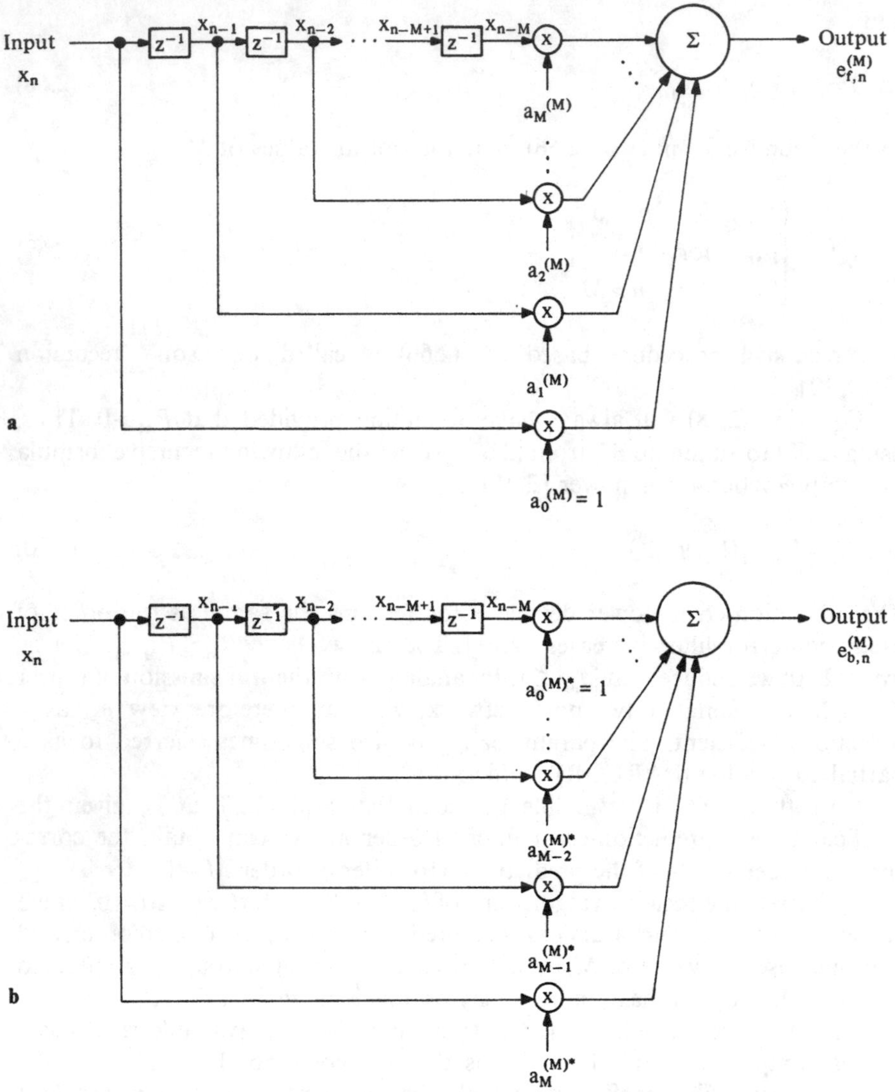

Fig. 2.4a, b. A prediction-error filter operating in: (**a**) forward direction, and (**b**) backward direction

coefficients $\{a_m^{(M)}\}$, $m=1,2,\ldots,M$. We therefore find that the set of reflection coefficients $\{\varrho_m\}$ plays a central role in the design of prediction-error filters.

There are several procedures that can be used to calculate the reflection coefficient ϱ_M [2.22]. For example, ϱ_M may be obtained by minimizing the forward error power defined by

$$P_{f,M} = \mathrm{E}[(e_{f,n}^{(M)})(e_{f,n}^{(M)})^*], \tag{2.71}$$

where $e_{f,n}^{(M)}$ is the error that results at stage M, when the prediction-error filter is operated in the forward direction, as in Fig. 2.4a. That is, $e_{f,n}^{(M)}$ is defined by

$$e_{f,n}^{(M)} = \sum_{m=0}^{M} a_m^{(M)} x_{n-m}. \tag{2.72}$$

Note that the error term $e_{f,n}^{(M)}$ is the same as the e_n defined by (2.26). We have added the superscript M and the subscript f so as to emphasize the filter order and the fact that the filter is operated in the forward direction. We refer to the procedure based on minimization of $P_{f,M}$ as the forward method.

Another procedure is to minimize the backward error power defined by

$$P_{b,M} = \mathrm{E}[(e_{b,n}^{(M)})(e_{b,n}^{(M)})^*], \tag{2.73}$$

where $e_{b,n}^{(M)}$ is the error that results at stage M, when the prediction-error filter is operated in the backward direction, as in Fig. 2.4b. That is, $e_{b,n}^{(M)}$ is defined by

$$e_{b,n}^{(M)} = \sum_{m=0}^{M} a_{M-m}^{(M)*} x_{n-m}. \tag{2.74}$$

Note that in Fig. 2.4b, compared to Fig. 2.4a, the prediction-error filter coefficients are complex conjugated and appear in reversed order. We refer to the procedure based on minimization of $P_{b,M}$ as the backward method. Note that for a given filter order M, the backward error power $P_{b,M}$ has the same value as the forward error power $P_{f,M}$.

The main problem in using the forward method and backward method, however, is that the computed reflection coefficients are not always guaranteed to be less than one in magnitude, that is, the stability of the resulting prediction-error filter is not always guaranteed. One solution to this problem is to minimize the prediction-error power taken as the average for forward and backward prediction over the entire time interval, as shown by

$$P_M = \tfrac{1}{2}(P_{f,M} + P_{b,M}). \tag{2.75}$$

This formulation is originally due to *Burg* [2.21].

Burg's procedure begins by first considering the system for $M=0$. For this value of M, we find from (2.32) that the prediction-error power P_0 is equal to $R_x(0)$, where $R_x(0)$ is the zero-lag value of the autocorrelation function. We assume that the time series $\{x_n\}$ is the sample function of an ergodic process, so that we may substitute time averages for ensemble averages. We may thus write

$$\begin{aligned} P_0 &= R_x(0) \\ &= \frac{1}{N} \sum_{n=1}^{N} x_n x_n^*. \end{aligned} \tag{2.76}$$

We next proceed to the value $M=1$ for which we use (2.75) to determine the prediction-error power P_1, and then determine the value of ϱ_1 for which P_1 is a minimum by solving $\partial P_1/\partial \varrho_1 = 0$. We continue to proceed in this way for higher integer values of M, and thus deduce a general expression for the reflection coefficient ϱ_M at stage M of the iteration. Specifically, we may express ϱ_M as follows [2.12]

$$\varrho_M = \frac{-2 \sum\limits_{n=M+1}^{N} e_{f,n}^{(M-1)} e_{b,n-1}^{(M-1)*}}{\sum\limits_{n=M+1}^{N} [|e_{f,n}^{(M-1)}|^2 + |e_{b,n-1}^{(M-1)}|^2]}, \quad M=1,2,\ldots . \tag{2.77}$$

The forward and backward prediction-error terms for filter order M and time index n are defined by (2.72) and (2.74), respectively. These definitions may be readily adapted to compute the prediction-error terms in the right-hand side of (2.77) for the required values of filter order and time index. Note that in (2.77) the backward prediction-error term is delayed by one time unit.

Equation (2.77) is called Burg's formula [2.14]. According to this formula, except for the minus sign, the reflection coefficient ϱ_M at stage M is equal to the normalized value of the cross-correlation of the forward prediction error $e_{f,n}^{(M-1)}$ and the delayed backward prediction error $e_{b,n-1}^{(M-1)}$, with both errors referring to the preceding stage $M-1$, and the normalizing factor being equal to the average value of the sum of the power due to $e_{f,n}^{(M-1)}$ and that due to $e_{b,n-1}^{(M-1)}$.

Equation (2.77) applies to the general case of complex-valued data. Accordingly, the reflection coefficient ϱ_M has in general a complex value. However, for the case of real-valued data, the prediction-error terms and, therefore, the reflection coefficient ϱ_M, assume real values. For a further discussion of this issue, see Sect. 3.3.2.

The two-by-two matrix

$$\begin{bmatrix} (e_{f,n}^{(M-1)} e_{f,n}^{(M-1)*} + e_{b,n-1}^{(M-1)} e_{b,n-1}^{(M-1)*}) & 2e_{f,n}^{(M-1)} e_{b,n-1}^{(M-1)*} \\ 2e_{f,n}^{(M-1)*} e_{b,n-1}^{(M-1)} & (e_{f,n}^{(M-1)} e_{f,n}^{(M-1)*} + e_{b,n-1}^{(M-1)} e_{b,n-1}^{(M-1)*}) \end{bmatrix}$$

is nonnegative definite. We therefore find that

$$\left| 2 \sum_{n=M+1}^{N} e_{f,n}^{(M-1)} e_{b,n-1}^{(M-1)*} \right| \leq \sum_{n=M+1}^{N} [|e_{f,n}^{(M-1)}|^2 + |e_{b,n-1}^{(M-1)}|^2] \tag{2.78}$$

which means that the reflection coefficient ϱ_M has a magnitude equal to or less than one, independently of the values of the incoming data damples. Accordingly, the use of Burg's formula for calculating the reflection coefficients always assures the stability of the resulting prediction-error filter. The flow graph included in Appendix C illustrates the steps involved in the use of Burg's formula to design a prediction-error filter.

Note that according to Burg's formula, the optimum value of the reflection coefficient ϱ_M is equal to the harmonic mean of the following two values

$$\frac{-\sum_{n=M+1}^{N} e_{\mathrm{f},n}^{(M-1)} e_{\mathrm{b},n-1}^{(M-1)*}}{\sum_{n=M+1}^{N} |e_{\mathrm{b},n}^{(M-1)}|^2}$$

and

$$\frac{-\sum_{n=M+1}^{N} e_{\mathrm{f},n}^{(M-1)} e_{\mathrm{b},n-1}^{(M-1)*}}{\sum_{n=M+1}^{N} |e_{\mathrm{f},n-1}^{(M-1)}|^2}$$

which are obtained by using the forward and backward methods, respectively [2.22].

Another method which guarantees the stability of the prediction-error filter is the geometric mean method [2.23]. This method uses the value of ϱ_M which, except for a minus sign, is equal to the geometric mean of the values obtained by the forward and backward methods. Specifically, according to this method, the reflection coefficient is defined by [2.22].

$$\varrho_M' = \frac{-\sum_{n=M+1}^{N} e_{\mathrm{f},n}^{(M-1)} e_{\mathrm{b},n-1}^{(M-1)*}}{\left[\sum_{n=M+1}^{N} |e_{\mathrm{f},n}^{(M-1)}|^2 \sum_{n=M+1}^{N} |e_{\mathrm{b},n-1}^{(M-1)}|^2\right]^{1/2}}, \tag{2.79}$$

where the prime is used to distinguish this value of the reflection coefficient from that computed by using Burg's formula. Equation (2.79) is commonly used by workers in speech processing. Note, however, that the geometric mean method does not directly involve minimization of an error criterion.

2.6 Lattice Equivalent Model

As previously described, the quantity $e_{\mathrm{f},n}^{(M)}$ represents the error produced by forward operation of a prediction-error filter of order M, as in Fig. 2.4a. On the other hand, the quantity $e_{\mathrm{b},n}^{(M)}$ represents the error produced by backward operation of the filter, as in Fig. 2.4b where the filter coefficients are complex conjugated and appear in reversed order. In this section, we wish to develop recursive formulas for the efficient computation of $e_{\mathrm{f},n}^{(M)}$ and $e_{\mathrm{b},n}^{(M)}$ [2.12, 24, 25, 78].

Consider first the forward error $e_{\mathrm{f},n}^{(M)}$. Substituting the recursive formula of (2.66) in (2.72), we get

$$\begin{aligned} e_{\mathrm{f},n}^{(M)} &= \sum_{m=0}^{M} a_m^{(M-1)} x_{n-m} + \varrho_M \sum_{m=0}^{M} a_{M-m}^{(M-1)*} x_{n-m} \\ &= \sum_{m=0}^{M-1} a_m^{(M-1)} x_{n-m} + \varrho_M \sum_{m=1}^{M} a_{M-m}^{(M-1)*} x_{n-m}, \end{aligned} \tag{2.80}$$

where we have made use of the fact that $a_M^{(M-1)}$ is zero for all M. The first summation term on the right-hand side of (2.80) is recognized as the forward error for a prediction-error filter of order $M-1$. For the second summation term, we substitute m for $m-1$, and so find that this term is equal to the backward error for a prediction-error filter of order $M-1$, but delayed by one time unit. We may thus rewrite the expression for the forward error $e_{\mathrm{f},n}^{(M)}$ as follows

$$e_{\mathrm{f},n}^{(M)} = e_{\mathrm{f},n}^{(M-1)} + \varrho_M e_{\mathrm{b},n-1}^{(M-1)}. \tag{2.81}$$

Consider next the backward error $e_{\mathrm{b},n}^{(M)}$. We first substitute m for $M-m$ in (2.74), obtaining

$$e_{\mathrm{b},n}^{(M)} = \sum_{m=0}^{M} a_m^{(M)*} x_{n+m-M}. \tag{2.82}$$

Next, substituting the recursive formula of (2.66) in (2.82), we get

$$\begin{aligned} e_{\mathrm{b},n}^{(M)} &= \sum_{m=0}^{M} a_m^{(M-1)*} x_{n+m-M} + \varrho_M^* \sum_{m=0}^{M} a_{M-m}^{(M-1)} x_{n+m-M} \\ &= \sum_{m=0}^{M-1} a_m^{(M-1)*} x_{n+m-M} + \varrho_M^* \sum_{m=1}^{M} a_{M-m}^{(M-1)} x_{n+m-M}, \end{aligned} \tag{2.83}$$

where we have used the fact that $a_M^{(M-1)}$ is zero. By substituting m for $M-m-1$ in the first summation term on the right-hand side of (2.83), we find that this term is the backward error for a prediction-error filter of order $M-1$, delayed by one time unit. Next, by substituting m for $M-m$ in the second summation term, we find that this term is the forward error for a prediction-error filter of order $M-1$. We may thus rewrite the expression for the backward error $e_{\mathrm{b},n}^{(M)}$ as follows

$$e_{\mathrm{b},n}^{(M)} = e_{\mathrm{b},n-1}^{(M-1)} + \varrho_M^* e_{\mathrm{f},n}^{(M-1)}. \tag{2.84}$$

A signal-flow graph representation of the recursive pair of relations in (2.81, 84), in the form of a lattice structure, is presented in Fig. 2.5a.

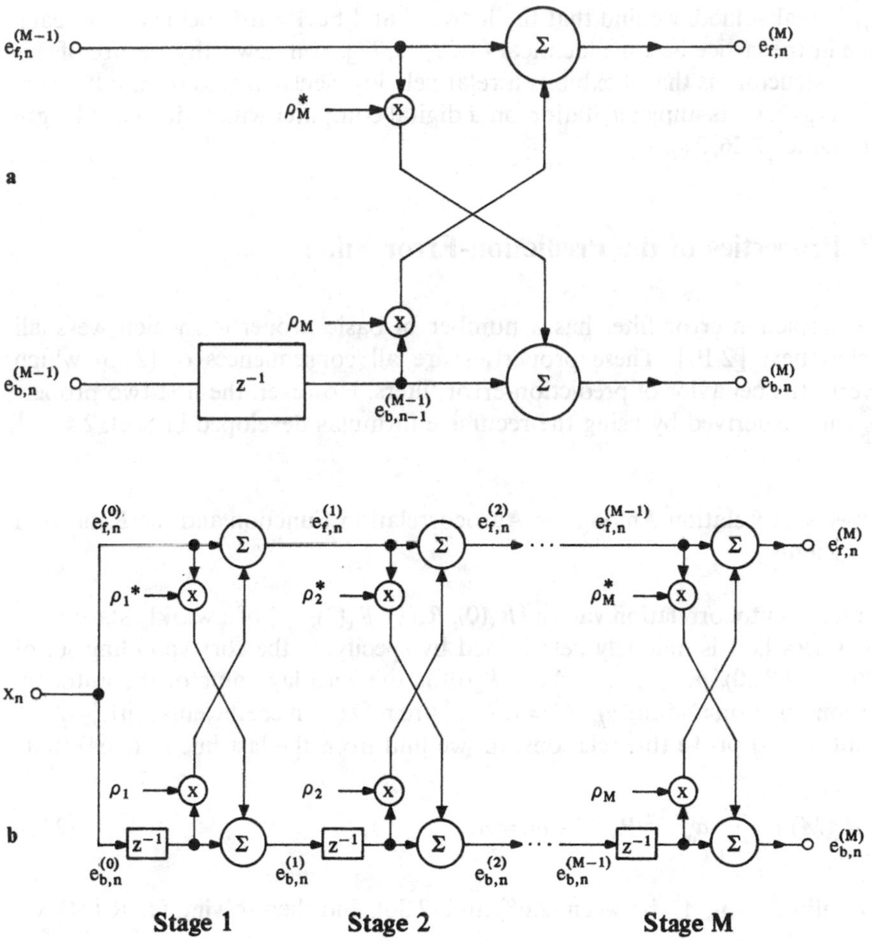

Fig. 2.5. a Signal-flow graph representation of (2.81) and (2.84); **(b)** Lattice equivalent model of a prediction-error filter of order M

We note that for the elementary case of $M=0$, the definitions of (2.72, 74) reduce to

$$e_{f,n}^{(0)}=e_{b,n}^{(0)}=x_n\,. \tag{2.85}$$

Therefore, starting with $M=0$, and incrementing the filter order by one at a time, we obtain the lattice equivalent model shown in Fig. 2.5b for a prediction-error filter of order M. Note that this equivalent model consists of a number of stages equal to the order of the prediction-error filter.

The lattice equivalent model of Fig. 2.5b assumes that the incoming time series $\{x_n\}$ is complex valued, so that all the reflection coefficients of the process are, in general, complex valued too. When, however, the incoming time series

$\{x_n\}$ is real valued, we find that the forward and backward coefficients of each stage in the lattice become identical [2.22, 23, 77]. A noteworthy feature of this lattice structure is that it exhibits a relatively low sensitivity to roundoff errors resulting from its implementation on a digital computer with finite word length arithmetic [2.26, 27].

2.7 Properties of the Prediction-Error Filter

The prediction-error filter has a number of basic properties, which we shall discuss next [2.10]. These properties are all consequences of (2.33) which governs the behavior of prediction-error filters. However, the first two properties will be derived by using the recursive formulas developed in Sect. 2.4.

Property 1: Relation Among the Autocorrelation Function and the Reflection Coefficients

The set of autocorrelation values $\{R_x(0), R_x(1), R_x(2), \ldots\}$ of a weakly stationary time series $\{x_n\}$ is uniquely determined by specifying the corresponding set of numbers $\{R_x(0), \varrho_1, \varrho_2, \ldots\}$, where $R_x(0)$ is the zero-lag value of the autocorrelation function and the ϱ_M, $M=1,2,\ldots$, are reflection coefficients with $|\varrho_M| \leqq 1$ for all M. To prove this relationship, we find from the last line of (2.64) that

$$R_x(M) + \sum_{m=1}^{M-1} a_m^{(M-1)} R_x(M-m) = \Delta_M \,. \tag{2.86}$$

Next, eliminating Δ_M between (2.68) and (2.86), and then solving for $R_x(M)$, we get

$$R_x(M) = -\varrho_M P_{M-1} - \sum_{m=1}^{M-1} a_m^{(M-1)} R_x(M-m) \,, \tag{2.87}$$

where we have used the identity $a_M^{(M)} = \varrho_M$. Therefore, if we are given the set of numbers $\{R_x(0), \varrho_1, \varrho_2, \ldots\}$ then we can recursively generate the unique set of numbers $\{R_x(0), R_x(1), R_x(2), \ldots\}$ by using (2.87). This follows from the fact that the values of P_{M-1} and $a_m^{(M-1)}$, $m=1,2,\ldots,M-1$ needed to find $R_x(M)$ are uniquely determined by the set of numbers $\{R_x(0), R_x(1), \ldots, R_x(M-1)\}$.

Conversely, if we are given a set of numbers $\{R_x(0), R_x(1), R_x(2), \ldots\}$, then we can recursively generate the unique set of numbers $\{R_x(0), \varrho_1, \varrho_2, \ldots\}$ again by using (2.87), provided that P_{M-1} is nonzero. If P_{M-1} is zero, this would have been the result of having $|\varrho_{M-1}| = 1$. In such a case, the sequence of reflection coefficients $\{\varrho_M\}$ can be terminated and, as we shall show later in Sect. 2.9, the values of $R_x(M), R_x(M+1), \ldots$, are rigidly determined by the condition that the

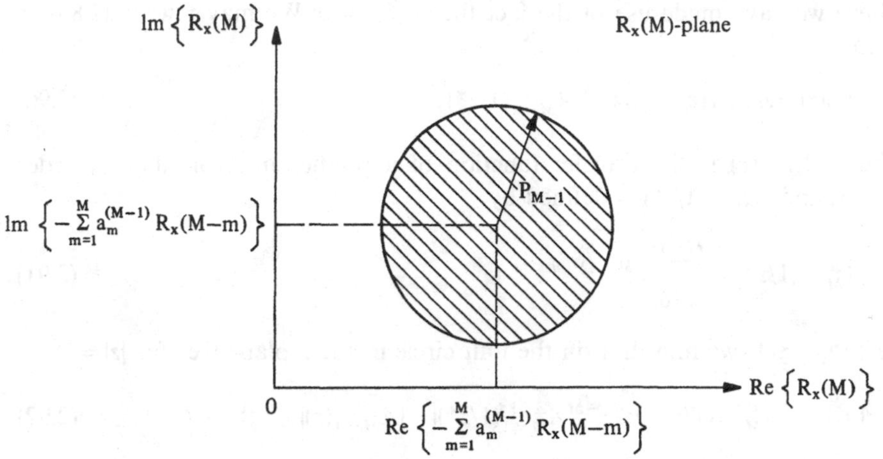

Fig. 2.6. Region (shown shaded) defining permissible values of $R_x(M)$

input set $\{R_x(0), R_x(1), R_x(2), \dots\}$ be representative of an autocorrelation function.

An important consequence of this property is that the set of numbers $\{R_x(0), \varrho_1, \varrho_2, \dots\}$ provides a new representation of the second-order statistics of a stationary time series, which is equivalent to specifying the autocorrelation function of the process [2.14].

As a corollary of the above property, we find that the permissible region for the Mth lag of the autocorrelation function [i.e., $R_x(M)$] is the interior (including circumference) of a circle (in the complex plane) of radius P_{M-1} and center at $-\sum_{m=1}^{M-1} a_m^{(M-1)} R_x(M-m)$, which follows directly from (2.87). This is illustrated in Fig. 2.6. Choosing $R_x(M)$ to be exactly at the center of the circle corresponds to the maximum-entropy spectrum, as will be shown in Sect. 2.8.

Property 2: Minimum-Phase Theorem

The prediction-error filter is minimum phase or, equivalently, minimum delay.

To prove this property, let the polynomial $A_M(z)$ denote the transfer function of a prediction-error filter of order M, as shown by

$$A_M(z) = \sum_{m=0}^{M} a_m^{(M)} z^{-m}, \tag{2.88}$$

where $a_0^{(M)} = 1$. Substitution of (2.66) in (2.88) yields

$$A_M(z) = \sum_{m=0}^{M-1} a_m^{(M-1)} z^{-m} + \varrho_M \sum_{m=1}^{M} a_{M-m}^{(M-1)*} z^{-m}, \tag{2.89}$$

where we have made use of the fact that $a_{M-1}^{(M)}=0$. We may rewrite (2.89) as follows

$$A_M(z)=A_{M-1}(z)+\varrho_M z^{-M}A_{M-1}^*(1/z^*), \tag{2.90}$$

where $A_{M-1}(z)$ is the transfer function of a prediction-error filter of order $M-1$, and $A_{M-1}^*(1/z^*)$ is defined by

$$A_{M-1}^*(1/z^*)=\sum_{m=0}^{M-1} a_m^{(M-1)*}z^m. \tag{2.91}$$

Since $|\varrho_M|\leqq 1$, we find that on the unit circle in the z plane (i.e., for $|z|=1$),

$$|\varrho_M z^{-M}A_{M-1}^*(1/z^*)|<|z^{-M}A_{M-1}^*(1/z^*)|=|A_{M-1}(z)|. \tag{2.92}$$

Now suppose that we know that $A_{M-1}(z)$ has no zeros outside the unit circle in the z plane. Then, from Rouche's theorem[2] we deduce that $A_M(z)$ also has no zeros outside the unit circle. This proof by induction is completed by noting that the prediction-error filter of order zero (i.e., $M=0$) has no zeros at all and is thus minimum phase. It follows therefore that a prediction-error filter of any order is minimum phase.

A characteristic of a minimum-phase filter is that its impulse response is delayed the least with respect to the instant of time when the impulse is applied to the filter input, i.e., the energy contained in the impulse response is concentrated as closely as possible at the front end of the response [2.28]. We therefore find that minimum-phase filters are also referred to as minimum-delay filters [2.29, 30].

The proof presented above for the minimum-phase or minimum-delay property of a prediction-error filter is based on the Levinson recursion. In Chap. 3, another procedure is presented for the proof of this property, which is based on an orthogonal transformation of the Toeplitz autocorrelation matrix of the given data.

Property 3: Principle of Orthogonality

The sequence of data samples $\{x_n\}$ applied to the input of a prediction-error filter of order M and the resulting sequence of error samples $\{e_n\}$ are orthogonal, that is,

$$\mathrm{E}[e_n x_i^*]=0, \quad i=n-1, n-2, \ldots, n-M, \tag{2.93}$$

2 Rouche's theorem [2.17] states that if the function $F(z)$ is analytic upon a contour C in the z plane and within the region enclosed by this contour, and if a second function $G(z)$, in addition to satisfying the same analyticity conditions, fulfills the relation, $|G(z)|<|F(z)|$ on the contour C, then the function $F(z)+G(z)$ has the same number of zeros within the region enclosed by C as does the function $F(z)$.

where

$$e_n = \sum_{m=0}^{M} a_m^{(M)} x_{n-m} . \tag{2.94}$$

The principle of orthogonality is a basic property of all linear mean-square estimation systems [2.31]. To prove it, we will show that if the set of filter coefficients is chosen so as to satisfy (2.93), then the mean-square value of the prediction error e_n is minimum.

Consider a linear filter with an arbitrary set of coefficients $\{A_0, A_1, \ldots, A_M\}$, which produces the following output

$$e'_n = \sum_{m=0}^{M} A_m x_{n-m} \tag{2.95}$$

in response to the input $\{x_n\}$. In terms of the prediction-error output e_n, we may rewrite (2.95) as follows

$$e'_n = e_n + \sum_{m=0}^{M} (A_m - a_m^{(M)}) x_{n-m} . \tag{2.96}$$

The mean-square value of the arbitrary filter output is therefore

$$\begin{aligned} \mathrm{E}[(e'_n)(e'_n)^*] &= \mathrm{E}\left\{\left[e_n + \sum_{m=0}^{M} (A_m - a_m^{(M)}) x_{n-m}\right] \right. \\ &\quad \left. \cdot \left[e_n^* + \sum_{m=0}^{M} (A_m^* - a_m^{(M)*}) x_{n-m}^*\right]\right\} \\ &= \mathrm{E}[|e_n|^2] + E\left[\left|\sum_{m=0}^{M} (A_m - a_m^{(M)}) x_{n-m}\right|^2\right] \\ &\geqq \mathrm{E}[|e_n|^2] , \end{aligned} \tag{2.97}$$

where we have made use of (2.93). Hence, the left-hand side of (2.97) is minimum if we choose $A_m = a_m^{(M)}$ for, $m = 0, 1, \ldots, M$, and the principle of orthogonality is thereby proved.

As a corrolary of the orthogonality property, we may state that the error time series produced by a prediction-error filter of order M, in response to an autoregressive process of order M, is a white noise process. This follows from the fact that in order for the difference equation defining the error time series $\{e_n\}$, namely, (2.94), to have a weakly stationary solution satisfying the condition of (2.93) for the case when e_n, $n = 0, \pm 1, \ldots$, is a white noise process, it is sufficient that all of the zeros of the transfer function of the prediction-error filter, represented by the set of coefficients $\{a_n^{(M)}\}$, lie inside the unit circle [2.32]. Since it is known that the prediction-error filter is minimum phase (see

Property 2), we therefore deduce that with an autoregressive process of order M as input, the error time series produced by the prediction-error filter of order M is a white noise process.

Indeed, virtually every time series encountered in practice (after appropriate preprocessing) can be approximated to any desired accuracy by a finite autoregressive model of sufficiently high degree [2.32]. It follows therefore that, in the limit, with an arbitrary time series as input, the error time series produced by a prediction-error filter of infinite length is a white noise process.

Property 4: Decoupling Between Successive Stages of the Lattice

The backward prediction errors in the equivalent lattice model are orthogonal to each other. To prove this property, we first rewrite (2.74), which defines the backward prediction error $e_{b,n}^{(M)}$ for any value of filter order M from zero and up, in matrix form as follows:

$$e_{b,n}^{T} = x^{T} A_l^{H}, \tag{2.98}$$

where the superscript T signifies transposition, and

$$e_{b,n} = \begin{bmatrix} e_{b,n}^{(0)} \\ e_{b,n}^{(1)} \\ e_{b,n}^{(2)} \\ \vdots \\ e_{b,n}^{(M)} \end{bmatrix}. \tag{2.99}$$

The A_l^{H} is an $(M+1)$ by $(M+1)$ upper diagonal matrix defined by

$$A_l^{H} = \begin{bmatrix} 1 & a_1^{(1)*} & a_2^{(2)*} & \dots & a_M^{(M)*} \\ & 1 & a_1^{(2)*} & \dots & a_{M-1}^{(M)*} \\ & & 1 & \dots & a_{M-2}^{(M)*} \\ & 0 & & \ddots & \vdots \\ & & & & 1 \end{bmatrix}. \tag{2.100}$$

The x^{T} is a row vector of length $M+1$ defined by

$$x^{T} = [x_n \; x_{n-1} \; x_{n-2} \; \dots \; x_{n-M}]. \tag{2.101}$$

Note that the elements of the mth column of the upper triangular matrix A_l^{H} represent the coefficients of the prediction-error filter operated in the backward direction, where $m = 0, 1, 2, \dots, M$. The autocorrelation matrix for the backward prediction errors is therefore given by

$$\begin{aligned} R_{e_b} &= \mathrm{E}[(e_{b,n}^{\mathrm{T}})^{\mathrm{H}} e_{b,n}^{\mathrm{T}}] \\ &= \mathrm{E}[(x^{\mathrm{T}}A_l^{\mathrm{H}})^{\mathrm{H}}(x^{\mathrm{T}}A_l^{\mathrm{H}})] \\ &= \mathrm{E}[A_l(x^{\mathrm{T}})^{\mathrm{H}}(x^{\mathrm{T}}A_l^{\mathrm{H}})] \\ &= A_l\,\mathrm{E}[(x^{\mathrm{T}})^{\mathrm{H}}(x^{\mathrm{T}})]A_l^{\mathrm{H}} \\ &= A_l R_x A_l^{\mathrm{H}}, \end{aligned} \tag{2.102}$$

where R_x is the autocorrelation matrix of the input data. However, the matrix product on the right-hand side of (2.102) is a diagonal matrix (see Appendix A), as shown by

$$A_l R_x A_l^{\mathrm{H}} = \begin{bmatrix} P_0 & & & & \\ & P_1 & & \boldsymbol{0} & \\ & & P_2 & & \\ & \boldsymbol{0} & & \ddots & \\ & & & & P_M \end{bmatrix}. \tag{2.103}$$

We may therefore rewrite (2.103) simply as follows:

$$\mathrm{E}[e_{b,n}^{(l)} e_{b,n}^{(m)*}] = \begin{cases} P_m, & l = m \\ 0, & l \neq m. \end{cases} \tag{2.104}$$

This means that the backward prediction errors in the equivalent lattice model of Fig. 2.5b are orthogonal to each other. That is, the backward prediction errors in the lattice are the result of an orthogonalization process (of the Gram-Schmidt type) on delayed versions of the input data [2.33, 34].

2.7.1 Consequences of the Orthogonality and Decoupling Properties

Based on the fact that the input data and the error time series appearing at the output of a prediction-error filter are orthogonal, and that successive stages of the equivalent lattice model of the filter are decoupled from each other, we may make the following observations with reference to this model [2.33–35]:

1) The reflection coefficient ϱ_m at stage m of the model may be computed independently of the reflection coefficients of those following stage m. In fact, the global minimization of the prediction-error power with respect to ϱ_m may be achieved as a sequence of local minimization problems, one at each stage.

2) To compute the forward prediction error $e_{f,n}^{(m)}$ at stage m, it is sufficient to update the forward prediction error $e_{f,n}^{(m-1)}$ with a constant (namely, the reflection coefficient ϱ_m) times the delayed backward prediction error $e_{b,n-1}^{(m-1)}$, with both $e_{f,n}^{(m-1)}$ and $e_{b,n-1}^{(m-1)}$ referring to the preceding stage $m-1$. The orthogonality of the data and error sequences ensures that this new forward

prediction error $e_{\mathrm{f},n}^{(m)}$ is smallest in the least-squares sense, and that previous coefficients need not be changed. Similar arguments hold for the backward prediction error $e_{\mathrm{b},n}^{(m)}$ at stage m.

3) Equation (2.70), reproduced here for convenience,

$$P_m = P_{m-1}(1 - |\varrho_m|^2), \quad m = 1, 2, \ldots, M \tag{2.105}$$

defines the minimum prediction-error power at stage m in terms of the minimum prediction-error power at the preceding stage $m-1$ and the reflection coefficient of stage m. This relation may be viewed as a manifestation of the decoupling between successive stages of the lattice.

2.8 Maximum-Entropy Extension of the Autocorrelation Function

Suppose that we are given the values of the autocorrelation function $R_x(m)$ for $m = 0, 1, 2, \ldots, M$, which make up a nonnegative definite $(M+1)$ by $(M+1)$ autocorrelation matrix $\boldsymbol{R}_x$. Then solving (2.33), we obtain the elements of the impulse response vector $\boldsymbol{a}$ of length $M+1$, which characterizes the corresponding prediction-error filter. For the special case when the set of reflection coefficients is extended so that ϱ_m is zero for $m > M$, we find from (2.87) that the corresponding extension of the autocorrelation function is defined by the convolution sum

$$R_x(m) = -\sum_{k=1}^{M} R_x(m-k)a_k^{(M)}, \quad m > M \tag{2.106}$$

or, equivalently,

$$\sum_{k=0}^{M} R_x(m-k)a_k^{(M)} = 0, \quad m > M. \tag{2.107}$$

Indeed, this special form of extension for the autocorrelation function corresponds to the maximum-entropy spectral estimate $\hat{S}_x(f)$ derived previously in Sect. 2.3.

In order to establish the validity of this statement, let the z transform of the autocorrelation function be denoted by the sum $\Psi_x(z) + \Psi_x^*(1/z^*)$, where function $\Psi_x(z)$ is defined by

$$\Psi_x(z) = \frac{R_x(0)}{2} + R_x(1)z^{-1} + R_x(2)z^{-2} + \ldots. \tag{2.108}$$

Then, recognizing that the convolution of two sequences of numbers is transformed into the product of their respective z transforms, we may write

$$\Psi_x(z)A_M(z) = \frac{R_x(0)}{2} D_M(z), \tag{2.109}$$

where $A_M(z)$ is the transfer function of the prediction-error filter [see (2.88)]. The polynomial $D_M(z)$ is defined by

$$D_M(z)=1+d_1z^{-1}+d_2z^{-2}+\dots, \tag{2.110}$$

where the coefficient d_m is zero for $m>M$ in accordance with (2.107). From (2.109), we thus have

$$\Psi_x(z)=\frac{R_x(0)}{2}\frac{D_M(z)}{A_M(z)}. \tag{2.111}$$

The expression on the right-hand side of (2.111) represents a legitimate z transform, that is, its Taylor series expansion in z^{-1} converges on the unit circle because the denominator polynomial $A_M(z)$ has no zeros outside the unit circle.

We next wish to derive a recursive formula for determining the polynomial $D_M(z)$. We do this by induction, adopting a procedure similar to that used in Sect. 2.4. We begin by considering a prediction-error filter of order $M-1$ for which we may write, based on (2.109),

$$\begin{bmatrix} R_x(0)/2 & 0 & 0 & \dots & 0 \\ R_x(1) & R_x(0)/2 & 0 & \dots & 0 \\ \vdots & \vdots & \vdots & \ddots & \vdots \\ R_x(M-1) & R_x(M-2) & . & \dots & R_x(0)/2 \end{bmatrix} \begin{bmatrix} 1 \\ a_1^{(M-1)} \\ \vdots \\ a_{M-1}^{(M-1)} \end{bmatrix} = \frac{R_x(0)}{2} \begin{bmatrix} 1 \\ d_1^{(M-1)} \\ \vdots \\ d_{M-1}^{(M-1)} \end{bmatrix}. \tag{2.112}$$

Equation (2.61) defines the corresponding set of prediction-error filter equations, reproduced here for convenience,

$$\begin{bmatrix} R_x(0) & R_x(-1) & \dots & R_x(1-M) \\ R_x(1) & R_x(0) & \dots & R_x(2-M) \\ \vdots & \vdots & \ddots & \vdots \\ R_x(M-1) & R_x(M-2) & \dots & R_x(0) \end{bmatrix} \begin{bmatrix} 1 \\ a_1^{(M-1)} \\ \vdots \\ a_{M-1}^{(M-1)} \end{bmatrix} = \begin{bmatrix} P_{M-1} \\ 0 \\ \vdots \\ 0 \end{bmatrix}. \tag{2.113}$$

Therefore, subtracting (2.112) from (2.113), we get

$$\begin{bmatrix} R_x(0)/2 & R_x(-1) & \dots & R_x(1-M) \\ 0 & R_x(0)/2 & \dots & R_x(2-M) \\ \vdots & \vdots & \ddots & \vdots \\ 0 & 0 & \dots & R_x(0)/2 \end{bmatrix} \begin{bmatrix} 1 \\ a_1^{(M-1)} \\ \vdots \\ a_{M-1}^{(M-1)} \end{bmatrix} = \begin{bmatrix} P_{M-1} \\ 0 \\ \vdots \\ 0 \end{bmatrix} - \frac{R_x(0)}{2} \begin{bmatrix} 1 \\ d_1^{(M-1)} \\ \vdots \\ d_{M-1}^{(M-1)} \end{bmatrix}. \tag{2.114}$$

Taking the complex conjugates of both sides of (2.114), and rearranging the order in which the entries are made, we may write

$$\begin{bmatrix} R_x(0)/2 & 0 & \dots & 0 \\ R_x(1) & R_x(0)/2 & \dots & 0 \\ \vdots & \vdots & \ddots & \vdots \\ R_x(M-1) & R_x(M-2) & \dots & R_x(0)/2 \end{bmatrix} \begin{bmatrix} a_{M-1}^{(M-1)*} \\ a_{M-2}^{(M-1)*} \\ \vdots \\ 1 \end{bmatrix}$$

$$= \begin{bmatrix} 0 \\ 0 \\ \vdots \\ P_{M-1} \end{bmatrix} - \frac{R_x(0)}{2} \begin{bmatrix} d_{M-1}^{(M-1)*} \\ d_{M-2}^{(M-1)*} \\ \vdots \\ 1 \end{bmatrix}. \tag{2.115}$$

We may now use (2.112, 115) in order to generate the corresponding set of relations for a prediction-error filter of order M, and so write

$$\begin{bmatrix} R_x(0)/2 & 0 & \dots & 0 & 0 \\ R_x(1) & R_x(0)/2 & \dots & 0 & 0 \\ \vdots & \vdots & \ddots & \vdots & \vdots \\ R_x(M-1) & R_x(M-2) & \dots & R_x(0)/2 & 0 \\ R_x(M) & R_x(M-1) & \dots & R_x(1) & R_x(0)/2 \end{bmatrix}$$

$$\cdot \left\{ \begin{bmatrix} 1 \\ a_1^{(M-1)} \\ \vdots \\ a_{M-1}^{(M-1)} \\ 0 \end{bmatrix} + \varrho_M \begin{bmatrix} 0 \\ a_{M-1}^{(M-1)*} \\ \vdots \\ a_1^{(M-1)*} \\ 1 \end{bmatrix} \right\}$$

$$= \frac{R_x(0)}{2} \begin{bmatrix} 1 \\ d_1^{(M-1)} \\ \vdots \\ d_{M-1}^{(M-1)} \\ 0 \end{bmatrix} + \begin{bmatrix} 0 \\ 0 \\ \vdots \\ 0 \\ \Delta_M \end{bmatrix} + \varrho_M \begin{bmatrix} 0 \\ 0 \\ \vdots \\ 0 \\ P_{M-1} \end{bmatrix} - \varrho_M \frac{R_x(0)}{2} \begin{bmatrix} 0 \\ d_{M-1}^{(M-1)*} \\ \vdots \\ d_1^{(M-1)*} \\ 1 \end{bmatrix}$$

$$= \frac{R_x(0)}{2} \left\{ \begin{bmatrix} 1 \\ d_1^{(M-1)} \\ \vdots \\ d_{M-1}^{(M-1)} \\ 0 \end{bmatrix} - \varrho_M \begin{bmatrix} 0 \\ d_{M-1}^{(M-1)*} \\ \vdots \\ d_1^{(M-1)*} \\ 1 \end{bmatrix} \right\}, \tag{2.116}$$

where the last equality follows from the fact that $\Delta_M + \varrho_M P_{M-1}$ is zero [see (2.68)].

From (2.116) we deduce the following recursive formula

$$d_m^{(M)} = d_m^{(M-1)} - \varrho_M d_{M-m}^{(M-1)*} \quad m = 1, 2, \ldots, M \tag{2.117}$$

which is rather similar to the corresponding formula for the prediction-error filter coefficients, except that the negatives of the pertinent reflection coefficients are used. Note that $d_0^{(M)} = 1$ for all M. Hence, the recursive formula for the polynomial $D_M(z)$ is as follows:

$$D_M(z) = D_{M-1}(z) - \varrho_M z^{-M} D_{M-1}^*(1/z^*). \tag{2.118}$$

Using (2.111), we find that the z transform of the autorcorrelation function is given by

$$\begin{aligned} \Psi_x(z) + \Psi_x^*(1/z^*) &= \frac{R_x(0)}{2}\left[\frac{D_M(z)}{A_M(z)} + \frac{D_M^*(1/z^*)}{A_M^*(1/z^*)}\right] \\ &= \frac{R_x(0)}{2}\,\frac{D_M(z)A_M^*(1/z^*) + A_M(z)D_M^*(1/z^*)}{A_M(z)A_M^*(1/z^*)}. \end{aligned} \tag{2.119}$$

Next, using the recursive formulas of (2.90, 118), we find that, after cancellation of terms, the numerator polynomial in the right hand (2.119) may be expressed as follows

$$\begin{aligned} &D_M(z)A_M^*(1/z^*) + A_M(z)D_M^*(1/z^*) \\ &= (1 - |\varrho_M|^2)\,[D_{M-1}(z)A_{M-1}^*(1/z^*) + A_{M-1}(z)D_{M-1}^*(1/z^*)]. \end{aligned} \tag{2.120}$$

By repeated application of this recursive formula, and starting with $M = 1$ and noting that $A_0(z) = D_0(z) = 1$, we ultimately obtain the result

$$D_M(z)A_M^*(1/z^*) + A_M(z)D_M^*(1/z^*) = 2\prod_{m=1}^{M}(1 - |\varrho_m|^2). \tag{2.121}$$

Hence, substituting (2.121) in (2.119), we get

$$\Psi_x(z) + \Psi_x^*(1/z^*) = \frac{R_x(0)\prod_{m=1}^{M}(1 - |\varrho_m|^2)}{A_M(z)A_M^*(1/z^*)}. \tag{2.122}$$

Next, we note that the spectral density estimate $\hat{S}_x(f)$ may be computed from the z transform of the autocorrelation function of (2.122) by using the formula

$$\hat{S}_x(f) = \Delta t[\Psi_x(z) + \Psi_x^*(1/z^*)]_{z = \exp(j2\pi f \Delta t)}, \tag{2.123}$$

where Δt is the sampling period. Thus, substituting (2.122) in (2.123), and using the recursive formula of (2.70), we obtain the same formula for the spectral density estimate $\hat{S}_x(f)$ which was derived previously in Sect. 2.4 using the concept of maximum entropy. We conclude, therefore, that the extension of the autocorrelation function defined by (2.106) does indeed correspond to the maximum-entropy spectrum [2.14].

2.9 Statistical Properties of the Maximum-Entropy Spectral Estimator

The value of a spectral estimate is dependent on the observations of a stochastic process in that as new sets of measurements are taken, the numerical value of the estimate changes. Since the spectral estimate, at a specified frequency, is itself a random variable, we may be interested in finding the mean and variance of the estimate, and the covariance between values of the estimate at two different frequencies. These properties determine the degree of statistical stability of the particular estimator.

Since the maximum-entropy spectral estimator is nonlinear, it is not possible to obtain general analytical expressions for its statistical properties. This is one of the major shortcomings of nonlinear spectral analysis methods. However, the asymptotic properties of the maximum-entropy spectral estimator have been determined [2.36–38] and compared with those of linear estimators based on the periodogram. It is found that the maximum-entropy estimate is asymptotically normal and asymptotically unbiased. In other words, when both the number of data samples N and order of the prediction-error filter M are sufficiently large, then

$$\mathrm{E}[\hat{S}_x(f)] = S_x(f), \tag{2.124}$$

where $S_x(f)$ is the true spectral density of the process. Furthermore, the variance of the estimate is given by

$$\mathrm{Var}[\hat{S}_x(f)] = \frac{2}{v} S_x^2(f), \tag{2.125}$$

where v is the number of degrees of freedom and is related to the order of the prediction-error filter M by $v = N/M$. Both (2.124) and (2.125) hold for large M and N and are valid when the true spectrum is reasonably smooth. The same asymptotic results are valid for periodogram-based estimators for which the number of degrees of freedom is given by $v = 2B_w T$, where T is the record length in seconds and B_w is the bandwidth of the spectral window in hertz [2.39].

It has been shown [2.37] that the covariance

$$C_{f_1, f_2} = \mathrm{Cov}[s_M(f_1) s_M(f_2)] \tag{2.126}$$

tends to zero with probability of one when $f_1 \neq f_2$, that is

$$\mathrm{Prob}[|C_{f_1,f_2}|<\varepsilon]\to 1-\eta \tag{2.127}$$

for arbitrarily small numbers ε and η. The function $s_M(f)$ in (2.126) is defined by

$$s_M(f)=\sqrt{\frac{N}{M}}\,[\hat{S}_x(f)-S_x(f)]\,, \tag{2.128}$$

where $\hat{S}_x(f)$ is the maximum-entropy spectral estimate associated with a prediction-error filter of order M.

Asymptotically then, the statistical properties of the maximum-entropy spectral estimator are approximately equivalent to those of periodogram-based estimators with a suitable chosen truncation length [2.40].

It is also of interest to note that the real and imaginary parts of the transfer function of a prediction-error filter have a joint-probability density which is a two-dimensional generalization of the noncentral student's t distribution [2.41].

2.10 Choice of Filter Order

In discussing the procedure for calculating the prediction-error filter coefficients in Sect. 2.4, 5, we did not consider the question of how long this filter should be, i.e., which value of M would be most appropriate for the design of the filter. We shall now discuss this important issue.

In order to determine the most suitable filter order M, one approach would be to try different values of M experimentally, and then choose the particular value which seems to be optimal in the sense that it satisfies some predetermined requirements for the particular situation. In general, it has been observed that, for a given value of record length N, small values of M yield spectral estimates with insufficient resolution, whereas for large values of M the estimates are statistically unstable with spurious details. Thus, it is expected that the value of the filter order M should be in the vicinity of some percentage of the record length N. Since the choice also depends on the statistical properties of the time series under analysis, it turns out that for the majority of practical measurements where the data can be considered short-term stationary, the optimum value of M lies in the range from 0.05 to $0.2N$. The experimental results included in the next section confirm the validity of this range.

However, for the proper choice of the optimal value of filter order, denoted by M_{opt}, we may use one of several objective criteria. These criteria are based on the sample moments of the given time series, and are evaluated iteratively at each value of M. When M attains the optimal value M_{opt}, the iteration process may be terminated automatically. We now discuss three specific criteria for choosing the order of a prediction-error filter:

1) Final Prediction-Error (FPE) Criterion. The first criterion by *Akaike* [2.42, 43] is defined as an estimate of the mean-square error in prediction (i.e., the prediction variance in the case of a zero-mean process) expected when a predictive filter, calculated from one observation of the process, is applied to another independent observation of the same process. For a filter of order M, the FPE is defined by, in the case of a zero-mean process,

$$\mathrm{FPE}(M)=\frac{N+M+1}{N-M-1}P_M, \tag{2.129}$$

where P_M is the output error power of the filter. Since P_M decreases with M, while the other term in (2.129) increases with M, the FPE(M) will have a minimum at some value $M=M_{\mathrm{opt}}$. This value defines the optimal value for the order of the prediction-error filter, based on the final prediction-error criterion.

2) Information Theoretic Criterion (AIC). Another criterion, also originally suggested by *Akaike* [2.44], is based on the minimization of the log-likelihood of the prediction-error variance as a function of the filter order M. This criterion, called the information theoretic criterion (AIC) is defined by

$$\mathrm{AIC}(M)=\ln(P_M)+\frac{2M}{N}. \tag{2.130}$$

Here, again, AIC(M) may be calculated at each recursion of the filter design procedure, and M_{opt} is the value for which the AIC(M), expressed as a function of M, is minimum. Note that FPE(M) and AIC(M) are asymptotically equivalent [2.45]; that is,

$$\lim_{N\to\infty}\left\{\ln[\mathrm{FPE}(M)]\right\}=AIC(M). \tag{2.131}$$

For a detailed treatment of the information theoretic criterion and its relationship to the final prediction-error criterion, see Sect. 3.6.

3) Autoregressive Transfer Function Criterion (CAT). A third criterion has been proposed by *Parzen* [2.46] and is known as the autoregressive transfer function criterion (CAT). According to this criterion, the optimal filter order M_{opt} is obtained when the estimate of the difference in the mean-square errors between the true filter, which exactly gives the prediction error, and the estimated filter, is minimum. *Parzen* [2.46] has shown that this difference can be calculated, without explicitly knowing the exact filter, by using the formula

$$\mathrm{CAT}(M)=\frac{1}{N}\sum_{m=1}^{M}\frac{N-m}{NP_m}-\frac{N-M}{NP_M}. \tag{2.132}$$

Any (or all) of the criteria described above may be calculated in each iteration of the algorithm used to design the prediction-error filter [see the flow graph of Appendix C]. The value thus obtained for the pertinent objective criterion is compared with the previously calculated value. The iteration terminates when the new value is larger than the previous one, and the optimal filter order M_{opt} is thereby obtained. The application of any of the above criteria yields a value for M_{opt} which is, in most cases, the best compromise between the variance and the bias of spectral estimate. It has been demonstrated [2.47] that for one period of a complex exponential signal, for example, all three criteria give the same value for M_{opt}.

Based on the experience gained by using these three criteria for the spectral analysis of radar signals, and the experimental results presented by *Landers* and *Lacoss* [2.47], we have adopted the use of the FPE criterion for subsequent analysis, since it gives, in general, slightly better results than the other two criteria.

To illustrate the applicability of the maximum-entropy method to the analysis of data records that can be approximated by an autoregressive model of finite order, we will consider three examples. The usefulness of the minimum FPE criterion will also be demonstrated in these examples.

Example 1: Real-Valued Autoregressive Process of Order Two

For the first example, consider a real-valued autoregressive process of order two, which is described by the difference equation

$$x_n = -a_1 x_{n-1} - a_2 x_{n-2} + w_n, \tag{2.133}$$

where $\{w_n\}$ is a Gaussian-distributed, white noise process with zero mean and variance σ_w^2. The theoretical spectral density of $\{x_n\}$ is given by

$$S_x(f) = \frac{\sigma_w^2}{|1 + a_1 \exp(-j2\pi f) + a_2 \exp(-j4\pi f)|^2}. \tag{2.134}$$

In (2.133, 134) we have put the sampling period $\Delta t = 1$ second for convenience. The values of the coefficients a_1 and a_2 are to be chosen such that the process has an autocorrelation function that is a damped sine-wave, with spectral peak at some frequency f_0, and with most of its power located at the lower frequency range. To meet these requirements the parameters a_1 and a_2 should satisfy the following relations [2.39]:

$$-(1 + a_2) < a_1 < 0, \tag{2.135}$$

$$|a_1(1 + a_2)| \leqq |4a_2|, \tag{2.136}$$

$$a_1^2 - 4a_2 < 0. \tag{2.137}$$

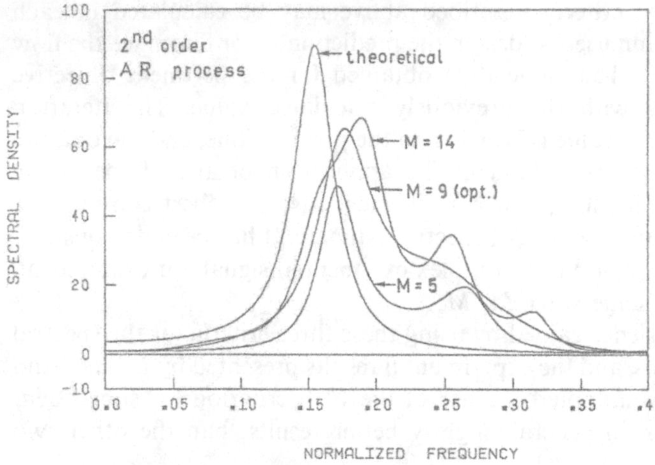

Fig. 2.7. Spectral density of a second-order real-valued autoregressive process and its maximum-entropy estimates for three values of M

The location of a spectral peak, f_0, is related to the parameters a_1 and a_2 by

$$\cos(2\pi f_0) = -\frac{a_1(1+a_2)}{4a_2}. \tag{2.138}$$

Choosing $\sigma_w^2 = 2.56$, $f_0 = 0.155$, and $a_1 = -1$, we find from (2.138) that $a_2 = 0.8$. This choice of parameters satisfies the inequalities of (2.135) to (2.137). The corresponding plot of (2.134) is shown in Fig. 2.7. This curve is labelled "theoretical". The peak value of the plot is 90 and occurs at $f_0 = 0.155$ as prescribed. The negative portion of the curve is symmetric with respect to the ordinate axis, and it was therefore not plotted.

The remaining curves represent the maximum-entropy spectral density estimates obtained from 128 samples of $\{x_n\}$ with $\{w_n\}$ obtained from a Gaussian-distributed random number generator. The first 100 samples of the time series so generated were ignored in order to reduce transient effects. The plot for $M=9$ corresponds to the optimum prediction-error filter, for which the FPE is a minimum. In Fig. 2.7 we see that for both $M=5$ and $M=14$ the maximum-entropy spectral density estimator exhibits minor peaks which do not appear in the plot for M_{opt}. The number of these peaks and their amplitudes increase with increasing filter order. Also, the location of the main peak shifts to the right and its amplitude also increases with increasing M. We may thus consider the curve corresponding to M_{opt} as an optimum estimate in the sense that it is a compromise between the fidelity of the estimate and the computation time for the filter coefficients. Note that, since the curve for FPE as a function of M usually has a relatively broad minimum, the resulting estimate is somewhat insensitive to the choice of M in the immediate vicinity of M_{opt}.

Example 2: Complex-Valued Autoregressive Process of Order Two

For the second example, consider the case of a complex-valued, second-order autoregressive process involving data generated in a manner similar to that described in the previous example. In this case, however, two different sets of random numbers are generated according to (2.133), which form the real and imaginary parts of the process [2.48]. The parameters of the process are $a_1 = -1$ and $a_2 = 0.8 - j0.2$, with σ_w^2 chosen to have the value

$$\sigma_w^2 = |1 + a_1 + a_2|^2 = 0.68 \tag{2.139}$$

so that the normalized spectral density at zero frequency is equal to unity. The theoretical spectral density of such a process, and its maximum-entropy estimates for 3 different data lengths are shown in Fig. 2.8. The theoretical spectrum exhibits assymetry in that the peak in the negative frequency region is broader, further away from the origin, and has a significantly smaller magnitude than the peak in the positive frequency region.

The estimates in Fig. 2.8a are obtained with 64 data samples of the process, with the first 50 samples ignored in order to reduce transient effects. It can be seen from Fig. 2.8a that the spectral estimates give a good representation of the theoretical spectrum in the positive frequency region, again with the closest approximation obtained for the optimal value of M. In the negative frequency region, however, the estimated peaks are shifted away from their actual

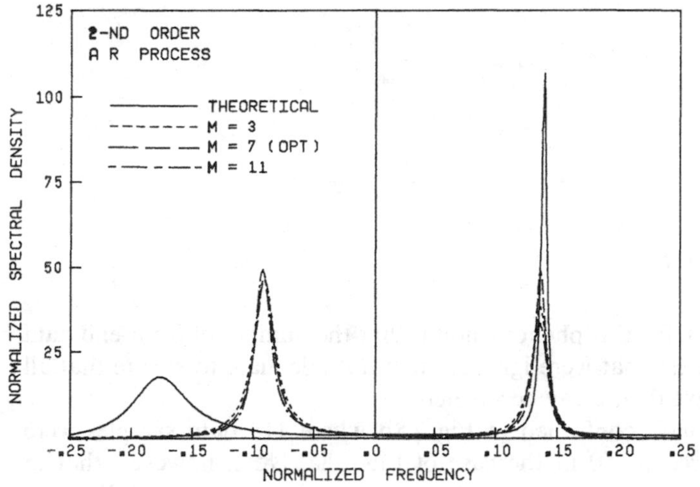

Fig. 2.8. a

Fig. 2.8. a Spectral density of a second-order complex-valued autoregressive process and its maximum-entropy estimates with the first 50 samples ignored and $N = 64$; **(b)** The same estimates as in **(a)** except that the first 114 data samples are ignored; **(c)** The same estimates with the first 114 data samples are ignored and $N = 192$

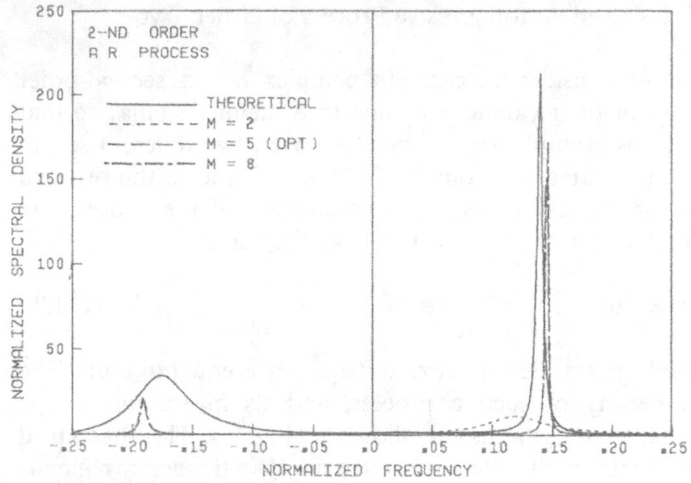

Fig. 2.8. b (Caption is on p. 45)

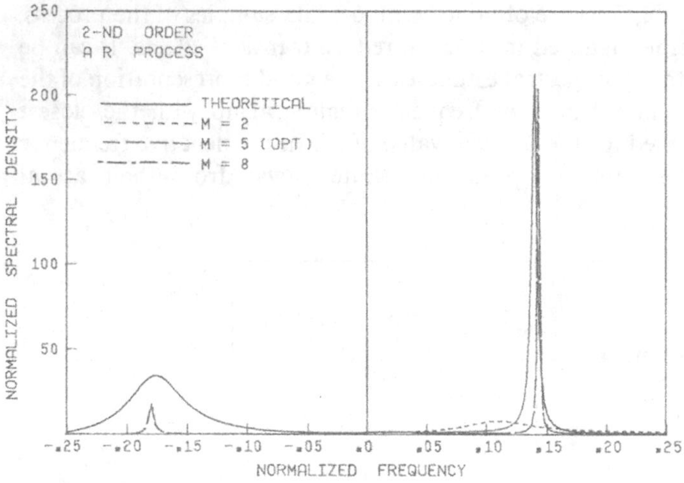

Fig. 2.8. c (Caption is on p. 45)

position. The reason for this phenomenon is that the number of front-end data samples of the process that were ignored were not adequate to ensure that all transient effects have died away completely.

This observation is confirmed in Fig. 2.8b where 114 data samples were ignored as compared to 50 in the case of Fig. 2.8a. Note, however, that in Fig. 2.8b the optimum length of the prediction-error filter is equal to 5. We thus find that in Fig. 2.8b the maximum-entropy estimates provide a closer approximation to the actual position of the peak for negative frequences, for both the optimal value of M and for $M=8$, than the previous estimates in Fig. 2.8a. Note also that the filter of order two gives a poor estimate of the peak in the positive

Table 2.1. Parameter values for the 4th-order autoregressive process

Parameter	a_1	a_2	a_3	a_4	σ_w	σ_w^2
Value	1.6	2.4	1.44	0.81	7.25	52.56

frequency region, and that the peak in the negative frequency region is not resolved at all.

Provided that an adequate number of front-end data samples are ignored so as to make transient effects negligible, the maximum-entropy estimates are relatively insensitive to the number of data samples used for driving the estimators. This is confirmed in Fig. 2.8c where the number of data samples ignored is the same as that in Fig. 2.8b, but the number of data samples used in the spectral analysis is three times as many.

Example 3: Real-Valued Autoregressive Process of Order Four

For the third example, consider a real-valued fourth-order autoregressive process described by the difference equation

$$x_n = -a_1 x_{n-1} - a_2 x_{n-2} - a_3 x_{n-3} - a_4 x_{n-4} + w_n \tag{2.140}$$

with a corresponding spectral density defined by

$$S_x(f) = \frac{\sigma_w^2}{|1 + a_1 \exp(-j2\pi f) + a_2 \exp(-j4\pi f) + a_3 \exp(-j6\pi f) + a_4 \exp(-j8\pi f)|^2}. \tag{2.141}$$

The parameters a_m, $m = 1, 2, 3, 4$ were chosen in such a way that we have two spectral peaks at the normalized frequences $f_1 = 0.3$ and $f_2 = 0.33$, with respective magnitudes of 8.3 and 10.0 [2.49]. The standard deviation of the white noise process $\{w_n\}$ was chosen to be $\sigma_w = |1 + a_1 + a_2 + a_3 + a_4|$ so that the value for $S_x(0)$ is normalized to unity. The values of the set of parameters $\{a_m\}$ are listed in Table 2.1.

The curve labelled "theoretical" in Fig. 2.9 is a plot of (2.141), while the remaining three curves represent the maximum-entropy estimates obtained from 128 samples of $\{x_n\}$. The random number generator used for generating $\{w_n\}$ was the same as in the previous two examples. Again, the plots are shown for positive frequencies only. In Fig. 2.9 we see that for $M = 3$ (the value which is less than the theoretical order of the autoregressive process), the existing peaks have not been resolved. Also, the single peak is shifted to the left from the larger peak by approximately ten percent of the full frequency range. With larger values of M, we see that both peaks can be resolved.

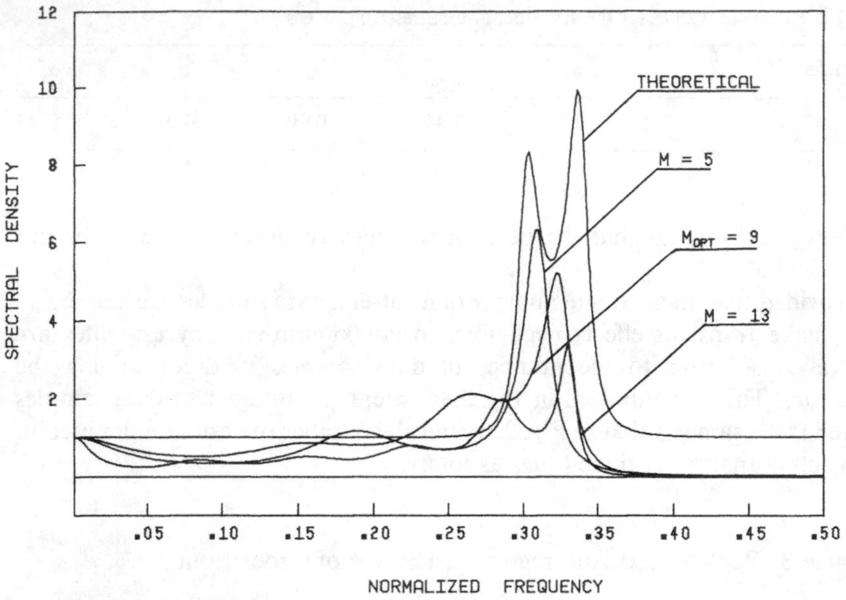

Fig. 2.9. Spectral density of a fourth-order real-valued autoregressive process and its maximum-entropy estimates for three values of M

Table 2.2. Locations and magnitudes of the spectral peaks for the theoretical 4th-order autoregressive spectrum and its maximum-entropy estimates with $M=5$, 9(opt), and 13

Parameter values	Peak I			Peak II		
Spectra	f_1	$\lvert f_1-f_{1t}\rvert$ [%]	$S_x(f_1)$	f_2	$\lvert f_2-f_{2t}\rvert$ [%]	$S_x(f_2)$
Theoretical	0.300	0.0	8.3	0.330	0.0	10.0
$M=5$	–	–	–	0.273	10.4	4.4
$M_{\text{opt}}=9$	0.257	8.4	1.8	0.321	1.8	4.9
$M=13$	0.280	4.0	2.2	0.327	0.6	4.0

Table 2.2 summarizes the locations and magnitudes of both peaks, as well as the percentage shifts from the true peak locations. We see that by increasing the filter order M, the positions of the maxima in the corresponding spectral estimates become closer to the exact positions. In this sense, the estimate corresponding to $M=13$ is better than the one for the optimal value of M. It should be noted, however, that by increasing M, the number of minor peaks in the left-hand portion of the diagram also increases.

2.11 Experimental Classification of Radar Clutter

In this section we will present some experimental results obtained by using the maximum-entropy spectral estimator to classify the different forms of clutter as encountered in an air traffic control radar environment. The term clutter is used to describe unwanted echoes on a radar display due to reflection of the transmitted wave from such objects as ground, weather disturbances, or migrating flocks of birds [2.50].

2.11.1 Modeling of Radar Clutter as an Autoregressive Process

According to *Barlow* [2.51], the spectral density of a continuous-time radar clutter process due to weather disturbances may be assumed to have a Gaussian shape, as shown by

$$S(f)=S(0)\exp\left(-\frac{f^2}{2\sigma_f^2}\right), \tag{2.142}$$

where $S(0)$ is the value of the spectral density at zero frequency, and σ_f is the spectral spread. This Gaussian-shaped spectral density, however, can be closely represented by an autoregressive model of relatively low order. For example, we may approximate a normalized form of the Gaussian function as follows [2.52]:

$$\frac{1}{\sqrt{2\pi}}\exp\left(-\frac{f^2}{2}\right)=\frac{1}{\alpha_0+\alpha_2 f^2+\alpha_4 f^4+\alpha_6 f^6}+\varepsilon(f), \tag{2.143}$$

where

$$\begin{aligned}\alpha_0 &= 2.490895\\ \alpha_2 &= 1.466003\\ \alpha_4 &= -0.024393\\ \alpha_6 &= 0.178257\end{aligned}$$

and

$$|\varepsilon(f)|\leq 2.7\times 10^{-3}.$$

Thus, ignoring the small error $\varepsilon(f)$, the remaining rational function on the right-hand side of (2.143) represents the spectral density of a continuous autoregressive process of order 3.

It appears therefore that the application of the maximum-entropy method to the spectral analysis of radar clutter is theoretically justified. Furthermore, the results obtained by using computer-simulated radar clutter data [2.48, 49, 53] and recorded clutter data [2.54, 55] confirm the practical usefulness of the method.

2.11.2 Experimental Results

The results reported in this section are based on clutter data obtained by using an airport surveillance radar ASR-8 located at Bagotville, Quebec [2.56, 57]. This radar is coherent, generating in-phase and quadrature components of the bipolar video signal.

Figure 2.10 shows the spectral density estimates of ground clutter obtained with the maximum-entropy method, using the minimum FPE criterion, for different numbers of data samples. The frequency values are normalized with respect to the average pulse-repetition frequency of the radar (equal to 1.04 kHz). Spectral density values are also normalized with respect to the maximum value, so that they extend from 0 dB down to the limiting value which is selected in advance. These normalizations are applied in all the figures that follow. A feature that is immediately obvious from these curves is the presence of a very narrow central peak at zero frequency. Furthermore, there is not much difference between the curves for $N=256$, 100 and 16. It is only when the number of data samples is low ($N=8$) that we find the spectral spread is relatively wide and sidelobe level high.

Figure 2.11 shows the maximum-entropy spectral density estimates of weather clutter for different numbers of data samples. Here, again, we see that the spectral change for different values of data samples is rather small. We also see that for weather clutter the spectral spread is larger and the sidelobe level is higher than the corresponding values for ground clutter.

Figure 2.12 shows the maximum-entropy spectral density estimates of clutter produced by migrating birds. Comparing this set of curves with those for ground and weather clutter, we see that: 1) the spectral spread is significantly wider, 2) the sidelobe levels are higher, and 3) some of the main peaks are shifted away from zero frequency, thereby resulting in a pronounced asymmetry.

The features which distinguish the maximum-entropy spectral density estimates of clutter due to ground, weather, and migrating flocks of birds are further illustrated in Fig. 2.13 for the case of 16 data samples. Figure 2.14 shows the corresponding spectral density estimates obtained by using the modified periodogram method [2.58] based on the fast Fourier transform. These two sets of curves clearly illustrate the superiority of the maximum-entropy method over the modified periodogram method in the spectral classification of different forms of radar clutter.

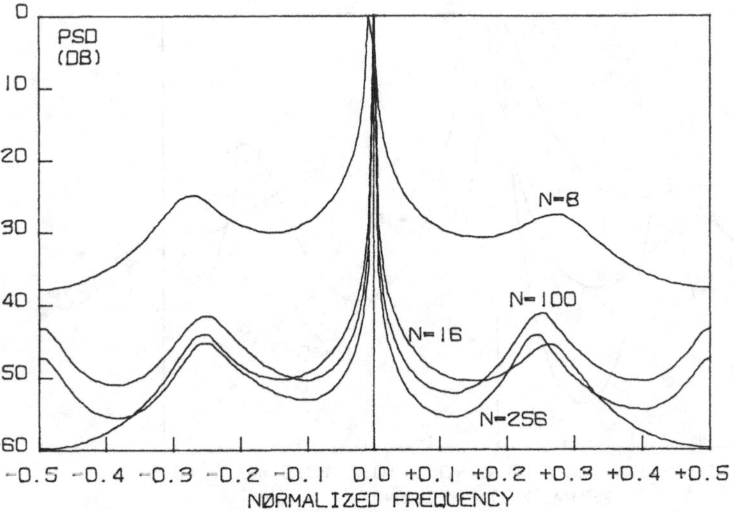

Fig. 2.10.
Maximum-entropy spectral estimates of ground clutter for varying values of data length N

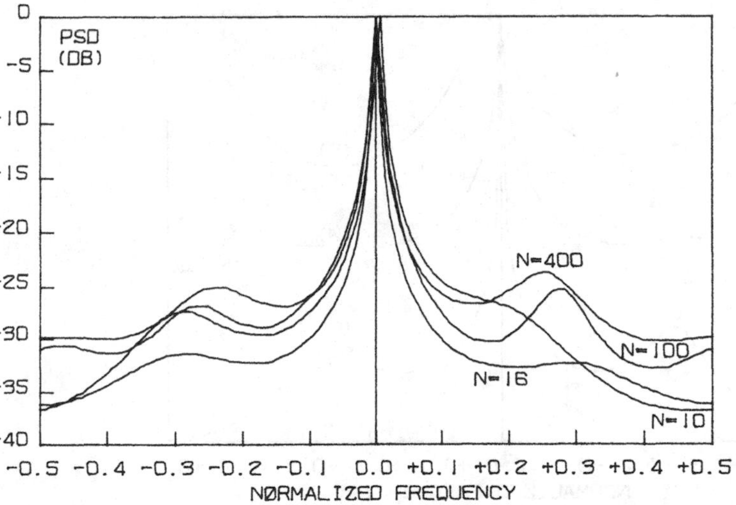

Fig. 2.11.
Maximum-entropy spectral estimates of weather clutter for varying values of data length N

The qualitative results described above indicate that it is possible to provide a practical means for the on-line classification of the various types of clutter, based on the maximum-entropy analysis of spectral density. In order to provide a feeling for the statistical variability of spectral spread for different clutter conditions and record lengths, a large number of spectra were calculated and the spectral spread, corresponding to the $-10\,\mathrm{dB}$ points, were extracted.

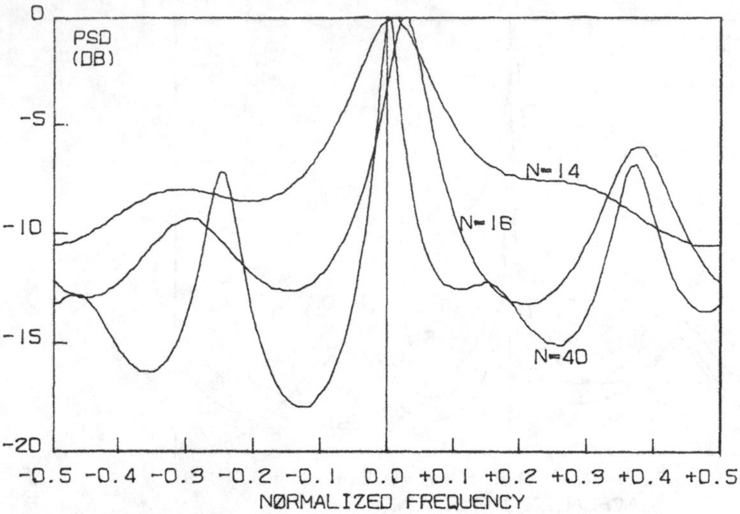

Fig. 2.12. Maximum-entropy spectral estimates of clutter due to birds for varying values of data length N

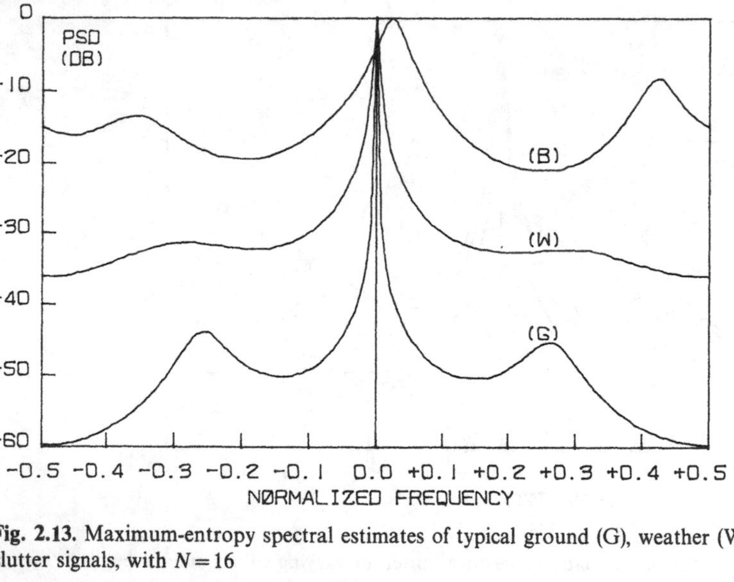

Fig. 2.13. Maximum-entropy spectral estimates of typical ground (G), weather (W), and bird (B) clutter signals, with $N = 16$

Figure 2.15 illustrates the differences in the statistical averages of spectral spread for the three types of clutter for 16 data samples. The solid vertical lines in the figure represent the sample mean, while the dotted lines represent the boundary values determined by standard diviations. Figure 2.15 further supports the usefulness of the maximum-entropy spectral estimator as a practical method for the classification of radar clutter.

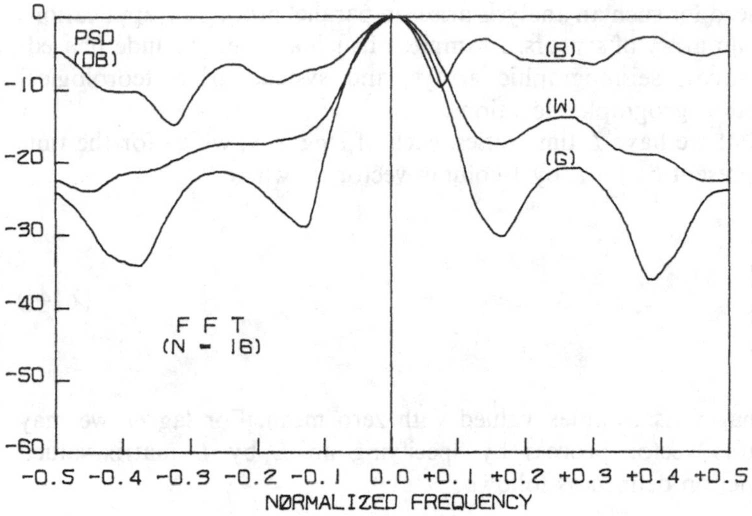

Fig. 2.14. Welch's spectral estimates of the same data as in Fig. 2.13

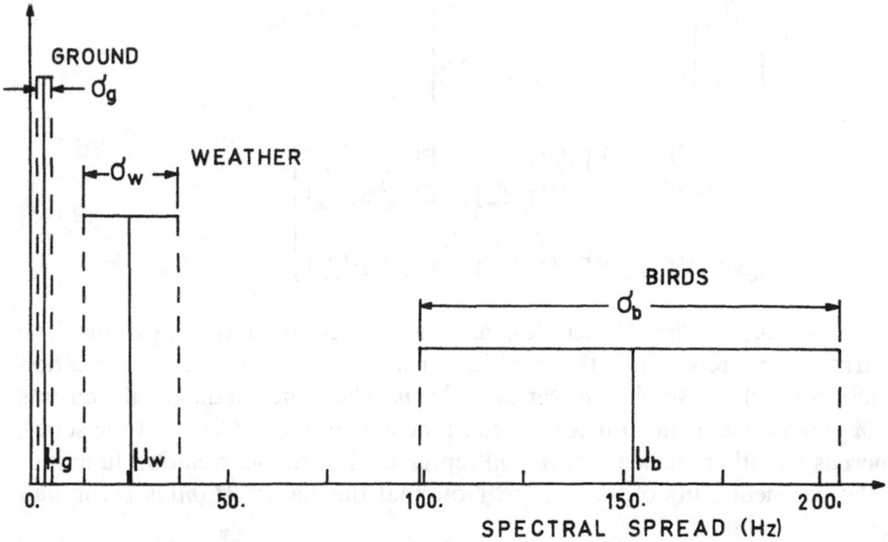

Fig. 2.15. Comparison of statistical averages of spectral spread for ground, weather and birds

2.12 Multichannel Maximum-Entropy Spectral Estimation

The material presented above applies to the spectral analysis of a scalar stationary discrete-time process using the maximum-entropy method. In this section we will describe the vector or multichannel extension of this method in order to evaluate the composite spectrum of a vector stationary discrete-time process. The need for such an analysis arises in parallel processing applications which involve an array of sensors. Examples of such a system include phased-array radar, sonar, seismographic arrays, and systems of meteorological stations at various geographic locations.

Suppose that we have L time series, each of length N, which for the time index n we represent by an L by 1 column vector $\boldsymbol{x}_n$, where

$$\boldsymbol{x}_n = \begin{bmatrix} x_n^{(1)} \\ x_n^{(2)} \\ \vdots \\ x_n^{(L)} \end{bmatrix}. \tag{2.144}$$

We assume that $\boldsymbol{x}_n$ is complex valued with zero mean. For lag m, we may characterize this vector process by specifying an L by L matrix-valued correlation function defined as follows:

$$\begin{aligned} \mathscr{R}_x(m) &= \mathrm{E}[\boldsymbol{x}_n \boldsymbol{x}_{n-m}^{\mathrm{H}}] \\ &= \mathrm{E}\left\{ \begin{bmatrix} x_n^{(1)} \\ x_n^{(2)} \\ \vdots \\ x_n^{(L)} \end{bmatrix} [x_{n-m}^{(1)*} \quad x_{n-m}^{(2)*} \ldots x_{n-m}^{(L)*}] \right\} \\ &= \begin{bmatrix} \mathrm{E}[x_n^{(1)} x_{n-m}^{(1)*}] & \mathrm{E}[x_n^{(1)} x_{n-m}^{(2)*}] \ldots & \mathrm{E}[x_n^{(1)} x_{n-m}^{(L)*}] \\ \mathrm{E}[x_n^{(2)} x_{n-m}^{(1)*}] & \mathrm{E}[x_n^{(2)} x_{n-m}^{(2)*}] \ldots & \mathrm{E}[x_n^{(2)} x_{n-m}^{(L)*}] \\ \vdots & \vdots \quad \ddots & \vdots \\ \mathrm{E}[x_n^{(L)} x_{n-m}^{(1)*}] & \mathrm{E}[x_n^{(L)} x_{n-m}^{(2)*}] \ldots & \mathrm{E}[x_n^{(L)} x_{n-m}^{(L)*}] \end{bmatrix}, \end{aligned} \tag{2.145}$$

where the superscript H denotes, as before, Hermitian transposition. The matrix $\mathscr{R}_x(m)$ represents the multichannel analog of the autocorrelation function $R_x(m)$ of a single time series for lag m. The principal diagonal elements of $\mathscr{R}_x(m)$ represent the autocorrelation functions of the individual time series, whereas the other elements of $\mathscr{R}_x(m)$ represent the cross-correlation functions of the pertinent pairs of time series. Note that the matrix $\mathscr{R}_x(m)$ is Hermitian but not Toeplitz.

The correlation matrix of the vector process of interest is, for the case of a multichannel prediction-error filter of order M, defined by the $M+1$ by $M+1$ block matrix:

$$\mathscr{R}_x = \begin{bmatrix} \mathscr{R}_x(0) & \mathscr{R}_x(-1) & \ldots & \mathscr{R}_x(-M) \\ \mathscr{R}_x(1) & \mathscr{R}_x(0) & \ldots & \mathscr{R}_x(1-M) \\ \vdots & \vdots & \ddots & \vdots \\ \mathscr{R}_x(M) & \mathscr{R}_x(M-1) & \ldots & \mathscr{R}_x(0) \end{bmatrix}. \tag{2.146}$$

Comparing this expression with the corresponding formula for the autocorrelation matrix of a scalar process, we find that in the right-hand side of (2.146) the entries are L by L matrices, whereas in the case of a scalar process the entries are scalar quantities. The correlation matrix $\mathcal{R}_x$ is clearly block-Toeplitz.

The multichannel version of a forward prediction-error filter of order M is defined by matrix-valued coefficients $\boldsymbol{A}_m^{(M)}$, $m=0,1,\ldots,M$, each of size L by L, with $\boldsymbol{A}_0^{(M)}=\boldsymbol{I}$, an identity matrix. Similarly, the multichannel backward prediction-error filter of order M is defined by a set of L by L matrices $\boldsymbol{B}_m^{(M)}$, $m-0,1,\ldots,M \Rightarrow m=0,1,\ldots,M$, with $\boldsymbol{B}_0^{(M)}=\boldsymbol{I}$. It is important to note that the multichannel forward and backward prediction-error filter coefficients are not related simply by complex conjugation and time reversal, as in the single-channel case. They consist of two separate sets of matrix-valued coefficients and need to be calculated simultaneously at each recursion step. The multichannel transfer functions of the forward and backward prediction-error filters are given by z transforms whose coefficients are made up of matrices as follows, respectively,

$$\boldsymbol{A}_M(z)=\boldsymbol{I}+\boldsymbol{A}_1^{(M)}z^{-1}+\ldots+\boldsymbol{A}_{M-1}^{(M)}z^{1-M}+\boldsymbol{A}_M^{(M)}z^{-M} \tag{2.147}$$

and

$$\boldsymbol{B}_M(z)=\boldsymbol{B}_M^{(M)}+\boldsymbol{B}_{M-1}^{(M)}z^{-1}+\ldots+\boldsymbol{B}_1^{(M)}z^{1-M}+\boldsymbol{I}z^{-M}, \tag{2.148}$$

where $\boldsymbol{A}_M(z)$ and $\boldsymbol{B}_M(z)$ are L by L matrices.

Based on the definitions of (2.72, 74) for the forward and backward prediction-error series in the single-channel case, we may next write the corresponding vector-valued expressions for a vector process as follows:

$$\boldsymbol{e}_{\mathrm{f},n}^{(M)}=\sum_{m=0}^{M}\boldsymbol{A}_m^{(M)}\boldsymbol{x}_{n-m} \tag{2.149}$$

and

$$\boldsymbol{e}_{\mathrm{b},n}^{(M)}=\sum_{m=0}^{M}\boldsymbol{B}_{M-m}^{(M)}\boldsymbol{x}_{n-m}, \tag{2.150}$$

where $\boldsymbol{e}_{\mathrm{f},n}^{(M)}$ and $\boldsymbol{e}_{\mathrm{b},n}^{(M)}$ are L by 1 column vectors. The corresponding expressions for the prediction-error powers are given by the L by L matrices

$$\boldsymbol{P}_{\mathrm{f},M}=\mathrm{E}[(\boldsymbol{e}_{\mathrm{f},n}^{(M)})(\boldsymbol{e}_{\mathrm{f},n}^{(M)})^{\mathrm{H}}] \tag{2.151}$$

and

$$\boldsymbol{P}_{\mathrm{b},M}=\mathrm{E}[(\boldsymbol{e}_{\mathrm{b},n}^{(M)})(\boldsymbol{e}_{\mathrm{b},n}^{(M)})^{\mathrm{H}}]. \tag{2.152}$$

Using the above definitions (2.146, 151, 152), we may now write the multichannel version of the prediction-error filter equations as follows [2.59]:

$$\mathscr{R}_x \begin{bmatrix} \boldsymbol{I} & \boldsymbol{B}_M^{(M)} \\ \boldsymbol{A}_1^{(M)} & \boldsymbol{B}_{M-1}^{(M)} \\ \vdots & \vdots \\ \boldsymbol{A}_M^{(M)} & \boldsymbol{I} \end{bmatrix} = \begin{bmatrix} \boldsymbol{P}_{\mathrm{f},M} & \boldsymbol{0} \\ \boldsymbol{0} & \boldsymbol{0} \\ \vdots & \\ \boldsymbol{0} & \boldsymbol{P}_{\mathrm{b},M} \end{bmatrix}, \tag{2.153}$$

where the submatrices $\boldsymbol{0}$ represent L by L null matrices. The maximum-entropy extension of the sequence $\mathscr{R}_x(m)$, for $|m| \leq M$, is defined by the matrix-valued z transform [2.14, 59]

$$\begin{aligned}\sum_{m=-\infty}^{\infty} \mathscr{R}_x(m) z^{-m} &= \boldsymbol{A}_M^{-1}(z) \boldsymbol{P}_{\mathrm{f},M} \boldsymbol{A}_M^{-\mathrm{H}}(1/z^*) \\ &= \boldsymbol{B}_M^{-1}(z) \boldsymbol{P}_{\mathrm{b},M} \boldsymbol{B}_M^{-\mathrm{H}}(1/z^*),\end{aligned} \tag{2.154}$$

where $\boldsymbol{A}^{-1}$ is the inverse of $\boldsymbol{A}$ and $\boldsymbol{A}^{-\mathrm{H}} = (\boldsymbol{A}^{-1})^{\mathrm{H}}$ denotes the Hermitian transpose of the inverse of A. The L by L matrices $\boldsymbol{A}_M(z)$, $\boldsymbol{B}_M(z)$, $\boldsymbol{P}_{\mathrm{f},M}$, and $\boldsymbol{P}_{\mathrm{b},M}$ are defined by (2.147, 148, 151, 152), respectively. Note that, for a given sampling period Δt, by substituting $z = \exp(\mathrm{j}2\pi\Delta t f)$ in (2.154), we obtain an L by L matrix $\hat{S}_x(f)$ which defines the maximum-entropy estimate of the composite power spectrum of a prescribed vector process consisting of L individual time series.

The set of normal equations (2.153) can be solved recursively by using the multichannel version of Levinson's algorithm [2.60]. We shall refer to this algorithm as the Levinson, Wiggins, Robinson (LWR) algorithm. Knowing the values for $\boldsymbol{A}_m^{(M-1)}$, $\boldsymbol{B}_m^{(M-1)}$, $\mathscr{R}_x(m)$, $\boldsymbol{P}_{f,M-1}$, and $\boldsymbol{P}_{b,M-1}$ from recursion $M-1$, we may use the LWR algorithm to calculate the corresponding values for recursion M as follows. We first define the L by L matrix

$$\boldsymbol{\Delta}_M = \sum_{m=0}^{M-1} \boldsymbol{A}_m^{(M-1)} \boldsymbol{R}_x(M-m) = \sum_{m=0}^{M-1} \boldsymbol{B}_m^{(M-1)} \mathscr{R}_x(m-M) \tag{2.155}$$

and then calculate the L by L matrix-valued coefficients for the forward and backward prediction-error filter by using the recursive formulas

$$\boldsymbol{A}_m^{(M)} = \boldsymbol{A}_m^{(M-1)} - \boldsymbol{\Delta}_M \boldsymbol{P}_{\mathrm{b},M-1}^{-1} \boldsymbol{B}_{M-m}^{(M-1)}, \quad m = 0, 1, \ldots, M \tag{2.156}$$

and

$$\boldsymbol{B}_m^{(M)} = \boldsymbol{B}_m^{(M-1)} - \boldsymbol{\Delta}_M^{\mathrm{H}} \boldsymbol{P}_{\mathrm{f},M-1}^{-1} \boldsymbol{A}_{M-m}^{(M-1)}, \quad m = 0, 1, \ldots, M, \tag{2.157}$$

where, $\boldsymbol{A}_0^{(M)} = \boldsymbol{B}_0^{(M)} = \boldsymbol{I}$ for all M, and $\boldsymbol{A}_m^{(M)} = \boldsymbol{B}_m^{(M)} = \boldsymbol{0}$ for $m > M$. Next, we update the L by L forward and backward prediction-error power matrices by using the recursive formulas

$$P_{f,M} = P_{f,M-1} - \Delta_M P_{b,M-1}^{-1} \Delta_M^H \tag{2.158}$$

and

$$P_{b,M} = P_{b,M-1} - \Delta_M^H P_{f,M-1}^{-1} \Delta_M . \tag{2.159}$$

The algorithm given by (2.155–159) is completed by introducing the initial values

$$A^{(0)} = B^{(0)} = I \tag{2.160}$$

and

$$P_{f,0} = P_{b,0} = \mathscr{R}_x(0) . \tag{2.161}$$

The main shortcoming of the LWR algorithm is that we have two sequences of matrix reflection coefficients: one corresponding to forward prediction, namely, $A_M^{(M)}$, and the other corresponding to backward prediction, namely, $B_M^{(M)}$. We may estimate these sequences separately but unfortunately they will yield different estimates of the power spectrum. This problem does not arise in the single-channel case because there is only a single set of reflection coefficients to deal with.

The above difficulty may be overcome by using a suitably normalized form of the LWR algorithm [2.61], by means of which we obtain a single sequence of matrix-valued reflection coefficients uniquely associated with a vector process. Each of the matrix-valued reflection coefficients obtained in this manner has singular values less than or equal to unity in magnitude. The singular values of a matrix A are defined as the positive square roots of the eigenvalues of AA^H. The normalized set of matrix reflection coefficients can be estimated directly from the given data, without having to compute the correlation matrix of the process. These estimates yield a unique multichannel minimum-phase prediction-error filter, and thereby a unique multichannel spectral estimate.

In order to outline the normalized form of the LWR algorithm we first define, for any positive-definite matrix R, the square-root matrix $R^{1/2}$ as a lower triangular matrix satisfying the condition

$$R^{1/2} R^{H/2} = R , \tag{2.162}$$

where $R^{H/2} = (R^{1/2})^H$ is the Hermitian transpose of $R^{1/2}$. For convenience we also let $R^{-1/2} = (R^{1/2})^{-1}$ and $R^{-H/2} = (R^{-1/2})^H = [(R^{1/2})^{-1}]^H$. The matrix $R^{1/2}$ may be made unique by requiring the diagonal elements to be positive.

The normalized form of the LWR algorithm involves the use of two L by L normalization matrices $\boldsymbol{P}_M^{1/2}$ and $\boldsymbol{Q}_M^{1/2}$ defined as follows, respectively:

$$\boldsymbol{P}_M^{1/2} = \boldsymbol{P}_{\mathrm{f},M-1}^{-1/2}\boldsymbol{P}_{\mathrm{f},M}^{1/2} \tag{2.163}$$

and

$$\boldsymbol{Q}_M^{1/2} = \boldsymbol{P}_{\mathrm{b},M-1}^{-1/2}\boldsymbol{P}_{\mathrm{b},M}^{1/2} \tag{2.164}$$

with the initial conditions

$$\boldsymbol{P}_0^{1/2} = \boldsymbol{Q}_0^{1/2} = \boldsymbol{\mathcal{R}}_x^{1/2}(0)\,. \tag{2.165}$$

For a multichannel prediction-error filter of order M, we may also define an L by L matrix of normalized reflection coefficients as follows:

$$\boldsymbol{\varrho}_M = \boldsymbol{P}_{\mathrm{f},M-1}^{-1/2}\boldsymbol{\Delta}_M\boldsymbol{P}_{\mathrm{b},M-1}^{-\mathrm{H}/2}\,, \tag{2.166}$$

where the L by L matrix $\boldsymbol{\Delta}_M$ is given by (2.155). Combining (2.155, 158, 159, 163, 164, 166), we thus find that

$$\boldsymbol{P}_M = \boldsymbol{I} - \boldsymbol{\varrho}_M\boldsymbol{\varrho}_M^{\mathrm{H}} \tag{2.167}$$

and

$$\boldsymbol{Q}_M = \boldsymbol{I} - \boldsymbol{\varrho}_M^{\mathrm{H}}\boldsymbol{\varrho}_M\,. \tag{2.168}$$

Using the matrices $\boldsymbol{P}_M^{1/2}$, $\boldsymbol{Q}_M^{1/2}$, and $\boldsymbol{\varrho}_M$, we may next define the following sets of normalized formulas:

a) For forward and backward prediction, the normalized matrix-valued prediction-error filter coefficients are respectively computed by using the recursive relations

$$\tilde{\boldsymbol{A}}_m^{(M)} = \boldsymbol{P}_M^{-1/2}[\tilde{\boldsymbol{A}}_m^{(M-1)} - \boldsymbol{\varrho}_M\tilde{\boldsymbol{B}}_{M-m}^{(M-1)}]\,, \quad m = 0, 1, \ldots, M-1 \tag{2.169}$$

and

$$\tilde{\boldsymbol{B}}_m^{(M)} = \boldsymbol{Q}_M^{-1/2}[\tilde{\boldsymbol{B}}_m^{(M-1)} - \boldsymbol{\varrho}_M^{\mathrm{H}}\tilde{\boldsymbol{A}}_{M-m}^{(M-1)}]\,, \quad m = 0, 1, \ldots, M-1\,, \tag{2.170}$$

where the tilde signifies that the pertinent quantities are normalized. For a multichannel prediction-error filter of order $M=0$, we have the initial conditions

$$(\tilde{\boldsymbol{A}}_0^{(0)})^{-1} = \left(\sum_{n=1}^{N} \boldsymbol{x}_n\boldsymbol{x}_n^{\mathrm{H}}\right)^{1/2} \tag{2.171}$$

and

$$(\tilde{\boldsymbol{B}}_0^{(0)})^{-1} = \left(\sum_{n=0}^{N-1} \boldsymbol{x}_n \boldsymbol{x}_n^{\mathrm{H}}\right)^{1/2}. \tag{2.172}$$

b) The normalized forms of L by L matrix-valued transfer functions of the forward and backward multichannel prediction-error filters are respectively computed by using the recursive relations

$$\tilde{\boldsymbol{A}}_M(z) = \boldsymbol{P}_M^{-1/2}[\tilde{\boldsymbol{A}}_{M-1}(z) - z^{-1}\boldsymbol{\varrho}_M \tilde{\boldsymbol{B}}_{M-1}(z)] \tag{2.173}$$

and

$$\tilde{\boldsymbol{B}}_M(z) = \boldsymbol{Q}_M^{-1/2}[z^{-1}\tilde{\boldsymbol{B}}_{M-1}(z) - \boldsymbol{\varrho}_M^{\mathrm{H}} \tilde{\boldsymbol{A}}_{M-1}(z)]. \tag{2.174}$$

For $M=0$, we have the initial conditions

$$\tilde{\boldsymbol{A}}_0(z) = \tilde{\boldsymbol{B}}_0(z) = \boldsymbol{\mathcal{R}}_x^{-1/2}(0). \tag{2.175}$$

c) The normalized forms of forward and backward prediction-error vectors (of dimensions L by 1) are respectively computed by using the relations

$$\tilde{\boldsymbol{e}}_{\mathrm{f},n}^{(M)} = \sum_{m=0}^{M} \tilde{\boldsymbol{A}}_m^{(M)} \boldsymbol{x}_{n-m} \tag{2.176}$$

and

$$\tilde{\boldsymbol{e}}_{\mathrm{b},n}^{(M)} = \sum_{m=0}^{M} \tilde{\boldsymbol{B}}_{M-m}^{(M)} \boldsymbol{x}_{n-m}. \tag{2.177}$$

d) The normalized forms of forward and backward error-power matrices as well as cross-power matrix (of dimensions L by L) are respectively computed by using the relations

$$\tilde{\boldsymbol{P}}_{\mathrm{f},M} = \sum_{n=M+1}^{N} \tilde{\boldsymbol{e}}_{\mathrm{f},n}^{(M)} (\tilde{\boldsymbol{e}}_{\mathrm{f},n}^{(M)})^{\mathrm{H}}, \tag{2.178}$$

$$\tilde{\boldsymbol{P}}_{\mathrm{b},M} = \sum_{n=M+1}^{N} \tilde{\boldsymbol{e}}_{\mathrm{b},n-1}^{(M)} (\tilde{\boldsymbol{e}}_{\mathrm{b},n-1}^{(M)})^{\mathrm{H}}, \tag{2.179}$$

and

$$\tilde{\boldsymbol{P}}_{\mathrm{fb},M} = \sum_{n=M+1}^{N} \tilde{\boldsymbol{e}}_{\mathrm{f},n}^{(M)} (\tilde{\boldsymbol{e}}_{\mathrm{b},n-1}^{(M)})^{\mathrm{H}}. \tag{2.180}$$

To estimate the normalized matrix-valued reflection coefficient, we may use a weighted geometric mean criterion, whereby in a particular recursion step the matrix-valued reflection coefficient is obtained by minimizing the geometric mean of the forward and backward prediction-error powers, weighted by their respective estimates from the previous step. It has been shown by *Morf* et al. [2.59] that the reflection coefficients obtained in this way are, except for a minus sign, equal to the geometric mean values of the forward and backward reflection coefficients obtained by minimizing the forward and backward prediction-error powers, respectively. That is, the weighted geometric mean criterion represents the multichannel version of the geometric mean method briefly described in Sect. 2.5. Thus, for an estimate of the normalized matrix-valued reflection coefficient ϱ_{M+1}, we may use the formula [2.59]

$$\varrho_{M+1} = \tilde{\boldsymbol{P}}_{\mathrm{f},M}^{-1/2} \tilde{\boldsymbol{P}}_{\mathrm{fb},M} \tilde{\boldsymbol{P}}_{\mathrm{b},M}^{-\mathrm{H}/2} . \tag{2.181}$$

As with the single-channel case, an important issue is to have a procedure for choosing the optimal filter order M_{opt}. Here again, we may use the multichannel version of the final prediction-error criterion. That is, we may define the optimal filter order, M_{opt}, as the particular value of filter order M for which the final prediction error

$$\mathrm{FPE}(M,L) = \left(\frac{N+LM+1}{N-LM-1}\right)^{L} \det\left[\tfrac{1}{2}(\boldsymbol{P}_{M+1} + \boldsymbol{Q}_{M+1})\right] \tag{2.182}$$

is minimum [2.62, 63].

Using the above definitions, and formulas, we may now outline the so-called square-root normalized MEM algorithm for computing the normalized matrix-valued reflection coefficients directly from the observed data [2.59]:

1) For a multichannel prediction-error filter of order $M=0$, use (2.171, 172) to compute $\tilde{\boldsymbol{A}}_0^{(0)}$ and $\tilde{\boldsymbol{B}}_0^{(0)}$. Then, use (2.176, 177) to compute $\tilde{\boldsymbol{e}}_{\mathrm{f},n}^{(0)}$ and $\tilde{\boldsymbol{e}}_{\mathrm{b},n}^{(0)}$. Next, use (2.178–180) to compute $\tilde{\boldsymbol{P}}_{\mathrm{f},0}$, $\tilde{\boldsymbol{P}}_{\mathrm{b},0}$, and $\tilde{\boldsymbol{P}}_{\mathrm{fb},0}$. Hence, use (2.181) to compute ϱ_1.

2) Substituting this value for ϱ_1 in (2.167, 168), compute $\boldsymbol{P}_1$ and $\boldsymbol{Q}_1$.

3) Use (2.182) to compute FPE$(0, L)$.

4) Increment the filter order M by 1, and use (2.169, 170) to compute $\tilde{\boldsymbol{A}}_m^{(M)}$ and $\tilde{\boldsymbol{B}}_m^{(M)}$ for $m=0, 1, \ldots, M-1$. Then, use (2.176, 177) to compute $\tilde{\boldsymbol{e}}_{\mathrm{f},n}^{(M)}$, and $\tilde{\boldsymbol{e}}_{\mathrm{b},n}^{(M)}$, and use these results in (2.178–180) to compute $\tilde{\boldsymbol{P}}_{\mathrm{f},M}$, $\tilde{\boldsymbol{P}}_{\mathrm{b},M}$, and $\tilde{\boldsymbol{P}}_{\mathrm{fb},M}$. Next, use (2.181) to compute ϱ_{M+1}, and (2.167, 168) to compute $\boldsymbol{P}_{M+1}$ and $\boldsymbol{Q}_{M+1}$.

5) Use (2.182) to compute FPE(M, L), and compare it with the previously computed value FPE(M', L), where $M'=M-1$. If FPE$(M,L) \leqq$ FPE(M',L); return to step 4). If, on the other hand, FPE$(M,L) >$ FPE(M',L) then $M_{\mathrm{opt}} = M$, and terminate the computation.

The normalized algorithm, outlined above, generates a unique spectral matrix estimate, and ensures stability by virtue of the fact that the singular values of the matrix reflection coefficient sequence are less than unity in magnitude in each recursion step, as shown by *Morf* et al. [2.59]. Moreover, they have shown [2.61] that the normalized LWR algorithm establishes a one-to-one correspondence between the matrix sequence $\{\boldsymbol{R}_x(0), \boldsymbol{R}_x(1), \boldsymbol{R}_x(2), \ldots\}$ and the normalized reflection coefficient sequence $\{\boldsymbol{R}_x(0), \varrho_1, \varrho_2, \ldots\}$. This property is proved for the single-channel case in Sect. 2.7.

Several other algorithms for the recursive estimation of multichannel reflection coefficients have also been proposed [2.64–70]. They are designed to generalize Burg's arithmetic mean method for the single-channel case, so that they reduce to it when the number of channels L becomes equal to 1. We shall briefly compare these methods with the method of *Morf* et al. described above.

The procedure proposed by *Jones* [2.64, 65] yields an estimate of the matrix reflection coefficient by minimizing the arithmetic mean of the forward and backward error powers, and may be viewed as a direct generalization of Burg's single-channel procedure. Unfortunately, this scheme does not produce positive-definite spectra, and does not guarantee stability, that is, the matrix reflection coefficients may have singular values greater than one in magnitude.

Nuttall [2.66–68], and *Strand* [2.69, 70] have applied a form of weighting to their error criteria, thereby obtaining a stable algorithm with positive-definite spectra. The current prediction-error powers are weighted by their respective estimates from the previous recursion step, and their arithmetic mean is then minimized with respect to the reflection coefficients. These schemes ensure that the singular values of the matrix reflection coefficient are less than unity in magnitude.

As far as uniqueness and stability of the spectral estimates are concerned, the weighted arithmetic mean algorithm developed by *Nuttall* and *Strand* is as good as the square-root normalized algorithm of *Morf* et al. However, preliminary results reported by *Morf* et al. [2.59] indicate that the square-root normalized algorithm gives a somewhat better spectral resolution than the weighted arithmetic mean algorithm. More investigation is needed in order to obtain more conclusive results.

2.13 Conclusions

The important features of the maximum-entropy method of spectral analysis may be summarized as follows:

a) The basic idea of the method is to extrapolate the autocorrelation function of a given time series by maximizing the entropy of the process.

b) Unlike conventional linear methods of spectral analysis, the maximum-entropy method does not require the use of a window function.

c) The method lends itself to an efficient recursive formulation whereby the computation, based on a set of prediction-error filter coefficients, may be carried out in a recursive manner without having to compute the autocorrelation function of the process. Furthermore, the filter so obtained is always stable.

d) The method is well-suited to the spectral analysis of relatively short data records. As such, the resolution of the method is usually superior to that obtained by using conventional linear methods.

e) The time needed to carry out the computation is of the same order of magnitude as that for procedures based on use of the periodogram.

The attributes mentioned above are well supported by experimental results obtained by applying the method to analyze the spectra of recorded radar clutter data. It has been demonstrated that, in spite of the relatively short data records, it is possible to classify radar clutter into one of three basic forms: ground clutter, weather clutter, and clutter due to migrating flocks of birds.

Finally, we briefly considered the multichannel version of the maximum-entropy spectral estimate.

Appendix A: Properties of the Autocorrelation Matrix

The equidiagonal autocorrelation matrix of coefficients

$$\boldsymbol{R}_x = \begin{bmatrix} R_x(0) & R_x(-1) & \dots & R_x(-M) \\ R_x(1) & R_x(0) & \dots & R_x(1-M) \\ \vdots & \vdots & \ddots & \vdots \\ R_x(M) & R_x(M-1) & \dots & R_x(0) \end{bmatrix} \tag{A.1}$$

is called a Toeplitz matrix in honor of the mathematician Toeplitz. Note that all the elements along any diagonal of a Toeplitz matrix have the same value. A great deal is known about the behavior of Toeplitz matrices, with the most complete references being *Grenander* and *Szegö* [2.71], and *Widom* [2.72]. For a more readable account of this subject, however, see *Gray* [2.73], and *Smylie* et al. [2.12]. In this appendix, we will summarize and prove four important properties of the autocorrelation matrix, which are relevant to the issues of prediction-error filtering and maximum-entropy spectral analysis.

Property 1: The autocorrelation matrix is nonnegative definite, i.e., its determinant is always a nonnegative quantity.

We show this by first writing the matrix characteristic equation

$$\boldsymbol{R}_x \boldsymbol{\varepsilon} = \lambda \boldsymbol{\varepsilon}\,, \tag{A.2}$$

where λ is an eigenvalue of the autocorrelation matrix $\boldsymbol{R}_x$, and $\boldsymbol{\varepsilon}$ is the corresponding eigenvector. Premultiplying (A.1) by the Hermitian transpose of $\boldsymbol{\varepsilon}$ we get

$$\boldsymbol{\varepsilon}^{\mathrm{H}} \boldsymbol{R}_x \boldsymbol{\varepsilon} = \lambda \boldsymbol{\varepsilon}^{\mathrm{H}} \boldsymbol{\varepsilon}\,. \tag{A.3}$$

The Hermitian form on the left-hand side of (A.3) may be expressed as follows:

$$\begin{aligned}\boldsymbol{\varepsilon}^{\mathrm{H}}\boldsymbol{R}_x\boldsymbol{\varepsilon} &= \sum_{m=0}^{M}\sum_{k=0}^{M} \varepsilon_m^* R_x(m-k)\varepsilon_k \\ &= \sum_{m=0}^{M}\sum_{k=0}^{M} \varepsilon_m^* \mathrm{E}[x_m x_k^*]\varepsilon_k \\ &= \mathrm{E}\left\{\left[\sum_{m=0}^{M} \varepsilon_m^* x_m\right]\left[\sum_{k=0}^{M} \varepsilon_k x_k^*\right]\right\} \\ &= \mathrm{E}\left[\left|\sum_{m=0}^{M} \varepsilon_m^* x_m\right|^2\right] \geqq 0 .\end{aligned} \tag{A.4}$$

Since

$$\boldsymbol{\varepsilon}^{\mathrm{H}}\boldsymbol{\varepsilon} = \sum_{m=0}^{M} \varepsilon_m^* \varepsilon_m = \sum_{m=0}^{M} |\varepsilon_m|^2 > 0 , \tag{A.5}$$

it follows from (A.3, 4) that

$$\lambda = \frac{\boldsymbol{\varepsilon}^{\mathrm{H}}\boldsymbol{R}_x\boldsymbol{\varepsilon}}{\boldsymbol{\varepsilon}^{\mathrm{H}}\boldsymbol{\varepsilon}} \geqq 0 . \tag{A.6}$$

Hence, the eigenvalues of autocorrelation matrix $\boldsymbol{R}_x$ are real and nonnegative quantities.

Let us now define an $(M+1)$ by $(M+1)$ matrix $\boldsymbol{E}_r$, consisting of the eigenvectors of $\boldsymbol{R}_x$ as columns, as shown by

$$\boldsymbol{E}_r = [\boldsymbol{\varepsilon}_0, \boldsymbol{\varepsilon}_1, \ldots, \boldsymbol{\varepsilon}_M] \tag{A.7}$$

and an $(M+1)$ by $(M+1)$ diagonal matrix $\boldsymbol{\Lambda}$, with the eigenvalues of $\boldsymbol{R}_x$ as elements, as shown by

$$\boldsymbol{\Lambda} = \begin{bmatrix} \lambda_0 & & & 0 \\ & \lambda_1 & & \\ & & \ddots & \\ 0 & & & \lambda_M \end{bmatrix} . \tag{A.8}$$

The following matrix equation then holds

$$\boldsymbol{R}_x\boldsymbol{E}_r = \boldsymbol{\Lambda}\boldsymbol{E}_r \tag{A.9}$$

which is actually the sum of the characteristic equations (A.2) for each of the $M+1$ eigenvectors. Taking the determinants of both sides of (A.9), we have

$$\det(\boldsymbol{R}_x)\det(\boldsymbol{E}_r) = \det(\boldsymbol{\Lambda})\det(\boldsymbol{E}_r) \tag{A.10}$$

or

$$\det(\boldsymbol{R}_x) = \det(\Lambda) = \prod_{m=0}^{M} \lambda_m \geqq 0, \tag{A.11}$$

where the last inequality follows from (A.6). It follows, therefore, that the autocorrelation matrix $\boldsymbol{R}_x$ is nonnegative definite.

Property 2: The spectral density $S_x(f)$ of a weakly stationary time series $\{x_n\}$ determines the limiting form of the determinant of the $(M+1)$ by $(M+1)$ autocorrelation matrix $\boldsymbol{R}_x$ of the time series as M becomes large.

This property is a special case of "Szegö's theorem" which may be stated as follows [2.12]: If $g(\cdot)$ is a continuous function, then

$$\lim_{M\to\infty} \frac{g(\lambda_0)+g(\lambda_1)+\ldots+g(\lambda_M)}{M+1} = \frac{1}{2B} \int_{-B}^{B} g[2BS_x(f)]df, \tag{A.12}$$

where it is assumed that the time series $\{x_n\}$ is limited to the frequency interval $-B \leqq f \leqq B$. If $g(\cdot)$ is the logarithmic function, the theorem states that

$$\lim_{M\to\infty} \frac{\ln\lambda_0+\ln\lambda_1+\ldots+\ln\lambda_M}{M+1} = \lim_{M\to\infty} \ln[(\lambda_0\lambda_1\ldots\lambda_M)^{1/M+1}]$$

$$= \frac{1}{2B} \int_{-B}^{B} \ln[2BS_x(f)]df. \tag{A.13}$$

Substituting (A.11) in (A.13), we get

$$\lim_{M\to\infty} [\det(\boldsymbol{R}_x)]^{1/M+1} = 2B\exp\left\{\frac{1}{2B} \int_{-B}^{B} \ln[S_x(f)]df\right\}, \tag{A.14}$$

where it is assumed that the determinant of the autocorrelation matrix $\boldsymbol{R}_x$ is nonzero. Equation (A.14) is the desired form of the relationship between the spectral density $S_x(f)$ of the time series $\{x_n\}$ and the limiting value of the determinant of the $(M+1)$ by $(M+1)$ autocorrelation matrix $\boldsymbol{R}_x$ when M approaches infinity.

Property 3: The determinant of the $(M+1)$ by $(M+1)$ autocorrelation matrix of a weakly stationary time series $\{x_n\}$ is related to the output power of all the pertinent prediction-error filters, up to order M, as follows:

$$\det(\boldsymbol{R}_x) = \prod_{m=0}^{M} P_m. \tag{A.15}$$

For the proof of this property, we will add a superscript to the prediction-error filter weights in order to identify the filter order. Thus, $a_m^{(M)}$ denotes the

mth coefficient of a prediction-error filter of order M, where $m=0,1,2,\ldots,M$. Using this notation, we may rewrite the system of Eq. (2.32) as follows:

$$\sum_{k=0}^{M} a_k^{(m)} R_x(m-k) = \begin{cases} P_M, & m=0 \\ 0, & m=1,2,\ldots,M. \end{cases} \tag{A.16}$$

Let A_l denote an $(M+1)$ by $(M+1)$ lower triangular matrix whose elements represent the prediction-error filter coefficients as indicated below:

$$A_l = \begin{bmatrix} 1 & & & & \boldsymbol{0} \\ a_1^{(1)} & 1 & & & \\ a_2^{(2)} & a_1^{(2)} & 1 & & \\ \vdots & \vdots & \vdots & \ddots & \\ a_M^{(M)} & a_{M-1}^{(M)} & a_{M-2}^{(M)} & \cdots & 1 \end{bmatrix}. \tag{A.17}$$

Since (A.16) applies for any value of filter order M, we find that by using the definitions for the matrices A_l and R_x, we may express the matrix product $A_l R_x$ in the form of an upper triangular matrix as follows:

$$A_l R_x = \begin{bmatrix} P_0 & & & U_1 \\ & P_1 & & \\ & & P_2 & \\ & & & \ddots \\ \boldsymbol{0} & & & P_M \end{bmatrix} \tag{A.18}$$

where the upper triangular part U_1 of the matrix is possibly nonzero. Since the autocorrelation matrix R_x is Hermitian, we may write

$$(A_l R_x)^{\mathrm{H}} = R_x A_l^{\mathrm{H}}, \tag{A.19}$$

where

$$A_l^{\mathrm{H}} = \begin{bmatrix} 1 & a_1^{(1)*} & a_2^{(2)*} & \cdots & a_M^{(M)*} \\ & 1 & a_1^{(2)*} & \cdots & a_{M-1}^{(M)*} \\ & & 1 & \cdots & a_{M-2}^{(M)*} \\ & & & \ddots & \vdots \\ \boldsymbol{0} & & & & 1 \end{bmatrix}. \tag{A.20}$$

Since both $A_l R_x$ and A_l^{H} are upper triangular matrices, their product is itself an upper triangular matrix whose elements along the main diagonal are equal to the products of the respective diagonal elements of $A_l R_x$ and A_l^{H}. We may therefore write

$$A_l R_x A_l^{\mathrm{H}} = \begin{bmatrix} P_0 & & & & U_2 \\ & P_1 & & & \\ & & P_2 & & \\ & & & \ddots & \\ 0 & & & & P_M \end{bmatrix}. \tag{A.21}$$

However, the matrix product $A_l R_x A_l^{\mathrm{H}}$ is a symmetric matrix because we have

$$(A_l R_x A_l^{\mathrm{H}})^{\mathrm{H}} = A_l R_x A_l^{\mathrm{H}}. \tag{A.22}$$

It follows therefore that the upper triangular part U_2 of the matrix on the right-hand side of (A.21) is zero, and so we may write

$$A_l R_x A_l^{\mathrm{H}} = \begin{bmatrix} P_0 & & & & 0 \\ & P_1 & & & \\ & & P_2 & & \\ & & & \ddots & \\ 0 & & & & P_M \end{bmatrix}. \tag{A.23}$$

Taking the determinants of both sides of (A.23), we have the desired result, namely,

$$\begin{aligned} \det(A_l R_x A_l^{\mathrm{H}}) &= \det(R_x) \\ &= \prod_{m=0}^{M} P_m . \end{aligned} \tag{A.24}$$

The first equality in (A.24) follows from the fact that

$$\det(A_l) = \det(A_l^{\mathrm{H}}) = 1 . \tag{A.25}$$

Property 4: The output power of a prediction-error power of infinite order is related to the spectral density $S_x(f)$ of the input time series $\{x_n\}$ by

$$P_\infty = 2B \exp\left\{\frac{1}{2B} \int_{-B}^{B} \ln[S_x(f)]\, df\right\}. \tag{A.26}$$

To prove this property, we take the logarithms of both sides of (A.15), and then divide the result by $M+1$, obtaining

$$\frac{1}{M+1} \ln[\det(R_x)] = \frac{1}{M+1} \sum_{m=0}^{M} \ln P_m . \tag{A.27}$$

Therefore,

$$\lim_{M\to\infty} \{\ln[\det(\boldsymbol{R}_x)]\}^{1/M+1} = \ln P_\infty , \tag{A.28}$$

where P_∞ is the output power of a prediction-error filter of infinite order. Combining (A.24, 28), we get the desired relationship between the spectral density of the input time series $\{x_n\}$ and the prediction-error power P_∞.

In the case of a deterministic (i.e., predictable) time series $\{x_n\}$, the prediction-error power P_∞ is, by definition, zero. On the other hand, if the time series $\{x_n\}$ is nondeterministic (i.e., random), then P_∞ is greater than zero. From (A.16) we therefore deduce that for a time series $\{x_n\}$ to be nondeterministic its spectral density $S_x(f)$ must satisfy the condition

$$\int_{-B}^{B} \ln[S_x(f)]df > -\infty \tag{A.29}$$

which is in accord with the "Paley-Wiener criterion" [2.74].

Appendix B: Entropy of a Gaussian Process

Suppose that we have a communication system which transmits M different messages $m_1, m_2, \ldots$, with probabilities of occurrence denoted by $p_1, p_2, \ldots$, respectively. Suppose further that during a long period of transmission, a sequence of L messages has been generated. If L is very large, then, on the average, we may expect to find in the sequence $p_1 L$ messages of m_1, $p_2 L$ messages of m_2 and so on. The total information in such a sequence is given by [2.75]

$$I_{\text{total}} = p_1 L \log_2\left(\frac{1}{p_1}\right) + p_2 L \log_2\left(\frac{1}{p_2}\right) + \ldots . \tag{B.1}$$

The average information per message interval, represented by the symbol H, is therefore

$$\begin{aligned} H &= \frac{I_{\text{total}}}{L} \\ &= p_1 \log_2\left(\frac{1}{p_1}\right) + p_2 \log_2\left(\frac{1}{p_2}\right) + \ldots \\ &= \sum_{k=1}^{M} p_k \log_2\left(\frac{1}{p_k}\right). \end{aligned} \tag{B.2}$$

The quantity H is called the entropy.

The base of the logarithm used in (B.2) is two. According to this definition, the entropy H is measured in binary digits or bits. The base of the logarithm used in the definition of entropy as a measure of average information content, however, depends on the encoding scheme used for the generation of the message. It is therefore quite arbitrary, and in general the entropy is defined by

$$H=\sum_{k} p_k \log_r\left(\frac{1}{p_k}\right), \quad r\text{-ary digits} \tag{B.3}$$

which, in terms of the natural logarithm, may be rewritten as

$$H=\frac{1}{\ln r}\sum_{k} p_k \ln\left(\frac{1}{p_k}\right), \quad r\text{-ary digits.} \tag{B.4}$$

The entropy measured in bits is equal to $\log_2 r$ times the entropy measured in r-ary digits.

When a given random variable takes on a continuum of possible values, the sum in the definition of the entropy H is replaced by an integral. Furthermore, when we deal with realizations of the time series $\{x_1, x_2, \ldots, x_N\}$, the probability is replaced by the joint-probability density function $p(x_1, x_2, \ldots, x_N)$, which, for convenience, we will write as $p(\boldsymbol{x})$. We may thus express the entropy of such a process in the form of an N-dimensional integral as follows:

$$H=-\int_{-\infty}^{\infty} p(\boldsymbol{x})\ln[c^{2N}p(\boldsymbol{x})]\,d\boldsymbol{x}, \tag{B.5}$$

where c is a constant that fixes the reference level of the entropy measure or the absolute value of the entropy.

If the time series $\{x_n\}$ is derived from a Gaussian process, we find that the joint-probability density function is given by [2.76]

$$p(\boldsymbol{x})=\frac{1}{[(2\pi)^N \det(\boldsymbol{C}_x)]^{1/2}}\exp[-\tfrac{1}{2}(\boldsymbol{x}-\boldsymbol{\mu}_x)^{\mathrm{H}}\boldsymbol{C}_x^{-1}(\boldsymbol{x}-\boldsymbol{\mu}_x)], \tag{B.6}$$

where $\boldsymbol{C}_x$ is the autocovariance matrix with elements

$$C_x(j-i)=\mathrm{E}[(x_i-\mu_x)(x_j^*-\mu_x^*)] \tag{B.7}$$

and μ_x is the mean value of x_i or x_j. The $\boldsymbol{\mu}_x$ denotes the vector of mean values. Substituting (B.6) in (B.5) yields the following expression for the entropy of a Gaussian process:

$$H=\tfrac{1}{2}\ln[\det(\boldsymbol{C}_x)], \tag{B.8}$$

where we have chosen the arbitrary constant $c=(2\pi)^{1/4}$.

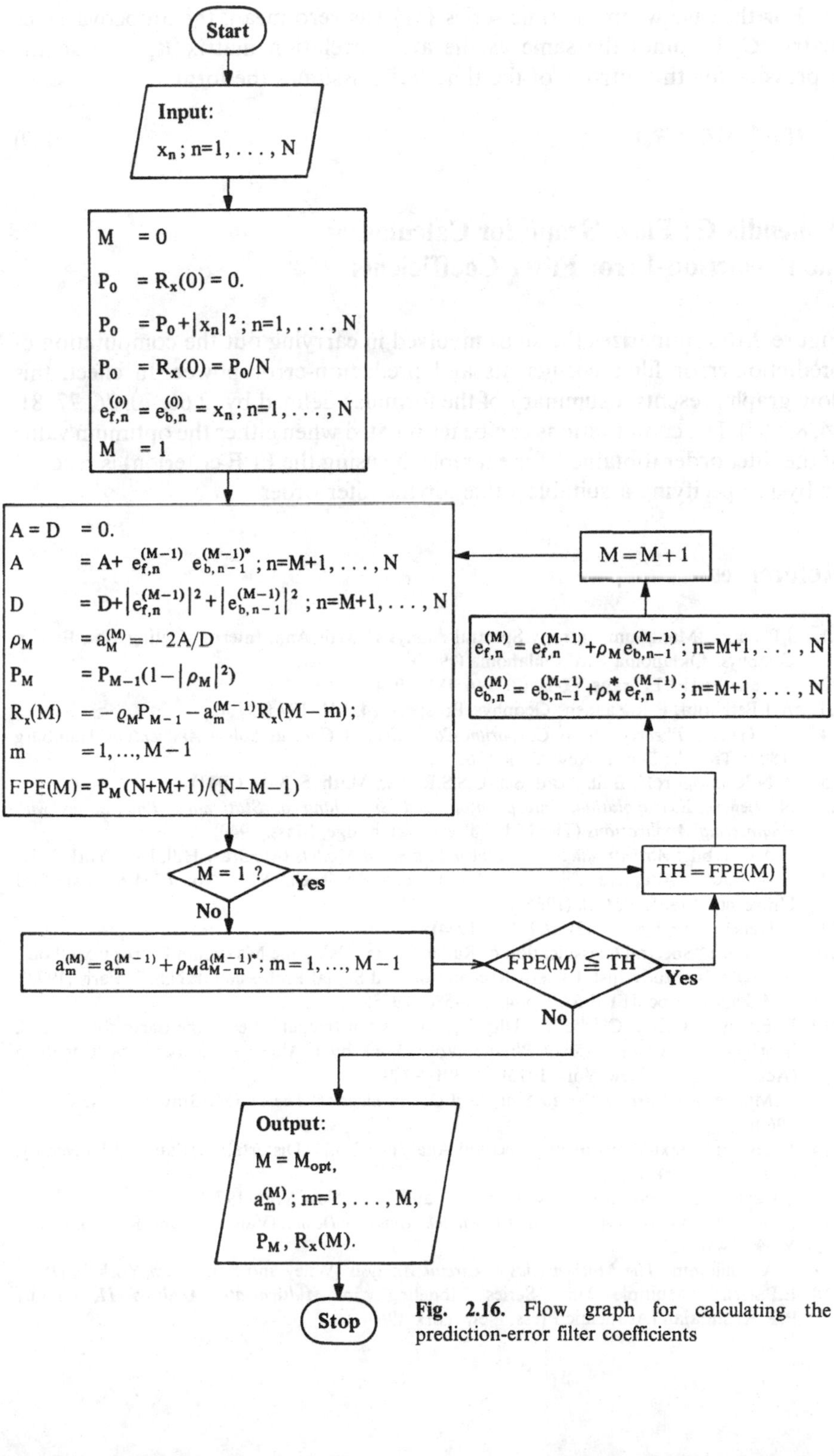

Fig. 2.16. Flow graph for calculating the prediction-error filter coefficients

For the case when the time series $\{x_n\}$ has zero mean, the autocovariance matrix C_x becomes the same as the autocorrelation matrix R_x, and so the expression for the entropy of the time series assumes the form

$$H = \tfrac{1}{2}\ln[\det(R_x)]\,. \tag{B.9}$$

Appendix C: Flow Graph for Calculating the Prediction-Error Filter Coefficients

Figure 2.16 summarizes the steps involved in carrying out the computation of prediction-error filter coefficients and prediction-error power. In effect, this flow graph presents a summary of the formulas defined by (2.66, 70, 76, 77, 81, 84, 85, 87). The computations can be terminated when either the optimum value of the filter order (obtained, for example, by using the FPE criterion) is reached or by prespecifying a suitable value for the filter order.

References

2.1 J.P.Burg: "Maximum Entropy Spectral Analysis", 37th Ann. Intern. Meeting, Soc. Explor. Geophys., Oklahoma City, Oklahoma (1967)

2.2 T.Kailath: IEEE Trans. IT-**20** (2), 146–181 (1974)

2.3 A.J.Berkhout, P.R.Zaanen: Geophys. Prospect. **24**, 141–197 (1976)

2.4 C.F.Gauss: *Theoria Motus Corporum Coelestium in Conicus Solem Ambientum*, Hamburg (1809) (Transl.: Dover, New York 1963)

2.5 A.N.Kolmogoroff: Bull. Acad. Sci. U.S.S.R., Ser. Math. **5**, 3–14 (1941)

2.6 N.Wiener: *Extrapolation, Interpolation, and Smoothing of Stationary Time Series with Engineering Applications* (The M.I.T. Press, Cambridge, Mass. 1949)

2.7 F.A.Graybill: *An Introduction to Linear Statistical Models* (McGraw-Hill, New York 1961)

2.8 B.Widrow: "Adaptive Filters 1: Fundamentals"; Tech. Rpt. No. 6764-6, Standford University, Stanford, Cal. (1966)

2.9 S.Treitel: Geophys. **39**, (2), 169–173 (1974)

2.10 S.Haykin: "Spectral Classification of Radar Clutter Using the Maximum Entropy Method", NATO Adv. Study Inst. Pattern Recognition and Signal Processing, E.N.S.T., Paris (1978)

2.11 J.Makhoul: Proc. IEEE **63**, No. 4, 561–580 (1975)

2.12 D.E.Smylie, G.K.C.Clarke, T.J.Ulrych; "Analysis of Irregularities in the Earth's Rotation", in *Methods in Computational Physics*, Vol. 13, ed. by B.Alder, S.Fernbach, M.Rotenberg (Academic Press, New York 1973) pp. 391—430

2.13 D.Middleton: *Introduction to Statistical Communication Theory* (McGraw-Hill, New York 1960)

2.14 J.P.Burg: "Maximum Entropy Spectral Analysis"; Ph.D. Dissertation, Stanford University, Stanford, Cal. (1975)

2.15 S.Haykin: *Communication Systems* (Wiley and Sons, New York 1978)

2.16 H.W.Bode: *Network Analysis and Feedback Amplifier Design* (Van Nostrand Reinhold, New York 1945)

2.17 E.A.Guillemin: *The Mathematics of Circuit Analysis* (Wiley and Sons, New York 1949)

2.18 E.Parzen: "Multiple Time Series Modeling", in *Multivariate Analysis II*, ed. by P.R.Krishnaiah (Academic Press, New York 1969)

2.19 N. Levinson: J. Math. Phys. **25**, 261–278 (1947)
2.20 J. Durbin: Rev. Inst. Int. Statist. **28**, (3), 233–243 (1960)
2.21 J.P. Burg: "A New Analysis Technique for Time Series Data", NATO Adv. Study Inst. Signal Processing, Enschede, Netherlands (1968)
2.22 J. Makhoul: IEEE Trans. ASSP-**25**, (5), 423–428 (1977)
2.23 F. Itakura, S. Saito: "Digital Filtering Techniques for Speech Analysis and Synthesis", 7th Intern. Conf. Acoust., Budapest, paper 25-C-1 (1971)
2.24 R.N. McDonough: Geophys. **39** (6), 843–851 (1974)
2.25 S. Haykin, S. Kesler: Proc. IEEE **64** (5), 822–823 (1976)
2.26 J.D. Markel, A.H. Gray, Jr.: IEEE Trans. ASSP-**23** (5), 473–486 (1975)
2.27 C.T. Mullis, R.A. Roberts: IEEE Trans. ASSP-**24** (6), 538–550 (1976)
2.28 A.V. Oppenheim, R.W. Schafer: *Digital Signal Processing* (Prentice Hall, Engelwood Cliffs, N.J. 1975)
2.29 E.A. Robinson: *Statistical Communication and Detection* (Hafner Press, New York 1967)
2.30 S. Treitel, E.A. Robinson: IEEE Trans. GE-**4** (1), 25–38 (1966)
2.31 A. Papoulis: *Probability, Random Variables, and Stochastic Processes* (McGraw-Hill, New York 1965)
2.32 L.H. Koopmans: *The Spectral Analysis of Time Series* (Academic Press, New York 1974)
2.33 L.J. Griffiths: "A Continuous-Adaptive Filter Implemented as a Lattice Structure", IEEE Intern. Conf. Acoust., Speech, Signal Processing, Hartford, Conn. (1977) pp. 683–686
2.34 J. Makhoul: IEEE Trans. ASSP-**26** (4), 304–314 (1978)
2.35 J. Makhoul, R. Viswanathan: "Adaptive Lattice Methods for Linear Prediction", IEEE Intern. Conf. Acoust., Speech, Signal Processing, Tulsa, Ok. (1978) pp. 83–86
2.36 R.E. Kromer: "Asymptotic Properties of the Autoregressive Spectral Estimator"; Ph.D. dissertation, Dept. of Statistics, Tech. Rpt. No. 13, Stanford University, Stanford, Cal. (1970)
2.37 K.N. Berk: Ann. Statistics **2** (3), 489–502 (1974)
2.38 T.J. Ulrych, T.N. Bishop: Rev. Geophys. Space Phys. **13**, 183–200 (1975)
2.39 G.M. Jenkins, D.G. Watts: *Spectral Analysis and Its Application* (Holden-Day, San Francisco, Cal. (1968)
2.40 W. Gersch, D.R. Sharpe: IEEE Trans. AC-**18**, (6), 367–369 (1973)
2.41 A.B. Baggeroer: IEEE Trans. IT-**22** (5), 534–545 (1976)
2.42 H. Akaike: Ann. Inst. Statist. Math. **21**, 243–247 (1969)
2.43 H. Akaike: Ann. Inst. Statist. Math. **22**, 203–217 (1970)
2.44 H. Akaike: IEEE Trans. AC-**19** (6), 716–723 (1974)
2.45 T.J. Ulrych, R.W. Clayton: Phys. Earth Planet. Inter. **12**, 188–200 (1976)
2.46 E. Parzen: "An Approach to Time Series Modeling and Forecasting Illustrated by Hourly Electricity Demands", Statistical Sci. Div., State University of New York in Buffalo, Tech. Rpt. 37 (1976)
2.47 T.E. Landers, R.T. Lacoss: IEEE Trans. GE-**15** (1), 26–32 (1977)
2.48 S.B. Kesler, S. Haykin: Can. Elec. Eng. J. **3** (1), 11–16 (1978)
2.49 S.B. Kesler, S. Haykin: IEEE Trans. IT-**24**, (2), 269–272 (1978)
2.50 M.I. Skolnik: *Radar Handbook* (McGraw-Hill, New York 1970)
2.51 E.J. Barlow: Proc. IRE **37**, 340–355 (1949)
2.52 M. Abramowitz, I.A. Stegun: *Handbook of Mathematical Functions* (Dover, New York 1965)
2.53 S.B. Kesler, S. Haykin: "Maximum Entropy Estimation of Radar Clutter Spectra", Nat. Telecommun. Conf., Birmingham, Ala. (1978)
2.54 S. Haykin, S. Kesler, B. Currie: "Autoregressive Spectral Analysis and Its Application to Radar Clutter Classification", Intern. Conf. on Radar, Paris, France (1978)
2.55 S. Haykin, S. Kesler, B. Currie: Proc. IEEE **67**, 332–333 (1979)
2.56 S.B. Kesler: "Nonlinear Spectral Analysis of Radar Clutter": Ph.D. dissertation, Commun. Res. Lab. Rpt. No. CRL-**51**, McMaster University, Hamilton, Ontario (1977)
2.57 B.W. Currie: "Design and Hardware Implementation of Radar Video Display and Sampler/Digitizer for Use in Radar Clutter Studies"; M. Eng. Thesis, McMaster University, Hamilton, Ontario (1976)

2.58 P.D.Welch: IEEE Trans. AU-**15** (3), 70–73 (1967)
2.59 M.Morf, A.Vieira, D.T.L.Lee, T.Kailath: IEEE Trans. GE-**16** (2), 85–94 (1978)
2.60 R.A.Wiggins, E.A.Robinson: J. Geophys. Res. **70**, 1885–1891 (1965)
2.61 M.Morf, A.Vieira, T.Kailath: Ann. Statistics **6** (3), 643–648 (1978)
2.62 H.Akaike: Ann. Inst. Statist. Math. **23**, 163–180 (1971)
2.63 G.J.Fryer, M.E.Odegard, G.H.Sutton: Geophys. **40**, (3), 411–425 (1975)
2.64 R.H.Jones: "Multivariate Maximum Entropy Spectral Analysis", Appl. Time Series Anal. Symp., Tulsa, Ok., May (1976)
2.65 R.H.Jones: "Multivariate Autoregression Estimation Using Residuals", Appl. Time Series Anal. Symp., Tulsa, Ok., May (1976)
2.66 A.H.Nuttall: "Fortran Program for Multivariate Linear Predictive Spectral Analysis Employing Forward and Backward Averaging", Naval Underwater System Center, Tech. Doc. 5419, New London, Conn. May (1976)
2.67 A.H.Nuttall: "Multivariate Linear Predictive Spectral Analysis Employing Weighted Forward and Backward Averaging: A Generalization of Burg's Algorithm", Naval Underwater System Center, Tech. Doc. 5501, New London, Conn., Oct. (1976)
2.68 A.H.Nuttall: "Positive Definite Spectral Estimate and Stable Correlation Recursion for Multivariate Linear Predictive Spectral Analysis, Naval Underwater System Center, Tech. Doc. 5729, New London, Conn., Nov. (1976)
2.69 O.N.Strand: "Computer Programs for Maximum Entropy Spectral Analysis of Real and Complex Multichannel Time Series (with Microfilm Plots)", National Oceanic and Atmospheric Administration, Tech. Ucmo ERL WPL-**22**, Boulder, Colo., April (1977)
2.70 O.N.Strand: IEEE Trans. AC-**22** (4), 634–640 (1977)
2.71 U.Grenander, G.Szegö: *Toeplitz Forms and Their Applications* (University of California Press, Berkeley and Los Anglees 1958)
2.72 H.Widom: "Toeplitz Matrices", in *Studies in Real and Complex Analysis, MAA Studies in Mathematics*, ed. by I.I.Hirschman, Jr. (Prentice Hall, Englewood Cliffs, N.J. 1965)
2.73 R.M.Gray: "Toeplitz and Circulant Matrices: A Review", Tech. Rpt. No. 6502-1, Stanford University, Cal. (1971)
2.74 R.E.A.C.Paley, N.Wiener: Amer. Math. Soc. Colloquium Publ. **19**, 16–17 (1934)
2.75 C.E.Shannon: Bell Syst. Tech. J. **27**, 379–423, 623–656 (1948)
2.76 H.L.Van Trees: *Detection, Estimation, and Modulation Theory*, Part I (Wiley and Sons, New York 1968)
2.77 J.D.Markel, A.H.Gray, Jr.: *Linear Prediction of Speech*, Communications and Cybernetics, Vol. 12 (Springer, Berlin, Heidelberg, New York 1976)
2.78 N.O.Andersen: Geophys. **39**, 69–72 (1974)

3. Autoregressive and Mixed Autoregressive-Moving Average Models and Spectra

T. J. Ulrych and M. Ooe

With 13 Figures

In recent years, the autoregressive (AR) representation of a time series has received considerable attention in a variety of disciplines. There are a number of very good reasons why this is so. In the first place the fitting of an AR model to an observed time series is a linear process which can be handled using well tried and highly efficient computing algorithms. Secondly, it has recently been shown by *van den Bos* [3.1] that the AR representation is equivalent to a model of a time series which has maximum entropy. The principle of maximum entropy has been applied to spectral analysis by *Burg* [3.2–4] and represents a different and important viewpoint in the analysis of time series, particularly when the data segment is short compared to the period of interest. Finally, the AR model is a predictive model; it tells us something about the time series outside the known interval and as such has wide application to many diverse problems.

Of course, not all processes can be modeled as AR with equal success. In fact, the fundamental theorem in the decomposition of stationary time series proposed by *Wold* [3.5] tells us that the most general decomposition of a stochastic process is a moving average (MA) one. Although all MA processes can be modeled as AR from a spectral point of view, a finite-order MA process theoretically requires an infinite-order AR representation. The principle of parsimony is, we feel, an important one in data analysis and consequently we look for models with the least number of parameters. An excellent example of such an approach is the case of harmonic processes corrupted by additive white noise. It turns out that the optimal and most parsimonious representation of such a process is a mixed autoregressive-moving average (ARMA) model for which the AR and MA coefficients are the same.

The identification problem in time series modeling, in other words choosing an AR, MA or ARMA representation for a particular process, is a very complex one. The importance of specifying the correct model in spectral analysis is very graphically demonstrated by *Treitel* et al. [3.6]. These authors make the point that the identification problem referred to above remains an unresolved issue.

We would like to stress in this section that we consider the problems of spectral analysis and time series modeling to be intimately connected. This connection is emphasised by the parameteric approach which we take in describing a process. The periodogram [3.7] or the autocovariance [3.8] estimates are nonparametric in the sense that they do not rely on a model of the process. The AR or ARMA approach, on the other hand, is to describe the available data, which represent a realization of the process to be modeled by a suitable set of parameters, and this set is then representative of the process itself. Consequently, if the order of the particular model has been correctly determined, the spectral estimate based on this set of parameters has the extremely important property that it is optimally smooth. The parametric approach can also be used to determine the AR (or maximum entropy) spectra of nonstationary signals [3.9]. This approach has found an important application in the analysis and synthesis of strong ground motion accelerograms used in studies of structural response [3.10].

3.1 Overview

We do not intend, in this chapter, to review the considerable body of literature available in the field of AR and ARMA modeling and spectral analysis. Rather, we intend to present those aspects of the problem which we consider to be centrally important and to illustrate these with synthetic and real examples. This chapter is divided into two main parts. The first part deals with AR models and in particular with various methods of determining the AR, or equivalently, the prediction coefficients. The second part is devoted to ARMA models and to a discussion of the Akaike AIC procedure for determining the appropriate model order. The complex ARMA model and identification procedure recently developed by *Ooe* [3.11] is applied to the analysis of the Chandler wobble. We introduce an unbiased autocovariance estimator, based on the principle of maximum likelihood, which we use in the identification of the ARMA model.

3.2 The General Time Series Model

The reason why the AR model finds considerable application in a variety of disciplines lies in its relationship to a fundamental theorem in the decomposition of time series which was proposed by *Wold* [3.5]. This theorem states that any real valued, stationary stochastic process allows the decomposition

$$y_t = u_t + v_t, \qquad (3.1)$$

where u_t and v_t are stationary, mutually uncorrelated and have the following properties.

1) u_t is deterministic. This simply means that $u_{t+\alpha}$ can be predicted for any α from its past values u_{t-1}, u_{t-2}, etc., with full accuracy.

2) v_t is nondeterministic with an absolutely continuous spectral distribution function and has the one-sided MA representation

$$v_t = \sum_{i=0}^{\infty} \psi_i q_{t-i}, \tag{3.2}$$

where $\psi_0 = 1$ and q_t has the properties

$$E[q_t] = 0 \quad \text{and} \quad E[q_t q_s] = \sigma_q^2 \delta_{ts} \tag{3.3}$$

where $E[\cdot]$ is the expectation operator and δ_{ts} is the Kronecker delta function. The time series $\{q_t\}$ is a white noise process.

In many cases of interest the process y_t either does not contain a deterministic component u_t, or this component can be removed prior to analysis. For example, in seismology, a reflection seismogram, x_t, can be modeled as the convolution of a wavelet with a random series of reflection coefficients denoted by q_t. For a finite length wavelet, b_i, $i = 0, 1, \ldots, m$ we can write

$$x_t = \sum_{i=0}^{m} b_i q_{t-i}. \tag{3.4}$$

3.2.1 The MA Approximation

Equation (3.4) describes the MA model of a stochastic process x_t which does not contain a deterministic component. Since the Wold theorem specifies the generality of this model, let us briefly examine the problem of determining the MA coefficients b_i, $i = 0, \ldots, m$ assuming that the order, m, of the MA model is known. Equation (3.4) may be written in vector form as

$$x_t = \boldsymbol{q}_t^T \boldsymbol{b}, \tag{3.5}$$

where

$$\boldsymbol{q}_t^T = [q_t, q_{t-1}, \ldots, q_{t-m}], \tag{3.6}$$

$$\boldsymbol{b}^T = [b_0, b_1, \ldots, b_m], \tag{3.7}$$

and the superscript T denotes transpose. If $\boldsymbol{q}_t$ were known we could determine $\boldsymbol{b}$ in the usual manner by forming the regressive residual

$$e_t = x_t - \boldsymbol{q}_t^T \boldsymbol{b} \tag{3.8}$$

from which we obtain the variance of e_t as

$$\mathrm{Var}[e_t] = E[e_t^2] = E[(x_t - \boldsymbol{q}_t^T \boldsymbol{b})^2]. \tag{3.9}$$

Using the method of least squares we minimize the gradient of the regressive variance. Thus

$$\nabla(\mathrm{E}[e_t^2]) = -2(\mathrm{E}[x_t - \boldsymbol{q}_t^{\mathrm{T}}\boldsymbol{b}]\boldsymbol{q}_t) = 0\,, \tag{3.10}$$

where $\nabla[(e_t^2)]$ denotes the gradient of (e_t^2) with respect to $\boldsymbol{b}$ and is an $m+1$ dimensional column vector with elements

$$\frac{\partial(e_t^2)}{\partial b_i}\,, \qquad i=0,1,\ldots,m\,. \tag{3.11}$$

We therefore obtain the $m+1$ dimensional column vector

$$\mathrm{E}[e_t\boldsymbol{q}_t] = 0 \tag{3.12}$$

which are the normal equations for this problem.

Using (3.3), namely, the fact that

$$\mathrm{E}[q_t q_s] = \sigma_q^2 \delta_{ts} \tag{3.13}$$

we have

$$\boldsymbol{b} = \frac{1}{\sigma_q^2}\mathrm{E}[x_t\boldsymbol{q}_t]\,. \tag{3.14}$$

However, the problem is that we do not know $\mathrm{E}[x_t q_t]$ and so (3.14) cannot be used to determine $\boldsymbol{b}$. The information which is available to us comes from the covariance structure of x_t. Taking expectations in (3.4) we get

$$\mathrm{E}[x_t x_{t-k}] = \begin{cases} \sigma_q^2 \sum\limits_{i=0}^{m-|k|} b_i b_{i+|k|} & \text{for} \quad |k| \leq m \\ 0 & \text{for} \quad |k| > m\,. \end{cases} \tag{3.15}$$

The solution of (3.15) such that all the roots of $B(z)$, where

$$B(z) = \sum_{n=0}^{\infty} b_n z^{-n} \tag{3.16}$$

is the z transform of $\boldsymbol{b}$, lie inside the unit circle $|z|=1$, is an iterative one using the Newton-Raphson method [3.12]. We note, therefore, that this approach forces the coefficient vector $\boldsymbol{b}$ to assume a minimum delay form. We will emphasize this important point in a later section. The minimum delay vector $\boldsymbol{b}$ may also be extracted by the method of spectral factorization [3.41].

3.2.2 The AR Approximation

Although the iterative solution which we have considered in the previous section forces $\boldsymbol{b}$ to be minimum delay, in practice $\boldsymbol{b}$ need not, of course, have this form. If however we have a priori information that $\boldsymbol{b}$ is in fact minimum delay, a much more convenient approach can be adopted.

Taking z transforms in (3.4), we get

$$X(z)=B(z)Q(z). \tag{3.17}$$

An inverse of $B(z)$, $G(z)=B^{-1}(z)$ exists owing to the assumed minimum-delay property of $\boldsymbol{b}$, consequently

$$X(z)G(z)=Q(z). \tag{3.18}$$

Since $g_0 b_0=1$, the inverse transform yields

$$x_t(1+g_1 z+\ldots)=q_t \tag{3.19}$$

or if $a_i=-g_i$, $i=1,2,\ldots,\infty$

$$x_t=a_1 x_{t-1}+a_2 x_{t-2}+\ldots+q_t. \tag{3.20}$$

Equation (3.20) is the AR representation of a process x_t. As we will see later, the determination of the AR parameters $a_i, i=1,2,\ldots$ is a linear problem and is consequently much simpler than the corresponding determination of the MA parameters.

It is important to point out that in cases where $\boldsymbol{b}$ is not minimum delay, so that $G(z)=B^{-1}(z)$ no longer holds, a minimum delay inverse can always be found by the method of least squares, so that an AR model with the same first- and second-order statistics as the original MA process can be identified [3.13].

3.2.3 The ARMA Approximation

We have seen that the general time series model contains both a deterministic and a nondeterministic part. In the description of the MA and AR models we have conveniently chosen to ignore the deterministic component. However, in general, many processes of interest are a combination of both. For example, in the case of a harmonic process with colored noise, we can represent the deterministic component as an AR process of order p with zero innovation [3.14],

$$u_t=\sum_{i=1}^{p} a_i u_{t-i}$$

and the nondeterministic part as an MA process of order q,

$$v_t = e_t + \sum_{i=1}^{q} b_i e_{t-i}.$$

A general model for the process is therefore

$$x_t = \sum_{i=1}^{p} a_i x_{t-i} + e_t - \sum_{i=1}^{p+q-1} c_i e_{t-i}, \tag{3.21}$$

where $c_i = \sum_{j=0}^{i} a_j b_{i-j}$ with $a_0 = -b_0 = -1$.

Equation (3.21) represents a mixed ARMA process of order $(p, p+q-1)$ where p and q, in this case, depend on the number of harmonics and on the spectral properties of the noise, respectively.

The problems which arise in the determination of MA parameters are also, of course, encountered in the determination of ARMA parameters. Various schemes which are in use for this purpose are described in Sect. 3.4.

3.2.4 Remarks

We have attempted, in this section, to give a rationale for representing processes in terms of AR and ARMA models. In summary, we make the following points:

a) The AR representation for a stochastic process is the most convenient in the sense that the computation of the AR parameters is a linear and robust procedure. An exact AR representation of a MA process can be found only providing that the MA coefficients have the minimum delay property. In this case however the AR representation is not a parsimonious one since the AR order is, theoretically, infinite.

b) The most general representation of a linear, stationary, stochastic process which contains a deterministic component is an ARMA model. (This point is discussed further in Sect. 3.5.) The determination of the parameters of this model can be accomplished in a number of ways, all of which, however, require an iterative approach. Although an AR representation of an ARMA process can in general be found, the importance of an ARMA representation lies in the fact that in certain cases it optimally and most parsimoniously models the process in question.

c) The determination of pure MA models is not considered in this chapter. The reason for this is twofold. In the first place the computation of the minimum delay operator is considered as a part of the determination of ARMA parameters in Sect. 3.5. Secondly, it is the delay or phase property of the MA coefficients which are of considerable interest in many instances, for example, in the identification of the so-called seismic wavelet. The determination of this property is a subject in itself and is outside the scope of the present discussion.

3.3 Autoregressive Processes

The AR model has become widely used in many different fields, notably in geophysics and in speech processing. A number of reviews of this particular decomposition have recently appeared [3.14–16, 18] and we refer the reader to these for details. In addition to the above an excellent treatment of the subject may be found in [3.17, 19].

There are some important aspects to the AR representation which will be discussed in some detail in this section. Specifically, certain assumptions which are implicit in the algorithms used to compute the AR parameters are not always fully appreciated. It turns out that, for short data lengths, these assumptions affect the final result quite considerably.

We begin the discussion of AR processes by considering the spectral representation associated with this model. The important equivalence of this spectral estimate with the maximum-entropy (ME) estimate is presented only heuristically in this section.

3.3.1 AR Spectra and Maximum Entropy

The spectral representation of an AR process can easily be deduced from a finite-order form of (3.20). We consider an AR process of order p

$$x_t = a_1 x_{t-1} + a_2 x_{t-2} + \ldots + a_p x_{t-p} + q_t . \tag{3.22}$$

Taking z transforms in (3.22) we obtain

$$X(z)\{1 - a_1 z^{-1} - a_2 z^{-2} - \ldots - a_p z^{-p}\} = Q(z) \tag{3.23}$$

and consequently, since $\mathrm{E}[q_t q_s] = \sigma_q^2 \delta_{ts}$, we may write

$$|X(z)|^2 = \frac{\sigma_q^2}{|1 - a_1 z^{-1} - a_2 z^{-2} - \ldots - a_p z^{-p}|^2} . \tag{3.24}$$

Evaluating the z transform on the unit circle, $|z| = 1$, where $z = \exp(\mathrm{j}2\pi f)$, we obtain half the power spectrum and consequently the AR spectral estimate $S_{\mathrm{AR}}(f)$ is given by:

$$S_{\mathrm{AR}}(f) = \frac{2\sigma_q^2}{\left|1 - \sum_{k=1}^{p} a_k \exp(-\mathrm{j}2\pi f k)\right|^2} . \tag{3.25}$$

Actually, although the AR model was used many years ago by *Yule* [3.20], the importance of (3.25) as a spectral estimate is due to the work of *Burg* [3.2, 3] who introduced the concept of maximum entropy into the field of spectral analysis. It turns out that the spectral estimate derived from this concept, the ME estimate, is equivalent to the AR estimate. This duality was shown rigorously by *Van den Bos* [3.1] and we will only sketch the relevant points here.

The object of the *Burg* ME approach is to get away from the assumption, inherent in conventional spectral estimates, that the data outside of the available record are zero. *Burg* [3.3, 4] assumed that these data are nonzero and that they are consistent with the assumption that all the information about the system is contained in the finite record. Since entropy is a measure of the average information, this assumption is equivalent to the requirement that the value of the entropy for the observed process be maximum subject to the constraint that it be consistent with the known second-order statistics. (A full discussion of this point is presented by *Ulrych* and *Bishop* [3.14].)

The entropy density h of a Gaussian process is

$$h=\frac{1}{4B}\int_{-B}^{B}\log S(f)df+\log(2\pi e)^{1/2}, \qquad (3.26)$$

where $S(f)$ is the spectral density of the process, and $B=1/2\Delta t$ is the bandwidth of the process. This assumes the use of Nyquist rate sampling.

Expressing $S(f)$ in terms of the autocovariance of the process $c(k)$ we have

$$h==\frac{1}{4B}\int_{-B}^{B}\log\left[\sum_{k=-\infty}^{\infty}c(k)\exp(-\mathrm{j}2\pi fk\Delta t)\right]df+\log(2\pi e)^{1/2}. \qquad (3.27)$$

Following the principle of maximum entropy, h is now maximized subject to the constraint that $S(f)$ is consistent with the known, or at least assumed known, autocovariances $c(k)$ $(-m\leqq k\leqq m)$.

Introducing the Lagrange multipliers, λ_k, the variational problem to be solved is

$$\delta\int_{-B}^{B}\left[\log S(f)-\sum_{k=-m}^{m}\lambda_k\{S(f)\exp(\mathrm{j}2\pi fk\Delta t)-c(k)\}\right]df=0. \qquad (3.28)$$

The solution of (3.28) gives the ME estimate $S_{\mathrm{ME}}(f)$ as

$$S_{\mathrm{ME}}(f)=\frac{P_m}{B\left|1+\sum_{k=1}^{m}g_k\exp(-\mathrm{j}2\pi fk\Delta t)\right|^2}. \qquad (3.29)$$

In (3.29) P_m is a constant and the coefficients g_k are determined from the data by solving

$$\boldsymbol{C}\boldsymbol{g}=P_m\boldsymbol{i}, \qquad (3.30)$$

where C is the estimate of the Toeplitz autocovariance matrix of the process, $g^T=(1,g_1,g_2,\ldots,g_m)$ is the so-called prediction-error operator (or PEO) and $i^T=(1,0,0,\ldots,0)$. We will discuss the normal equations (3.30) in detail in the following sections. A detailed derivation of (3.29) is given by *Smylie* et al. [3.21], and *Edward* and *Fitelson* [3.22] (see also Sect. 2.3).

A comparison of (3.25) and (3.29) establishes heuristically the equivalence of the spectral estimates $S_{AR}(f)$ and $S_{ME}(f)$. For $m=p$ we can see that the coefficients g_k, $k=1,2,\ldots,p$ are just the negative of the AR parameters of the pth order process. Since, as we have seen, these are the prediction coefficients, the filter

$$g^T=(1,-a_1,-a_2,\ldots,-a_p) \tag{3.31}$$

is the prediction-error filter [the PEO of (3.30)].

The AR/ME spectral estimate has two properties which make it an extremely useful tool in time series analysis. These are the resolution and smoothing characteristics of this estimate. At the same time, however, the AR/ME method suffers from the difficulty, in certain cases, of obtaining a correct estimate of the required length of the PEO, and from the lack of a clearly defined variance estimate. The asymptotic distribution of the AR/ME estimate is discussed by *Kromer* [3.23], *Akaike* [3.24], and *Berk* [3.25]. *Baggeroer* [3.26] has recently presented a method of determining the confidence intervals for the AR/ME estimator based on a Wishart model for the estimated covariance. For a discussion of the statistical properties of the ME spectral estimator, see also Sect. 2.9.

Determination of the AR Order

A rather crucial point in estimating AR/ME spectra is the determination of the order of the AR process, or equivalently, the length of the PEO. This length determines not only the resolution of the estimate, but also the smoothness. We will be concerned, in this chapter, with the *Akaike* FPE [3.24, 27] and AIC [3.28, 29] criteria. These criteria were discussed in Sect. 2.10. Also in Sect. 3.6.1, the AIC is developed in some detail for the identification of ARMA processes and since $n\log(\text{FPE})\to\text{AIC}$ as $n\to\infty$, where n is the number of data points, this development also concerns the FPE.

The expression for the FPE (final prediction-error) criterion for an mth order fit is

$$(\text{FPE})_m=\left\{\frac{n+(m+1)}{n-(m+1)}\right\}S_m^2, \tag{3.32}$$

where S_m^2 is the residual sum of squares and is computed as outlined by *Ulrych* and *Bishop* [3.14]. The order m which gives the minimum FPE is chosen as the correct order of the process.

We have found that the FPE criterion gives excellent results in the case of AR processes, or processes which can be adequately modeled by an AR representation. An example of the latter is an MA process with minimum delay coefficients. For harmonic processes with noise the FPE and AIC criteria underestimate the AR order and we have adopted an empirical rule for such processes which constrains the order m to

$$n/3-1 \leqq m \leqq n/2-1 .$$

A fuller discussion of the FPE is presented by *Ulrych* and *Bishop* [3.14]. *Landers* and *Lacoss* [3.30] have recently studied the behavior of the FPE, AIC, and CAT criteria when applied to harmonic processes. The CAT criterion, or the autoregressive transfer function criterion, was introduced by *Parzen* [3.31]. The conclusion reached by *Landers* and *Lacoss* [3.30] is that all three criteria give orders which produce acceptable spectra in the case of low noise but underestimate the order for higher noise levels.

Resolution of the AR/ME Spectral Estimate

The resolution, or the ability to resolve two frequency components, of conventional spectral techniques is known to depend on the record length. Specifically, for two sinusoids separated by $\Delta f = |f_1 - f_2|$, resolution is approximately achieved for a record length of $L > 1/\Delta f$. The same restriction does not hold for the AR/ME estimate since this estimate is, in effect, based on an infinitely long autocovariance function. Of course, the actual resolution depends rather heavily on the signal to noise ratio (SNR). For small SNR the conventional estimate must be smoothed, decreasing the resolution. The resolution of the AR/ME estimate depends on the proximity of the roots of $G(z)$, the z transform of the PEO, to the unit circle $|z|=1$. With decreasing SNR, for a fixed order, these roots move away from the unit circle and the resolution is decreased.

The two-sinusoid resolution properties of various spectral estimators were discussed quantitatively by *Marple* [3.32], and *Frost* [3.33]. These authors defined the condition for two-sinusoid resolution as the separation, $\Delta f = |f_1 - f_2|$, at which the power spectral density, $S(f_c)$, evaluated at the center frequency, $f_c = (f_1 + f_2)/2$, is equal to the average of the spectral densities evaluated at f_1 and f_2, i.e.,

$$S(f_c) = \tfrac{1}{2}[S(f_1) + S(f_2)] .$$

This definition is a very convenient one in comparing the resolution properties of the periodogram and the AR/ME estimates in that the latter estimate does not have the mainlobe function which allows the determination of the 3 dB width.

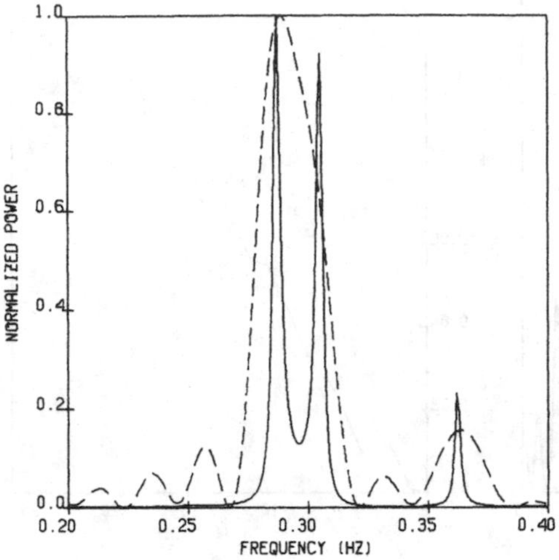

Fig. 3.1. Comparison of the resolution properties of the periodogram and the AR/ME estimates. Input consists of two sinusoids with frequencies 0.2875 and 0.3 Hz and a SNR of 6 dB

Marple [3.32], and *Frost* [3.33] defined a measure of the frequency resolution in terms of a dimensionless quantity, R, which is called the normalized resolution and is given by

$$R=2\pi M\Delta t\Delta f,$$

where M is either the number of points or the number of lags used, Δt is the sampling time interval, and Δf is the frequency separation at the point where the two frequencies are just resolved. A plot of R against M for various SNR allows the determination of the resolution, Δf, in Hz from

$$\Delta f=R/2\pi M\Delta t.$$

Using these definitions, *Marple* [3.32] presented detailed results of the comparison of the resolution of conventional and AR/ME spectral estimates based on the assumptions of perfectly known autocovariances and computer simulations in the case of imperfectly known autocovariances. It may be concluded that the resolution of the AR/ME estimate is superior to that of the conventional estimates. The degree of improvement is dependent on such factors as length of record, SNR, initial phases of the sinusoids and the location of the frequencies in the Nyquist bandwidth. Certainly a twofold increase in resolution is not uncommon when high SNR signals are analyzed. An example of a "typical" comparison of the resolution properties of the periodogram and AR/ME estimates is shown in Fig. 3.1. The input data consisted of two sinusoids of frequencies 0.2875 and 0.3 Hz with additive white noise. The SNR was 6 dB where we define the SNR by

$$\mathrm{SNR}=10\log_{10}\left(\frac{\text{signal power}}{\text{noise power}}\right). \tag{3.33}$$

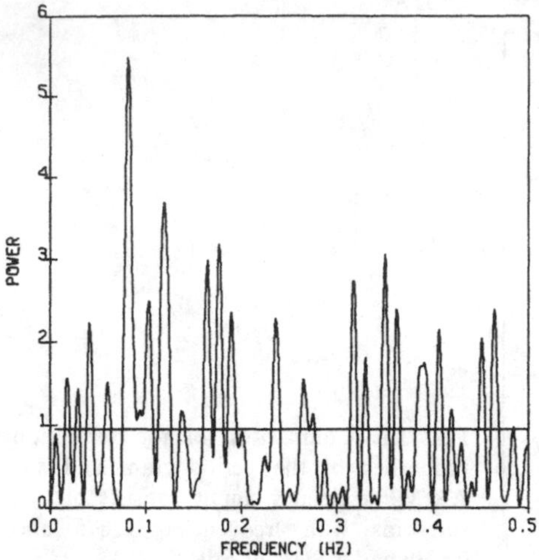

Fig. 3.2. Unsmoothed periodogram of 100 samples of white noise. The full line is the power spectrum computed from (3.25) with $k=0$

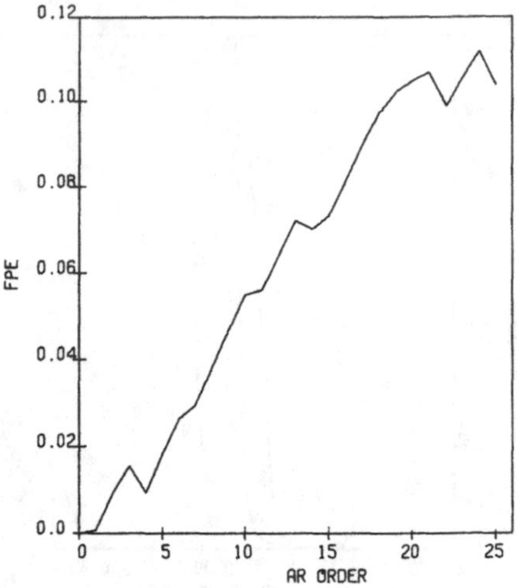

Fig. 3.3. Variation of log(FPE) for a white noise process

The length of the PEO, which was computed using the Burg technique (Sect. 3.3.2) was 25 s for a record length of 50 s with a sampling interval of one s.

A further look at the resolution problem is presented in Sect. 3.5.1 where we discuss the special ARMA representation of harmonics with noise.

The Smoothness of the AR/ME Estimator

It is well known that a trade-off exists between resolution and variance for conventional estimators [3.34]. Although an analytical relationship between these two properties has not been developed for the AR/ME estimator, this estimator does have the important property that, given a correct determination of the AR order, it is in a least-squares sense optimally smooth. This property can best be demonstrated by a very simple example. Fig. 3.2 shows the unsmoothed periodogram of 100 samples of white noise distributed as $N(0,1)$. This curve emphasizes not only the necessity of smoothing but also the difficulty in estimating the degree of smoothing required. Figure 3.3 illustrates the variation of the logarithm of the normalized FPE. The minimum value occurs at $m=0$, which indicates that the process is indeed white noise. The power spectrum computed using $p=0$ in (3.25) is shown in Fig. 3.2 as the heavy solid line, and is optimally smooth.

3.3.2 Estimation of the AR Parameters

We consider the problem of the determination of the AR parameters of a pth order AR process

$$x_t = a_1 x_{t-1} + a_2 x_{t-2} + \ldots + a_p x_{t-p} + e_t . \tag{3.34}$$

At this point we will assume that the order p is known, or it has been determined, and we consider two solutions to the problem. The first we call the Toeplitz solution and the second we call the least-squares (LS) solution. In speech analysis these two approaches are known as the autocorrelation and covariance methods, respectively, [3.15].

We have previously mentioned the minimum delay property of the PEO $\boldsymbol{g}$ which is related to the AR parameters by (3.31). This property is an important constraint for the algorithms which we will consider. Another important property is the relationship of the PEO to the eigenvectors of the autocorrelation matrix

The Eigenvector Decomposition of the Unit PEO

We consider the system of $(p+1)$ normal equations given by (3.30). It is well known that the matrix $\boldsymbol{C}$ can be subjected to an orthogonal transformation

$$\boldsymbol{C}=\boldsymbol{V}\boldsymbol{\Lambda}\boldsymbol{V}^{\mathrm{T}}, \tag{3.35}$$

where $\boldsymbol{V}$ is the $(p+1)$ by $(p+1)$ orthogonal matrix of eigenvectors $\boldsymbol{v}_i^{\mathrm{T}}=(v_{1i}, v_{2i}, \ldots, v_{p+1\,i})$ and $\boldsymbol{\Lambda}$ is the $(p+1)$ by $(p+1)$ diagonal matrix of corresponding eigenvalues, λ_i. Since $\boldsymbol{C}$ is symmetric and positive definite, the λ_i's are real and positive; repeated eigenvalues, if any, are assumed to be numbered separately.

Combining (3.30, 35), we get

$$\boldsymbol{g}=\boldsymbol{C}^{-1}\sigma^2\boldsymbol{i}=\boldsymbol{V}\boldsymbol{\Lambda}^{-1}\boldsymbol{V}^{\mathrm{T}}\sigma^2\boldsymbol{i} \tag{3.36}$$

where $\sigma^2=P_m$ is the prediction-error variance for the PEO of length $(p+1)$. Therefore,

$$\boldsymbol{g}=\sigma^2\left(\frac{v_{11}}{\lambda_1}\boldsymbol{v}_1+\frac{v_{12}}{\lambda_2}\boldsymbol{v}_2+\ldots+\frac{v_{1p+1}}{\lambda_{p+1}}\boldsymbol{v}_{p+1}\right). \tag{3.37}$$

If we let

$$\frac{v_{11}}{\lambda_1}=k_1,\ \frac{v_{12}}{\lambda_2}=k_2,\ \ldots,\ \frac{v_{1p+1}}{\lambda_{p+1}}=k_{p+1} \tag{3.38}$$

and if we order the numbers k_i such that

$$|k_1|\geqq|k_2|\geqq\ldots\geqq|k_{p+1}|. \tag{3.39}$$

(3.37) can be written as

$$\boldsymbol{g}=k_1\sigma^2\left(\boldsymbol{v}_1+\frac{k_2}{k_1}\boldsymbol{v}_2+\ldots+\frac{k_{p+1}}{k_1}\boldsymbol{v}_{p+1}\right), \tag{3.40}$$

where the v_i's have been suitably relabelled.

The PEO is composed of all of the eigenvectors of the matrix $\boldsymbol{c}$ weighted by the inverses of the eigenvalues. Thus, the eigenvector corresponding to the minimum eigenvalue dominates the estimate of the PEO. Since the unit PEO is minimum delay, the addition of the other eigenvectors has the effect of forcing the zeros of $\boldsymbol{g}$ off the unit circle. The particular decomposition used in this section is important for a later discussion of ARMA processes.

The Toeplitz Method

The expression for the AR process, (3.34), in vector notation is

$$\begin{aligned} x_t &= \sum_{k=1}^{p} a_k x_{t-k} + e_t \\ &= \boldsymbol{a}^{\mathrm{T}} \boldsymbol{x}_{t-1} + e_t \\ &= \boldsymbol{x}_{t-1}^{\mathrm{T}} \boldsymbol{a} + e_t \,, \end{aligned} \tag{3.41}$$

where

$$\boldsymbol{a}^{\mathrm{T}} = (a_1, a_2, \ldots, a_p) \tag{3.42}$$

and

$$\boldsymbol{x}_{t-1}^{\mathrm{T}} = (x_{t-1}, x_{t-2}, \ldots, x_{t-p}) \,. \tag{3.43}$$

The determination of the AR parameters follows the principle of least squares. In other words, the second moment of the regression residual, e_t, is minimized with respect to $\boldsymbol{a}$.

The second moment is given in terms of the expectation of e_t^2 by

$$\begin{aligned} \sigma_e^2 &= \mathrm{E}[e_t^2] \\ &= \mathrm{E}[(x_t - \boldsymbol{x}_{t-1}^{\mathrm{T}} \boldsymbol{a})^2] \\ &= \mathrm{E}[x_t^2] - 2\boldsymbol{c}^{\mathrm{T}} \boldsymbol{a} + \boldsymbol{a}^{\mathrm{T}} \boldsymbol{C} \boldsymbol{a} \,, \end{aligned} \tag{3.44}$$

where

$$\boldsymbol{c} = (c_1, c_2, \ldots, c_p)^{\mathrm{T}} = \mathrm{E}[x_t \boldsymbol{x}_{t-1}] \tag{3.45}$$

is the autocovariance of the process starting with lag one, and

$$\boldsymbol{C} = \mathrm{E}[\boldsymbol{x}_{t-1} \boldsymbol{x}_{t-1}^{\mathrm{T}}] \tag{3.46}$$

is the p by p autocovariance matrix which has the well-known Toeplitz form.

Since σ_e^2 is quadratic with respect to the coefficient vector $\boldsymbol{a}$ it has a unique minimum which occurs when $\partial\sigma_e^2/\partial\boldsymbol{a}=0$. Differentiating (3.44) with respect to $\boldsymbol{a}$ and equating the result to zero, we obtain

$$2\boldsymbol{c}=2\boldsymbol{C}\boldsymbol{a}$$

or

$$\boldsymbol{C}\boldsymbol{a}=\boldsymbol{c}, \tag{3.47}$$

where $\boldsymbol{a}$ now represents the vector of estimated coefficients.

Equation (3.47) represents the so-called normal equations, mentioned previously, which are solved to give $\boldsymbol{a}$. We will consider two different approaches to this solution, but first of all, let us remark on the assumptions inherent in deducing (3.47).

1) The first assumption which we have made is that the process x_t is, at least, weakly stationary. In other words that the mean and autocovariance of x_t are independent of time.

2) We have introduced the expectation operator $\mathrm{E}[\cdot]$, which is generally defined as the sum of *all* values, weighted by the appropriate probability distribution, that a stochastic variable can take. In other words, implicit in (3.47) is the assumption that, either we have an infinite number of ensembles, or, for an ergodic process, we have an infinitely long record. The reason why this unrealistic assumption is made is due to the structure of (3.47) which lends itself to a particularly efficient and robust solution.

The recursive solution of (3.47) is more conveniently developed if we obtain an alternate form of these normal equations.

Multiplying (3.34) by x_t and taking expectations, we obtain, since $\mathrm{E}[x_t e_t]=\mathrm{E}[e_t^2]=\sigma_e^2$

$$\mathrm{E}[x_t x_t]=c_0=a_1c_1+a_2c_2+\ldots+a_pc_p+\sigma_e^2 \tag{3.48}$$

or

$$c_0-\boldsymbol{c}^{\mathrm{T}}\boldsymbol{a}=\sigma_e^2\,. \tag{3.49}$$

Augmenting (3.47) by (3.49), redefining $\boldsymbol{C}$ to be of order $(p+1)$ by $(p+1)$, and introducing the PEO

$$\boldsymbol{g}^{\mathrm{T}}=(1,g_1,g_2,\ldots,g_p), \tag{3.50}$$

we obtain

$$\boldsymbol{C}\boldsymbol{g}=\sigma_e^2\boldsymbol{i}=P_p\boldsymbol{i}\,. \tag{3.51}$$

This equation is identical to the Wiener formulation for obtaining the optimum prediction-error filter [3.39].

The Yule-Walker Solution

The Yule-Walker (YW) solution takes advantage of the Toeplitz form of $\boldsymbol{C}$ in (3.51) and uses the Levinson recursion [3.40], which has been often described in the literature. In keeping with the spirit of this chapter we will discuss those features of the YW solution which, we feel, require particular emphasis.

The first point to note is that the YW solution requires an estimated autocovariance vector $\boldsymbol{c}$. The particular estimate plays an important role, especially in the determination of the AR/ME spectrum and we discuss the properties of different estimators in Sect. 3.6.4. The second point of importance arises in the recursion itself. Consider the recursive determination of the PEO $\boldsymbol{g}_3$ from a known PEO $\boldsymbol{g}_2$, where we introduce the convention

$$\boldsymbol{g}_m^{\mathrm{T}} = (1, g_{1m}, g_{2m}, \ldots, g_{mm}) \tag{3.52}$$

and $\boldsymbol{C}_m$ is the Toeplitz matrix generated from the vector

$$\boldsymbol{c}_m^{\mathrm{T}} = (c_0, c_1, \ldots, c_m) \,. \tag{3.53}$$

Thus, (3.51) for $\boldsymbol{g}_3$ becomes $\boldsymbol{C}_3 \boldsymbol{g}_3 = P_3 \boldsymbol{i}$ or

$$\begin{bmatrix} c_0 & c_1 & c_2 & c_3 \\ c_1 & c_0 & c_1 & c_2 \\ c_2 & c_1 & c_0 & c_1 \\ c_3 & c_2 & c_1 & c_0 \end{bmatrix} \begin{bmatrix} 1 \\ g_{13} \\ g_{23} \\ g_{33} \end{bmatrix} = \begin{bmatrix} P_3 \\ 0 \\ 0 \\ 0 \end{bmatrix} . \tag{3.54}$$

Using a little algebra, the middle two equations of (3.54) become

$$\begin{bmatrix} g_{13} \\ g_{23} \end{bmatrix} = \begin{bmatrix} g_{12} \\ g_{22} \end{bmatrix} + g_{33} \begin{bmatrix} g_{22} \\ g_{12} \end{bmatrix} . \tag{3.55}$$

We now rewrite (3.54) using the form suggested by (3.55)

$$\boldsymbol{C}_3 \begin{bmatrix} 1 \\ g_{13} \\ g_{23} \\ g_{33} \end{bmatrix} = \boldsymbol{C}_3 \left[\begin{pmatrix} 1 \\ g_{12} \\ g_{22} \\ 0 \end{pmatrix} + g_{33} \begin{pmatrix} 0 \\ g_{22} \\ g_{12} \\ 1 \end{pmatrix} \right] = \left[\begin{pmatrix} P_2 \\ 0 \\ 0 \\ \Delta_2 \end{pmatrix} + g_{33} \begin{pmatrix} \Delta_2 \\ 0 \\ 0 \\ P_2 \end{pmatrix} \right] = \begin{bmatrix} P_3 \\ 0 \\ 0 \\ 0 \end{bmatrix} , \tag{3.56}$$

where Δ_2 is defined by $\boldsymbol{C}_3$ and $\boldsymbol{g}_2$.

Equation (3.56) expresses the crux of the Levinson recursion. The trick, here, is to write the normal equations not only in terms of the forward system

$$C_3 \begin{bmatrix} 1 \\ g_{12} \\ g_{22} \\ 0 \end{bmatrix} = \begin{bmatrix} P_2 \\ 0 \\ 0 \\ \Delta_2 \end{bmatrix} \tag{3.57}$$

but also in terms of the *backward* system [3.41],

$$C_3 \begin{bmatrix} 0 \\ g_{22} \\ g_{12} \\ 1 \end{bmatrix} = \begin{bmatrix} \Delta_2 \\ 0 \\ 0 \\ P_2 \end{bmatrix}. \tag{3.58}$$

This split into a forward and backward system of equations is also pivotal in the Burg and the LS solutions of the normal equations which we will be discussing later.

For completeness, the recursion from $k=2$ to $k=3$ is, from (5.56, 57)

$$\Delta_2 \leftarrow \sum_{k=0}^{2} g_{k2} c_{3-k} \quad \text{where} \quad g_{02}=1, \tag{3.59}$$

$$g_{33} = -\Delta_2/P_2, \tag{3.60}$$

$$\begin{bmatrix} 1 \\ g_{13} \\ g_{23} \\ g_{33} \end{bmatrix} \leftarrow \begin{bmatrix} 1 \\ g_{12} \\ g_{22} \\ 0 \end{bmatrix} + g_{33} \begin{bmatrix} 0 \\ g_{22} \\ g_{12} \\ 1 \end{bmatrix}, \tag{3.61}$$

and

$$P_3 = P_2[1-(\Delta_2/P_2)^2]. \tag{3.62}$$

Computer programs to do the Levinson recursion may be found in [3.37, 41].

One final comment is concerned with the phase properties of $\boldsymbol{g}$. We showed previously that the PEO is minimum delay. This imposes a constraint on any scheme used to solve the normal equations. That the Levinson recursion does in fact ensure this property for the computed PEO may be easily seen from the recursion itself [3.41].

The Burg Solution

In order to obviate the requirement of obtaining an a priori estimate of the autocovariance function, which is demanded by the YW solution, *Burg* [3.2, 3] conceived an ingenious approach to the solution of (3.51) which, in effect, generates its own autocovariance estimate which is consistent with the semipositive definite requirement of the autocovariance function [3.21]. The motivation behind Burg's solution is the fact that Levinson's trick of considering both a forward and backward scheme of equations leads to a minimum delay PEO.

A brief account of Burg's approach was presented in Sect. 2.6. Here we will illustrate the procedure for a recursion from second to third order for the case of real-valued data. Burg's approach is to use (3.61) to update the filter, but to compute g_{33} from the data themselves. We can write the prediction-error power P_3 as

$$P_3 = \frac{1}{2(n-3)} \sum_{k=4}^{n} [(x_k + g_{13}x_{k-1} + g_{23}x_{k-2} + g_{33}x_{k-3})^2 + (x_{k-3} + g_{13}x_{k-2} + g_{23}x_{k-1} + g_{33}x_k)^2]. \tag{3.63}$$

Using (3.61) in (3.63), P_3 becomes

$$\begin{aligned} P_3 = \frac{1}{2(n-3)} \sum_{k=4}^{n} \{ & [(x_k + g_{12}x_{k-1} + g_{22}x_{k-2}) \\ & + g_{33}(g_{22}x_{k-1} + g_{12}x_{k-2} + x_{k-3})]^2 \\ & + [(x_{k-3} + g_{12}x_{k-2} + g_{22}x_{k-1}) + g_{33}(x_k + g_{12}x_{k-1} + g_{22}x_{k-2})]^2 \} \end{aligned} \tag{3.64}$$

which can be conveniently written in terms of the forward and backward second-order prediction errors e_{f2} and e_{b2} as

$$P_3 = \frac{1}{2(n-3)} \sum_{k=4}^{n} [(e_{f2} + g_{33}e_{b2})^2 + (e_{b2} + g_{33}e_{f2})^2], \tag{3.65}$$

where the definitions of e_{f2} and e_{b2} are contained in (3.64).

The coefficient g_{33} is now determined from $\frac{\partial P_3}{\partial g_{33}} = 0$, and is given by

$$g_{33} = \frac{-2 \sum_{k=4}^{n} e_{f2}e_{b2}}{\sum_{k=4}^{n} (e_{f2})^2 + (e_{b2})^2}. \tag{3.66}$$

Burg's solution guarantees that $|g_{33}|<1$ [3.41] and therefore yields the required minimum delay PEO.

The complete recursion may be found in [3.21, 42]. Computer programs for the Burg algorithm have been published by *Ulrych* and *Bishop* [3.14], *Claerbout* [3.41], and *Nuttal* et al. [3.43].

The Least-Squares Method

The Toeplitz formulation arises by virtue of the assumption that an infinite ensemble of records is available for analysis. In the least-squares (LS) approach we recognize the fact that only one finite length record is available. The solution is obtained by minimizing the prediction-error power with respect to all the AR coefficients at a particular AR order. The prediction-error power is taken as the sum of the variances of the innovations of both the forward and backward AR models.

We define the forward and backward errors for a pth order AR process at time t as

$$e_{f,t}=x_t-\sum_{j=1}^{p} a_j x_{t-j} \qquad t=p+1,p+2,\ldots,n \tag{3.67}$$

$$e_{b,t}=x_{t-p}-\sum_{j=1}^{p} a_j x_{t-p+j} \qquad t=p+1,p+2,\ldots,n. \tag{3.68}$$

The total prediction-error power $\mathscr{E}$ is

$$\mathscr{E}=\sum_{k=p+1}^{n}(e_{f,k}^2+e_{b,k}^2). \tag{3.69}$$

In convenient matrix form (3.69) may be written as

$$\begin{aligned}\mathscr{E}&=\boldsymbol{e}_f^T\boldsymbol{e}_f+\boldsymbol{e}_b^T\boldsymbol{e}_b\\&=(\boldsymbol{x}_{p+1}-\boldsymbol{X}_f\boldsymbol{a})^T(\boldsymbol{x}_{p+1}-\boldsymbol{X}_f\boldsymbol{a})+(\boldsymbol{x}_1-\boldsymbol{X}_b\boldsymbol{a})^T(\boldsymbol{x}_1-\boldsymbol{X}_b\boldsymbol{a}),\end{aligned} \tag{3.70}$$

where the definitions are as follows.

$$\boldsymbol{x}_{p+1}^T=(x_{p+1},x_{p+2},\ldots,x_n), \tag{3.71}$$

$$\boldsymbol{x}_1^T=(x_1,x_2,\ldots,x_{n-p}), \tag{3.72}$$

$$\boldsymbol{e}_f=\boldsymbol{x}_{p+1}-\boldsymbol{X}_f\boldsymbol{a}, \tag{3.73}$$

$$\boldsymbol{e}_b=\boldsymbol{x}_1-\boldsymbol{X}_b\boldsymbol{a}, \tag{3.74}$$

and X_f and X_b are the forward and backward propagating matrices for the problem:

$$X_f = \begin{bmatrix} x_p & x_{p-1} & \dots & x_1 \\ \vdots & & & \vdots \\ x_{2p-1} & & & x_p \\ \vdots & & & \vdots \\ x_{n-1} & \dots & \dots & x_{n-p} \end{bmatrix}; \quad X_b = \begin{bmatrix} x_2 & x_3 & \dots & x_{p+1} \\ \vdots & & & \vdots \\ x_{p+1} & & & x_{2p} \\ \vdots & & & \vdots \\ x_{n-p+1} & \dots & \dots & x_n \end{bmatrix}. \tag{3.75}$$

Minimizing (3.70) with respect to a we obtain the normal equations for this case.

$$(X_f^T X_f + X_b^T X_b) a = X_f^T x_{p+1} + X_b^T x_1 \tag{3.76}$$

which we can write as

$$\tilde{C} a = s . \tag{3.77}$$

The elements of the p by p matrix $\tilde{C}$ are given by

$$\tilde{c}_{jk} = \sum_{i=p+1}^{n} x_{i-j} x_{i-k} + \sum_{i=p+1}^{n} x_{i-p+j} x_{i-p+k}$$
$$j, k = 1, 2, \dots, p . \tag{3.78}$$

The s is a p length column vector with elements:

$$s_j = \sum_{i=p+1}^{n} x_{i-j} x_i + \sum_{i=p+1}^{n} x_{i-p+j} x_{i-p}$$
$$j = 1, 2, \dots, p . \tag{3.79}$$

It is immediately apparent that $\tilde{C}$ is a symmetrical but not necessarily a Toeplitz matrix, although it approaches the Toeplitz form as $n \to \infty$. Clearly the Levinson scheme cannot be used in solving (3.77). Recursive schemes are however available [3.44, 45] and have recently been described in the geophysical literature by *Berkhout* and *Zaanen* [3.46].

Remarks Concerning AR Parameter Determination

Chen and *Stegen* [3.47] have noted that, for short data segments of a harmonic process, the AR/ME spectral estimate is very sensitive to the initial phase of the sinusoid. In fact these authors have shown that frequency estimation errors of up to 30% may occur. The dashed curve in Fig. 3.4 shows such a variation for a 1 Hz signal obtained using the Burg power estimate. Although *Chen* and *Stegen*

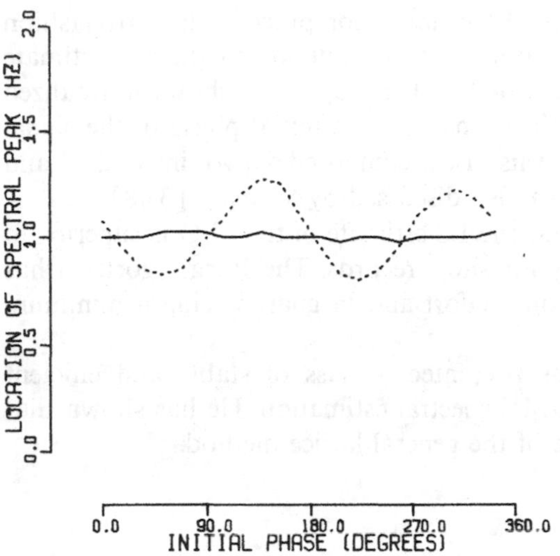

Fig. 3.4. Location of the spectral peak of 1 Hz sinusoid as a function of initial phase. LS estimate (———); Burg estimate (- - - -)

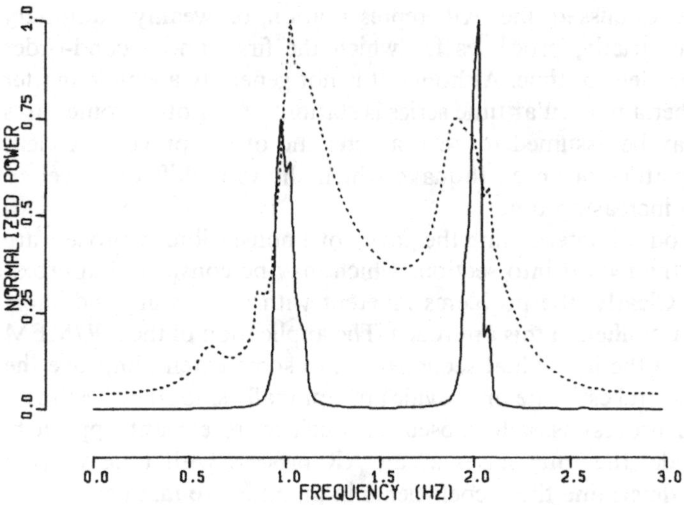

Fig. 3.5. Variation of the spectrum of two sinusoids as a function of initial phase computed using the forward-backward LS estimate (———) and the forward LS estimate (- - - -)

[3.47] interpreted this phase dependence as being due to the interference between positive and negative frequency spectral contributions, *Ulrych* and *Clayton* [3.16] suggested that the dependence was due to the assumption of a Toeplitz autocovariance matrix used in the Burg estimate, an assumption which is not correct for short records. The backward and forward LS scheme introduced by *Ulrych* and *Clayton* [3.16] and expressed by (3.76) leads to an AR/ME estimate which is far less sensitive to initial phase than the Toeplitz AR/ME estimate. This point is demonstrated by the solid line in Fig. 3.4.

It is of interest to point out that a LS estimate computed in standard fashion by minimizing the forward error only is much inferior to the LS estimate discussed here. This point is illustrated in Fig. 3.5, which shows normalized spectra which are averages of realizations with the initial phase of the signal varying from 0 to 2π. The signal in this case is composed of two sinusoids (1 and 2 Hz) with 10% noise. This point is also discussed by *Swingler* [3.48].

We wish to emphasize here that the LS estimate of the PEO is superior to the YW and Burg estimates only for short records. The Burg algorithm has distinct advantages in computational effort and in guaranteeing a minimum delay PEO.

Recently, *Makhoul* [3.49] has presented a class of stable and efficient recursive lattice methods for AR/ME spectral estimation. He has shown that Burg's procedure is a special case of the general lattice method.

3.4 Adaptive AR Processes

Thus far we have discussed the AR representation of weakly stationary processes, or more strictly, processes for which the first- and second-order statistics are independent of time. Although it is not generally a simple matter to determine whether a particular time series is stationary or not, in some cases nonstationarity may be assumed to be characteristic of the process. A clear example is the signature of an earthquake which shows a shift of power to longer period with increasing time.

The usual methods of determining the power of a nonstationary process are based on dividing the record into sections which may be considered approximately stationary. Clearly, the problems inherent with windowing and short record lengths are amplified in this approach. The application of the AR/MEM spectral technique to the individual sections can, to some extent, improve the resolution of the spectral estimate or provide an "optimally smoothed" estimate if the order of the process is well chosen. A much more elegant approach, however, is to model the time series as an AR process with time varying coefficients and to determine these coefficients in an adaptive manner.

In this section we will consider the adaptive algorithm of *Widrow* and *Hoff* [3.50] and we will illustrate the discussion with examples of adaptive spectral analysis of harmonic processes and adaptive modeling of strong ground motion. These examples are chosen to illustrate the resolution and smoothing properties of the AR/MEM spectral estimator as well as the modeling potential of the parametric approach.

3.4.1 The Adaptive LMS Algorithm

In this section, we consider an adaptive algorithm which was proposed by *Widrow* and *Hoff* [3.50] and has been applied to AR processes by *Griffiths*

[3.9], and *Griffiths* and *Prieto-Diaz* [3.51]. By way of illustration let us consider the adaptation of a first-order AR process in the manner of [3.41]. We have

$$x_t = a_1 x_{t-1} + e_t . \tag{3.80}$$

The sum of the squared prediction error is

$$E = \sum_t e_t^2 = \sum_t (x_t - a_1 x_{t-1})^2 . \tag{3.81}$$

The standard least-mean-squared (LMS) approach in the stationary case is to determine a_1 from $\partial E/\partial a_1 = 0$. Since the AR parameters are time dependent in the nonstationary case, however, we wish to redetermine these parameters as each new data point is considered. In this case $\partial E/\partial a_1$ will depart from zero and the adaptive algorithm is designed to minimize this departure. Thus, from (3.81) we write

$$E = e_t^2 = (x_t - a_1 x_{t-1})^2 \tag{3.82}$$

and

$$\begin{aligned} \frac{\partial E}{\partial a_1} &= \nabla[e_t^2] = -2(x_t - a_1 x_{t-1}) x_{t-1} \\ &= -2 e_t x_{t-1} . \end{aligned} \tag{3.83}$$

Thus the addition of a new data point x_t causes a change in $\partial E/\partial a_1$ of $-2e_t x_{t-1}$. This observation leads to the motivation of iteratively correcting the coefficient a_1 to be

$$a_1 \to a_1 + \mu e_t x_{t-1} \to a_1 + \mu \nabla[e_t^2] \tag{3.84}$$

which is equivalent to seeking the minimum mean-square error by the method of steepest descent. The constant μ is a convergence factor which adjusts the amount of correction which we are going to make. The choice of μ is considered later.

The simple discussion above illustrates the concept of the LMS technique. It is important to realize that this adaptive approach, like all stochastic approximation methods, [3.52], does not necessarily yield optimal estimates. It does however yield recursive estimates which do have well-defined convergence properties.

We now consider the LMS algorithm in more detail. Let $\boldsymbol{a}_k$ represent the column vector of AR coefficients for an lth order process at time k. Thus

$$\boldsymbol{a}_k^{\mathrm{T}} = (a_{1k}, a_{2k}, \ldots, a_{lk}) . \tag{3.85}$$

The LMS adaptive algorithm is described by

$$a_{k+1} = a_k + \eta \hat{\nabla}[E(e_k^2)], \tag{3.86}$$

where e_k is the prediction error at time k and $\hat{\nabla}[E(e_k^2)]$ indicates the estimate of the gradient of the mean-square error with respect to $\boldsymbol{a}_k$, and is a column vector. The practical noisy gradient algorithm proposed by *Widrow* and *Hoff* [3.50] (see also [3.53]) replaces $\hat{\nabla}[E(e_k^2)]$ by the gradient of a single time sample of the squared error. The rationale for this follows from the fact that, as shown below, this estimate of the gradient is unbiased [3.53, 54].

The prediction error, e_k, associated with the kth sample point is given by

$$\begin{aligned} e_k &= x_k - \sum_{j=1}^{l} a_{jk} x_{k-j} \\ &= x_k - \boldsymbol{a}_k^{\mathrm{T}} \boldsymbol{x}_{k-1}, \end{aligned} \tag{3.87}$$

where $\boldsymbol{x}_{k-1}^{\mathrm{T}} = (x_{k-1}, x_{k-2}, \ldots, x_{k-l})$ is the vector of data samples which enters into the prediction.

The estimated gradient is obtained as

$$\hat{\nabla}[E(e_k^2)] = \nabla[e_k^2] = 2e_k \nabla[e_k]. \tag{3.88}$$

From (3.87)

$$\nabla[e_k] = \nabla[x_k - \boldsymbol{a}_k^{\mathrm{T}} \boldsymbol{x}_{k-1}] = -\boldsymbol{x}_{k-1}. \tag{3.89}$$

Therefore

$$\begin{aligned} \hat{\nabla}[E(e_k^2)] &= -2e_k \boldsymbol{x}_{k-1} \\ &= -2[x_k - \boldsymbol{a}_k^{\mathrm{T}} \boldsymbol{x}_{k-1}] \boldsymbol{x}_{k-1}. \end{aligned} \tag{3.90}$$

Taking expectations in (3.90)

$$E\hat{\nabla}[E(e_k^2)] = -2[\boldsymbol{c} - \boldsymbol{C}\boldsymbol{a}_k], \tag{3.91}$$

where $\boldsymbol{c}$ and $\boldsymbol{C}$ are as defined previously in (3.45, 46).

We return to the definition of the error in (3.87) and obtain its square,

$$e_k^2 = x_k^2 - 2x_k \boldsymbol{a}_k^{\mathrm{T}} \boldsymbol{x}_{k-1} + \boldsymbol{a}_k^{\mathrm{T}} \boldsymbol{x}_{k-1} \boldsymbol{x}_{k-1}^{\mathrm{T}} \boldsymbol{a}_k. \tag{3.92}$$

Differentiating the expectation of (3.92) with respect to $\boldsymbol{a}_k$, we obtain

$$\nabla[E(e_k^2)] = -2\boldsymbol{c} + 2\boldsymbol{C}\boldsymbol{a}_k. \tag{3.93}$$

Comparing (3.93, 91) establishes that $\nabla[e_k^2]$ is an unbiased estimator of the gradient of the mean-square error. Using (3.86, 90) we obtain the LMS algorithm

$$\boldsymbol{a}_{k+1} = \boldsymbol{a}_k + \mu e_k \boldsymbol{x}_{k-1}, \tag{3.94}$$

where

$$\mu = -2\eta. \tag{3.95}$$

It is important to note that once the term μe_k has been computed, only one multiply and add operation is required to obtain a_{ik+1} from a_{ik}. This leads to an economic algorithm which, to a large extent, is independent of the initializing filter, $\boldsymbol{a}_0$.

The choice of the constant μ in (3.94) is important in that this parameter controls the rate of convergence of the vector $\boldsymbol{a}_k$ to the stationary Wiener solution given by the normal equations (3.47). In other words, it may be shown that, providing μ satisfies certain bounds,

$$\lim_{k\to\infty} \mathrm{E}[\boldsymbol{a}_{k+1}] = \boldsymbol{C}^{-1}\boldsymbol{c}. \tag{3.96}$$

Equation (3.96) requires stationarity of the input series and the assumption that successive input samples are independent.

The bounds on μ may be determined by taking expectations in (3.94) and considering $k+1$ iterations with an initial vector $\boldsymbol{a}_0$ [3.53]. The development makes use of the eigenvector decomposition of $\boldsymbol{C}$ given by (3.35) and leads to the result that (3.96) is satisfied providing

$$2/\lambda_{\max} > \mu > 0, \tag{3.97}$$

where $\lambda_{\max}$ is the maximum eigenvalue of $\boldsymbol{C}$. These bounds on μ are required to ensure stability of the algorithm. The value of μ affects the adaptive time constant τ and, in addition, it influences what is known as the misadjustment noise. The latter is a consequence of the fact that the adaptive filter adapts on the basis of small sample size statistics. Clearly, the faster the adaptation proceeds, the poorer the performance will be. Based on the work of *Widrow* [3.53], Griffiths [3.9] has shown that;

a)

$$\mu = \frac{\alpha}{l\sigma^2} \tag{3.98}$$

where $0 < \alpha < 2$ and σ^2 is the average input power.

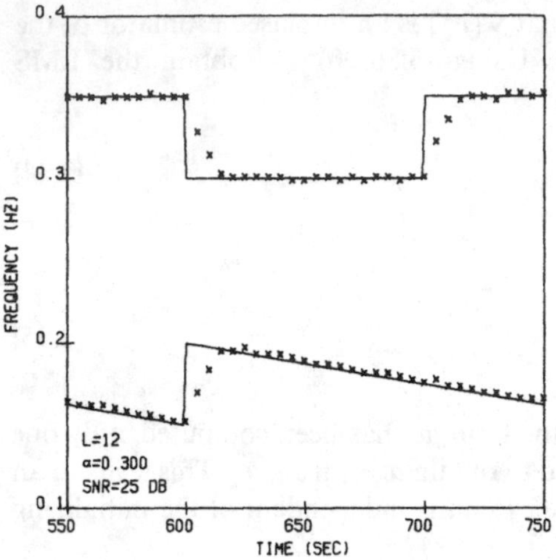

Fig. 3.6. Behavior of the adaptive LMS algorithm for a sinusoidal input (I). Actual frequencies (———); computed frequencies (×)

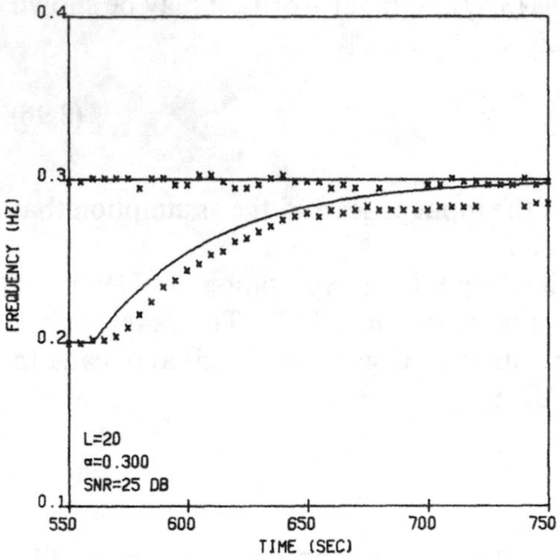

Fig. 3.7. Behavior of the adaptive LMS algorithm for a sinusoidal input (II). Actual frequencies (———); computed frequencies (×)

b)

$$\tau \approx \frac{\Delta t}{\log_e(1-\alpha/l)}, \tag{3.99}$$

where Δt is the sampling interval.

c) If M is the misadjustment due to gradient noise defined as [3.55]

$$M = \frac{\text{average excess mean-square error}}{\text{minimum mean-square error}} \tag{3.100}$$

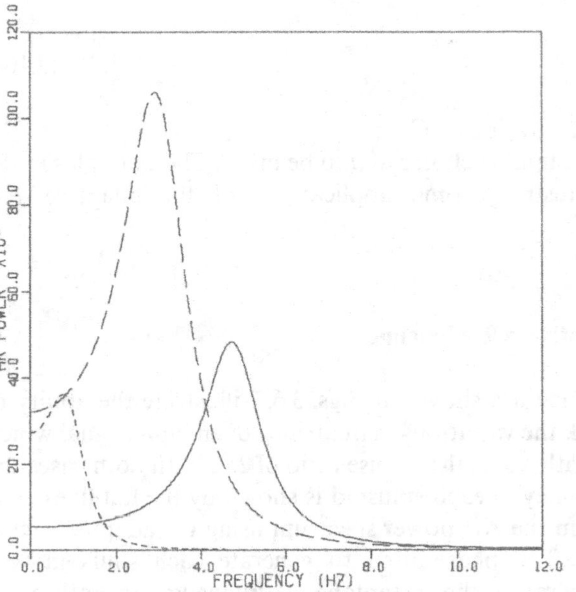

Fig. 3.8. Adaptive AR power spectra computed for the 1971 San Fernando earthquake recorded at Orion Blvd.

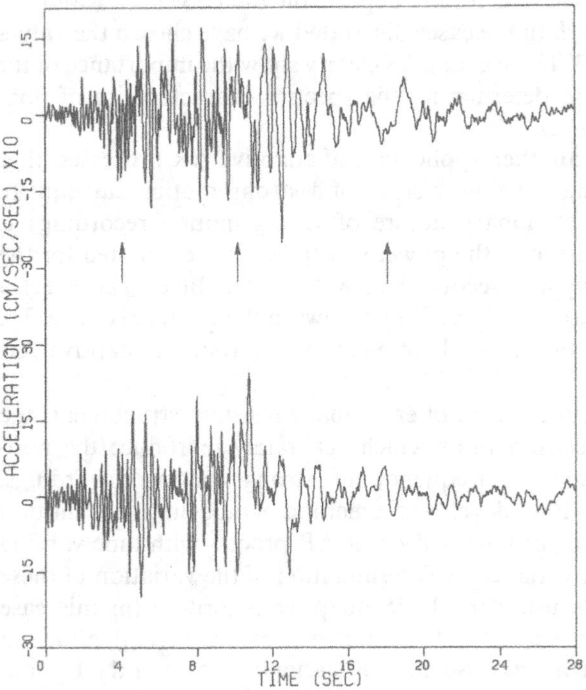

Fig. 3.9. The Orion recording of the 1971 San Fernando earthquake is shown in the top trace. The bottom trace illustrates the simulated earthquake using adaptive AR modeling [3.10]

then

$$M = \mu l \lambda_{av}, \qquad (3.101)$$

where λ_{av} is the average eigenvalue of $\boldsymbol{C}$.

These relations allow a suitable choice of μ to be made. The examples in the following section will illustrate some applications of the adaptive AR algorithm.

3.4.2 Applications of Adaptive AR Modeling

The first two examples which are shown in Figs. 3.6, 7 illustrate the ability of the LMS algorithm to track the variations in frequency of an input signal which consists of two sinusoids with a signal-to-noise ratio of 25 dB. In both cases the actual instantaneous frequency of each sinusoid is shown by the full lines. The frequencies as deduced from the AR power spectrum using the adapted vector $\boldsymbol{a}_k$ are shown as crosses. The phase used to generate each sinusoid was computed by digitally integrating the instantaneous frequency. In both cases the filter was initialized to zero and sufficient time was allowed for the filter to stabilize. As discussed above, the results depend on the choice of α and the length of the adaptive filter, l. In the cases illustrated we have chosen the values $l=12$ and $l=20$ with $\alpha=0.3$. These examples clearly show the importance of the LMS adaptive algorithm in determining the spectral characteristics of non-stationary harmonic processes.

Figures 3.8, 9 illustrate another application of adaptive AR processes, that of analyzing and simulating the time history of a strong motion earthquake recording [3.10]. The nonstationary nature of strong motion recordings is illustrated in Fig. 3.8 which shows the power spectral curves computed for the 1971 San Fernando earthquake recorded at 8244 Orion Blvd. The spectra correspond to the time locations indicated by arrows in the top trace of Fig. 3.9, which shows the Orion recording, and were computed from an adaptive AR filter of second order.

An important aspect in the design of earthquake-resistant structures is the computation of synthetic seismograms which accurately reproduce the non-stationarities in amplitude and frequency of strong motion recordings. *Jurkevics* and *Ulrych* [3.10] have developed a method of modeling such motion by representing the recording as a second-order AR process with time varying coefficients and innovation variance. A determination of the variation of these three parameters with time using the LMS adaptive algorithm (in this case extended to include backward as well as forward errors [3.10]) allows a simulation of the target earthquake. Such a simulation is shown in the bottom trace of Fig. 3.9.

3.5 Autoregressive-Moving Average Processes

We have already briefly discussed the ARMA model and its relationship to the Wold decomposition theorem. The motivation for representing a process by an ARMA model is the parsimony of the approach, such an approach often yields the fewest number of parameters, and the fact that, in some instances, the ARMA model has particular physical significance. For example, in seismic exploration, this model represents the impulse response of a perfectly elastic, horizontally stratified medium [3.6, 56]. The discrete ARMA representation is also of importance in the representation of continuous AR processes and in the estimation of models of AR processes plus white noise. Thus, *Bartlett* [3.57] has shown that a discrete process which is covariance-invariant [3.58] to a continuous AR process of order p is equivalent to an ARMA model of order $(p, p-1)$. Furthermore, it can be shown [3.59] that an AR process plus white noise is equivalent to an ARMA scheme.

In the last few years several different methods of determining ARMA parameters have been published. *Box* and *Jenkins* [3.19] have investigated both stationary and nonstationary ARMA models in a rigorous manner. Their algorithms do not automatically compute the order of the ARMA process, however, and their final least-squares solution, based on two separate computations, requires off-line decisions.

Mehra [3.60] identified a state space representation for an ARMA model subject to the restriction that the MA order is equal to or less than the AR order. This restriction may be rather severe in many cases of interest.

Krause and *Graupe* [3.61] have proposed two methods of determining the MA parameters of an ARMA process after the AR parameters have been estimated [3.62]. The first method is to identify the inverse of the MA parameters from the residual time series generated by the already computed AR coefficients. The second approach is identical to that of [3.19] and solves a system of nonlinear equations to obtain initial estimates.

Recently, *Treitel* et al. [3.6] have proposed a new iterative least-squares scheme which guarantees the minimum delay property of the AR parameters, a constraint for which there is considerable physical motivation. They also demonstrated that whereas the magnitude of the partial correlation coefficient may be used to determine the order of an AR process this approach breaks down for ARMA series.

In this section we will concentrate our discussion on a maximum-likelihood approach to the determination of ARMA parameters proposed by *Akaike* [3.63]. This approach, based on the assumption that both the AR and MA components are invertible, may be applied to processes of arbitrary orders and depends only on the estimated autocovariance function (see Sect. 3.6.4). The Akaike method is available as an algorithm written in Fortran [3.64] and has been applied to scalar data by *Gersch* et al. [3.65]. Recently *Ooe* [3.11] has extended the Akaike method to complex processes. An important advantage of the Akaike approach is that it incorporates an information criterion, called the AIC [3.29, 63, 66] which can be used to identify the optimum ARMA representation. The AIC has been successfully tested in practice by *Tong* [3.67–69], and *Ooe* [3.11].

3.5.1 The ARMA Model for Sinusoids Plus White Noise

Before considering the general ARMA model we will develop a model for a process which consists of $p/2$ harmonic components with additive white noise [3.16].

Thus

$$x_t = \sum_{k=1}^{p/2} q_k \sin(\omega_k t - \phi_k) + n_t , \tag{3.102}$$

where

$$\mathrm{E}[n_t n_s] = \sigma_n^2 \delta_{ts} \tag{3.103}$$

and q_i is the amplitude of the ith harmonic.

It is well-known that any harmonic process can be represented by a second-order difference equation

$$y_t = a_1 y_{t-1} + a_2 y_{t-2} , \tag{3.104}$$

where the roots of the polynomial $1 - a_1 z^{-1} - a_2 z^{-2} = 0$ lie on the unit circle $|z| = 1$ in the complex z plane. The angular frequency of the harmonic is given by the angle which the upper-half plane root subtends with the real axis.

For $p/2$ harmonics we have

$$y_t = \sum_{k=1}^{p} a_k y_{t-k} . \tag{3.105}$$

Although it may appear that the effect of the additive noise, n_t, is to make (3.105) into an AR process of order p, it must be realized that, whereas for the AR process given by (3.34) $\mathrm{E}[y_t e_t] = \sigma_e^2$, in this case $\mathrm{E}[y_t n_t] = 0$.

Therefore we write

$$\begin{aligned} x_t &= y_t + n_t \\ &= \sum_{k=1}^{p} a_k y_{t-k} + n_t . \end{aligned} \tag{3.106}$$

If we substitute $y_{t-k}=x_{t-k}-n_{t-k}$ into (3.106) we get

$$x_t=\sum_{k=1}^{p} a_k x_{t-k}+n_t-\sum_{k=1}^{p} a_k n_{t-k}. \tag{3.107}$$

We can see that (3.107) is a special case of the general ARMA (p,p) model. It is special in the sense that the AR and MA parameters are the same.

Expressing (3.107) in matrix form we have

$$\boldsymbol{x}_t^{\mathrm{T}}\boldsymbol{g}=\boldsymbol{n}_t^{\mathrm{T}}\boldsymbol{g}, \tag{3.108}$$

where

$$\boldsymbol{x}_t^{\mathrm{T}}=(x_t, x_{t-1}, \ldots, x_{t-p}), \tag{3.109}$$

$$\boldsymbol{n}_t^{\mathrm{T}}=(n_t, n_{t-1}, \ldots, n_{t-p}), \tag{3.110}$$

and as before

$$\boldsymbol{g}^{\mathrm{T}}=[1:-\boldsymbol{a}]. \tag{3.111}$$

We now premultiply (3.108) by $\boldsymbol{x}_t$ and take expectations:

$$\mathrm{E}[\boldsymbol{x}_t\boldsymbol{x}_t^{\mathrm{T}}]\boldsymbol{g}=\mathrm{E}[\boldsymbol{y}_t\boldsymbol{n}_t^{\mathrm{T}}]\boldsymbol{g}+\mathrm{E}[\boldsymbol{n}_t\boldsymbol{n}_t^{\mathrm{T}}]\boldsymbol{g}. \tag{3.112}$$

Therefore,

$$\boldsymbol{C}\boldsymbol{g}=\sigma_n^2\boldsymbol{g}, \tag{3.113}$$

where $\boldsymbol{C}$ is the Toeplitz autocovariance matrix of the process.

The solution of (3.113) is an eigenproblem in which the desired coefficient vector is the normalized eigenvector corresponding to the minimum eigenvalue. (See [3.32] for a proof.) This solution, in fact, corresponds to a method of spectral analysis suggested by *Pisarenko* [3.70], in which harmonics are retrieved from the autocovariance function. Accordingly, the spectral estimate obtained using $\boldsymbol{g}$ from (3.113) will be the Pisarenko spectral estimate.

The frequencies of the hamonics modeled by the ARMA process are determined from the roots of the polynomial $\boldsymbol{g}^{\mathrm{T}}\boldsymbol{z}=0$ which lie on the unit circle, where $\boldsymbol{z}^{\mathrm{T}}=(z^0, z^{-1}, \ldots, z^{-p})$.

Since

$$z^k=\exp(\mathrm{j}2\pi fk)$$

the fast Fourier transform (FFT) algorithm may be conveniently used to find the frequencies. Because the eigenvector does not contain amplitude information, the power associated with the harmonics must be determined by fitting the determined frequencies to the autocovariance function. Assuming that the noise affects only the zero lag of the autocovariance function c_τ, we can write

$$c_\tau = \sigma_n^2 \delta_\tau + \sum_{k=1}^{p/2} \frac{q_k^2}{2} \cos(2\pi f k \tau), \quad 0 \leqq \tau \leqq m, \tag{3.114}$$

where m is the number of lags.

The powers $p_k = q_k^2/2$, $k = 1, 2, \ldots, p/2$ are determined by least squares from the equation

$$\boldsymbol{D}\boldsymbol{p} = \boldsymbol{s}, \tag{3.115}$$

where $\boldsymbol{D}$ is a $(p-1)$ by $(p-1)$ matrix with terms

$$d_{ij} = \sum_{l=1}^{m} \cos(2\pi f i l) \cos(2\pi f j l) \quad i, j = 1, 2, \ldots, p, \tag{3.116}$$

$$\boldsymbol{s}^{\mathrm{T}} = (s_1, s_2, \ldots, s_p) \tag{3.117}$$

with

$$s_i = \sum_{l=1}^{m} c_l \cos(2\pi f i l) \quad i = 1, 2, \ldots, p \tag{3.118}$$

and

$$\boldsymbol{p}^{\mathrm{T}} = (p_1, p_2, \ldots, p_p). \tag{3.119}$$

Equation (3.115) excludes the zero lag, c_0, which contains the term σ_n^2.

3.5.2 Relationship Between the Pisarenko and the AR/ME Spectral Estimators

An interesting relationship which has been explored to some extent [3.16] exists between the AR/ME spectral estimator and the special ARMA or Pisarenko spectral estimate which we have described. Although an infinite order AR model is required to represent the discrete ARMA process of (3.107) [3.71] we will, for the sake of this discussion, choose an AR model of order p.

Thus we model x_t of (3.107) as

$$x_t = \sum_{k=1}^{p} \tilde{a}_k x_{t-k} + e_t. \tag{3.120}$$

As we have seen the normal equations (3.51), which are solved for the AR parameters, are

$$C\tilde{g}=\sigma_e^2 i\,, \tag{3.121}$$

where

$$\tilde{g}=[1:-\tilde{a}]\,. \tag{3.122}$$

From Sect. 3.3.2 and (3.36, 37) we see that

$$\tilde{g}=V\Lambda^{-1}V^{\mathrm{T}}\sigma_e^2 i$$

or

$$\tilde{g}=\sigma_e^2\sum_{k=1}^{p}\frac{v_{k1}}{\lambda_k}v_k\,. \tag{3.123}$$

In this particular case, the eigenvectors v_i in (3.123) are replaced by the eigenvectors g_i of which g_1 corresponds to the solution of the special ARMA model of (3.113).

Therefore

$$\tilde{g}=\sigma_e^2\sum_{k=1}^{p}\frac{g_{k1}}{\lambda_k}g_k\,. \tag{3.124}$$

Equation (3.124) is particularly interesting. It shows that the AR/ME estimate is composed of all the eigenvectors of (3.113) weighted by the inverses of the eigenvalues. The eigenvector corresponding to the minimum eigenvalue, which is the correct solution to (3.113), dominates $\tilde{g}$. The effect of the remaining eigenvectors is to force the zeros of $\tilde{g}$ off the unit circle. Since the proximitiy of the zeros of $\tilde{g}$ to the unit circle determines the resolution of the spectral estimator, the AR/ME estimator will always be a smoothed form of the Pisarenko estimator g_1 whose zeros lie on the unit circle.

3.5.3 Remarks Concerning the Pisarenko Estimate

As we have shown, the Pisarenko power estimate is the estimate with the highest resolution. However, when evaluating the performance of an estimator, the precision of the estimate is usually as important as the resolution. *Ulrych* and *Clayton* [3.16] have shown that the Pisarenko estimate is particularly sensitive to the estimate of the autocovariance matrix, and for short realizations of harmonic processes with noise, the frequencies of the components are determined less accurately than they are by using the LS estimate. For high

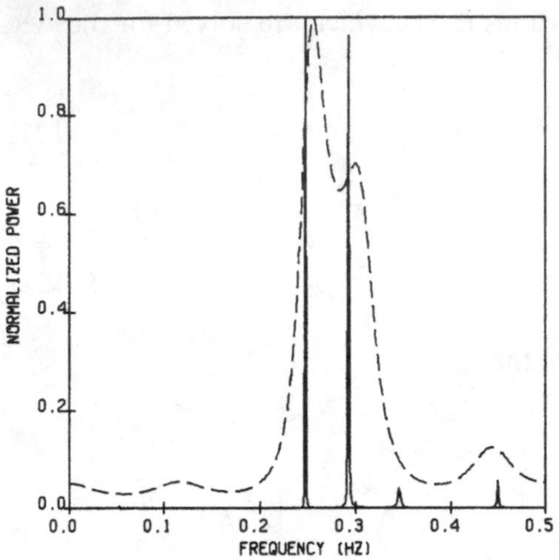

Fig. 3.10. Increase in resolution of normal AR/ME estimate (----) resulting from a decrease in the zero lag autocovariance

signal-to-noise ratios, however, the Pisarenko estimator is an important spectral estimator for harmonic processes. The expansion of the application of this estimator to sinusoids in correlated noise has recently been considered by *Satorius* and *Alexander* [3.72].

The resolution of the Pisarenko estimate can be approached using the AR/ME estimate under certain circumstances. Thus, assuming as before that we are dealing with a harmonic process with additive white noise, we can write the autocovariance function as

$$c_\tau = \sigma_n^2 \delta_\tau + \sum_{k=1}^{p} \frac{q_k^2}{2} \cos(2\pi f k \tau) \tag{3.125}$$

emphasizing that, theoretically, the noise power, σ_n^2, appears only in the zero lag autocovariance. *Frost* [3.33] suggested the procedure of estimating σ_n^2 by computing the smallest eigenvalue of $\boldsymbol{C}$ and subtracting a portion of it from c_0. This leads to the normal equations for computing the PEO given by

$$(\boldsymbol{C} - k\sigma_n^2 \boldsymbol{I})\boldsymbol{g} = (1-k)\boldsymbol{p}_{p+1}, \tag{3.126}$$

where k is a constant. We can see that, as $k \to 1$, the above equation becomes equivalent to (3.113) which gives the Pisarenko PEO.

An example of the procedure outlined above is presented in Fig. 3.10, which illustrates the power spectra for two sinusoids of equal amplitude with a signal-to-noise ratio of 6 dB. The resolved estimate was computed for $k\sigma_n^2$ equal to the residual sum of squares for a PEO of length ten. *Marple* [3.32] has proposed an iterative approach to the above procedure.

3.5.4 Determination of Parameters of the General ARMA Process

The general ARMA process is expressed by (3.21) which we rewrite here for convenience:

$$x_t = \sum_{i=1}^{p} a_i x_{t-i} + e_t - \sum_{i=1}^{q} b_i e_{t-1} \quad t = p+1, \ldots, n. \tag{3.127}$$

We repeat that the process e_t has the properties $\mathrm{E}[e_t]=0$ and $\mathrm{E}[e_t e_s]=\sigma_e^2 \delta_{ts}$.

We can rewrite (3.126) in terms of the z transforms of $\boldsymbol{a}$ and $\boldsymbol{b}$ remembering that z^{-1} is the unit-delay operator; thus

$$A(z)[x_t] = B(z)[e_t], \tag{3.128}$$

where $A(z)$ and $B(z)$ act as operators on x_t and e_t, respectively.

Since $A(z)[x_t]$ is correlated with $A(z)[x_{t+k}]$ for $0 \leq k \leq q$ but is not correlated with $A(z)[x_{t+k}]$ for $k \geqq q+1$, we can take expectations in (3.127) and arrive at the following matrix expressions:

$$\begin{bmatrix} c(q+1,1) & c(q+1,2) & \ldots & c(q+1,p) \\ c(q+2,1) & c(q+2,2) & \ldots & c(q+2,p) \\ \vdots & & & \\ c(q+p,1) & c(q+p,2) & \ldots & c(q+p,p) \end{bmatrix} \begin{bmatrix} a_1 \\ a_2 \\ \\ a_p \end{bmatrix} = \begin{bmatrix} c(q+1,0) \\ c(q+2,0) \\ \vdots \\ c(q+p,0) \end{bmatrix} \tag{3.129}$$

where

$$c(j,k) = \mathrm{E}[x_{t-j} x_{t-k}]. \tag{3.130}$$

Krause and *Graupe* [3.61] presented a least-squares recursive algorithm for solving (3.129). An estimate of $c(j,k)$ was taken to be

$$c_{jk} = \frac{1}{n-p-q} \left\{ \sum_{l=p+q+1}^{n} x_{l-k} x_{l-j} \right\}. \tag{3.131}$$

For a long realization of a stationary process, c_{jk} may be replaced by ${}_1c_{j-k}$ which is defined by

$$ {}_1c_{j-k} = {}_1c_l = \frac{1}{n} \sum_{k=1}^{n-l} x_k x_{k+l} . \tag{3.132} $$

A method to estimate $\boldsymbol{a}$ by using ${}_1c_l$ was presented by [Ref. 3.19, p. 202].

Once (3.129) is solved for the AR parameters, the MA parameters may be estimated by the first method of *Krause* and *Graupe* [3.61].

Let us define w_t by

$$ w_t = A(z)[x_t] . \tag{3.133} $$

We then solve the large order AR model

$$ D(z)[w_t] = n_t \tag{3.134} $$

by minimizing the variance of n_t in the usual manner. An estimate of the MA parameters is then obtained from

$$ B^{-1}(z) = D(z) , \tag{3.135} $$

where $b_0 = 1$.

The second approach used, [3.61], is to solve the following $q+1$ nonlinear equations with $q+1$ unknowns:

$$ \begin{aligned} &\frac{1}{n-p-k} \sum_{i=p+k+1}^{n} w_{i-k} w_i = \sigma_n^2 \sum_{j=0}^{q} b_{j-k} b_j \\ &k = 0, 1, \ldots, q \\ &b_0 = 1 \quad \text{and} \quad b_j = 0 \quad \text{for} \quad j < 0 . \end{aligned} \tag{3.136} $$

Equations (3.136) are solved by the linear convergence process described by [Ref. 3.19, pp. 202–205] which uses the Newton–Raphson algorithm [3.12].

The Method of *Treitel* et al. [3.6]

Treitel et al. [3.6] have developed an iterative least-squares procedure for determining the ARMA parameters which guarantees the physically important property of the minimum delay nature of the AR component. This procedure is treated in some detail in Chap. 4, and consequently we will only sketch the outline of the algorithm in this section. A comparison of the Akaike method and the *Treitel* et al. [3.6] method is presented later on in this section.

Our task is to represent the z transform $X(z)$ of a process x_t as a rational function

$$X(z)=\frac{V(z)}{W(z)}. \tag{3.137}$$

Comparing (3.137) with (3.128) we see that, if we define

$$V(z)=E(z)B(z) \tag{3.138}$$

and

$$W(z)=A(z) \tag{3.139}$$

then (3.137) is an ARMA representation of the process x_t with power spectral estimate

$$P_{\mathrm{ARMA}}(f)=\frac{|V(f)|^2}{|W(f)|^2}=\frac{\sigma_e^2|B(f)|^2}{|A(f)|^2}, \tag{3.140}$$

where we have substituted $z=\exp(\mathrm{j}2\pi f)$, assuming a sampling period of one s. In the time domain, we may write, with the usual definitions,

$$v_i=\boldsymbol{x}_i^{\mathrm{T}}\boldsymbol{w} \tag{3.141}$$

and we are faced with the problem of determining the vectors of coefficients $\boldsymbol{v}$ and $\boldsymbol{w}$. The *Treitel* et al. [3.6] procedure is to determine these iteratively, each iteration requiring the computation of three least-squares Wiener filters.

At iteration k we have, from (3.141), the following steps:

Step I Compute $\boldsymbol{w}^{(k)}$ with x_t as input and $\boldsymbol{v}^{(k-1)}$ as desired output.

Step II Compute $\boldsymbol{w}^{-1(k)}$ with $\boldsymbol{w}^{(k)}$ as input and a spike at zero delay as desired output.

Step III Compute $\boldsymbol{v}^{(k)}$ with $\boldsymbol{w}^{-1(k)}$ as input and x_t as desired output.

If convergence has occurred after m steps, we compute $\tilde{\boldsymbol{w}}^{(m)}$, the least-squares inverse of $\boldsymbol{w}^{-1(m)}$ and the ARMA spectral estimate is computed from

$$P_{\mathrm{ARMA}}(f)=\frac{|V^{(m)}(f)|^2}{|\tilde{W}^{(m)}(f)|^2}. \tag{3.142}$$

The minimum delay characteristic of $\tilde{W}^{(m)}(f)$ is guaranteed by the fact that the inverse is computed using the Toeplitz formulation which we discussed in Sect. 3.3.2. Of course, the orders of the polynomials $V(z)$ and $W(z)$ must be assigned for a given set of iterations. An example of the application of the above approach to the computation of an ARMA spectral estimate is given later on in this section.

The Method of *Akaike* [3.63]

Akaike [3.63] has proposed a method for the determination of the ARMA parameters which is based on a maximum-likelihood identification procedure and which requires only an estimate of the autocovariance of the process. An assumption required in this approach is that both the AR and MA parameters are invertible. As we have seen above, however, this assumption can be obviated by taking least-squares inverses.

We now outline the Akaike procedure for the scalar case.

Equation (3.128) may be written as

$$e_t = B^{-1}(z)A(z)[x_t]. \tag{3.143}$$

Let the order of $B^{-1}(z)A(z)$ be l where we write

$$B^{-1}(z)A(z) \simeq \sum_{i=0}^{l} u_i z^{-i} \tag{3.144}$$

with $u_0 = 1$.

We now form the prediction-error power, which, following the procedure outlined in Sect. 3.3 is taken as the average of the variances of both the forward and backward innovations.

The forward innovation for $t = l+1, l+2, \ldots, n$ is given by (3.143) and the backward innovation for $t = 1, 2, \ldots, n-l$ is given by $B^{-1}(z^{-1})A(z^{-1})[x_t]$.

Combining these two expressions for the forward and backward innovations, the expression for the prediction-error power is

$$\mathscr{E} = \boldsymbol{u}^{\mathrm{T}} \boldsymbol{C}_{0,0} \boldsymbol{u}, \tag{3.145}$$

where $\boldsymbol{u}^{\mathrm{T}} = (1, u_1, \ldots, u_l)$ is given by (3.144) and $\boldsymbol{C}_{0,0}$ is the matrix

$$\boldsymbol{C}_{0,0} = \begin{bmatrix} c_{00} & c_{01} & \cdots & c_{0l} \\ c_{10} & c_{11} & & c_{1l} \\ \vdots & & & \\ c_{l0} & \cdots & \cdots & c_{ll} \end{bmatrix} \tag{3.146}$$

with elements

$$c_{jk} = \frac{1}{2(n-l)} \left\{ \sum_{i=l+1}^{n} x_{i-j} x_{i-k} + \sum_{i=1}^{n-l} x_{i+j} x_{i+k} \right\}. \tag{3.147}$$

Let

$$\boldsymbol{\theta}^{\mathrm{T}} = [a_1, a_2, \ldots, a_p, b_1, b_2, \ldots, b_q] \tag{3.148}$$

represent the column vector of ARMA parameters. Then the condition for $\mathscr{E}$ to be minimum is

$$\left(\frac{\partial \mathscr{E}}{\partial \boldsymbol{\theta}}\right)_{\hat{\boldsymbol{\theta}}} = 0,$$

where $\hat{\boldsymbol{\theta}}$ is the column vector of estimated parameters. Using Taylor's expansion we write the above as

$$\left(\frac{\partial \mathscr{E}}{\partial \boldsymbol{\theta}}\right)_{\hat{\boldsymbol{\theta}}} \simeq \left(\frac{\partial \mathscr{E}}{\partial \boldsymbol{\theta}}\right)_{\boldsymbol{\theta}_0} + \left(\frac{\partial^2 \mathscr{E}}{\partial \boldsymbol{\theta}^{\mathrm{T}} \partial \boldsymbol{\theta}}\right)_{\boldsymbol{\theta}_0} (\hat{\boldsymbol{\theta}} - \boldsymbol{\theta}_0), \tag{3.149}$$

where $\boldsymbol{\theta}_0$ is close to $\hat{\boldsymbol{\theta}}$.

The minimization in (3.149) involves the computation of the gradient vector $\boldsymbol{g}'$ and the Hessian matrix $\boldsymbol{H}$. In order to define these quantities we must first of all define the column vectors $\boldsymbol{d}^{\mathrm{T}} = [d_0, d_1, \ldots, d_v]$ and $\boldsymbol{f}^{\mathrm{T}} = [f_0, f_1, \ldots, f_w]$ in terms of products of $\mathrm{B}^{-1}(z)$, $\mathrm{B}^{-2}(z)$, and $\mathrm{A}(z)$ which enter into the minimization expressed by (3.149).

Let

$$\sum_{i=0}^{v} d_i z^{-i} \simeq \mathrm{B}^{-1}(z) \tag{3.150}$$

and

$$\sum_{i=0}^{w} f_i z^{-i} \simeq \mathrm{B}^{-2}(z)\mathrm{A}(z). \tag{3.151}$$

The $\simeq$ sign in (3.144, 150, 151) indicates that we are concerned with approximate inverses, the orders of which are determined in reference to some specified truncation constant.

The $\boldsymbol{g}'$ and $\boldsymbol{H}$ are now defined as follows:

$$\boldsymbol{g}'^{\mathrm{T}} = [g'_{a1}, g'_{a2}, \ldots, g'_{ap}, g'_{b1}, g'_{b2}, \ldots, g'_{bq}] \tag{3.152}$$

with

$$g'_{ak} = \frac{\partial \mathscr{E}}{\partial a_k} = \boldsymbol{u}^{\mathrm{T}} \boldsymbol{C}_{0,k} \boldsymbol{d} \qquad k = 1, 2, \ldots, p, \tag{3.153}$$

$$g'_{bk} = \frac{\partial \mathscr{E}}{\partial b_k} = -\boldsymbol{u}^{\mathrm{T}} \boldsymbol{C}_{0,k} \boldsymbol{f} \qquad k = 1, 2, \ldots, q, \tag{3.154}$$

and

$$\boldsymbol{H} = \begin{bmatrix} \boldsymbol{H}^{aa} & \boldsymbol{H}^{ab} \\ \boldsymbol{H}^{ba} & \boldsymbol{H}^{bb} \end{bmatrix} \tag{3.155}$$

is a block matrix where the elements of the submatrices are defined by

$$H^{aa}=[h_{jk}^{aa}]; \quad h_{jk}^{aa}=d^{\mathrm{T}}C_{jk}d \quad j,k=1,2,\ldots,p, \tag{3.156}$$

$$H^{ab}=[h_{jk}^{ab}]; \quad h_{jk}^{ab}=-d^{\mathrm{T}}C_{jk}f \quad \begin{matrix} j=1,2,\ldots,p \\ k=1,2,\ldots,q, \end{matrix} \tag{3.157}$$

$$H^{ba}=[h_{jk}^{ba}]; \quad h_{jk}^{ba}=h_{kj}^{ab} \quad \begin{matrix} j=1,2,\ldots,q \\ k=1,2,\ldots,p, \end{matrix} \tag{3.158}$$

$$H^{bb}=[h_{jk}^{bb}]; \quad h_{jk}^{bb}=f^{\mathrm{T}}C_{jk}f \quad j,k=1,2,\ldots,q. \tag{3.159}$$

The covariance matrix C_{jk} is given by

$$C_{jk}=\begin{bmatrix} c_{j,k} & c_{j,k+1} & \cdots & c_{j,k+\max} \\ \vdots & & & \\ c_{j+\max,k} & \cdots & \cdots & c_{j+\max,k+\max} \end{bmatrix} \tag{3.160}$$

with elements

$$c_{jk}=\frac{1}{2(n-\max)}\left\{\sum_{i=\max+1}^{n} x_{i-j}x_{i-k}+\sum_{i=1}^{n-\max} x_{i+j}x_{i+k}\right\}, \tag{3.161}$$

where

$$\max=w+q \quad \text{if} \quad q\geqq p$$

and

$$\max=w+p \quad \text{if} \quad q<p.$$

In terms of the above definitions, the minimization expressed by (3.149) reduces to

$$g'=H(\theta_0-\hat{\theta}). \tag{3.162}$$

If H is invertible such that H^{-1} exists, we can write

$$\hat{\theta}=\theta_0-H^{-1}g'. \tag{3.163}$$

Thus, we see that the Newton-Raphson procedure for successive minimization is realized by correcting the current estimate of the ARMA parameters by an amount equal to the negative of the inverse of the Hessian times the gradient. In later work *Akaike* et al. [3.64] replaced the Newton-Raphson procedure by the variance minimization algorithm of *Devidon* [3.73, 74].

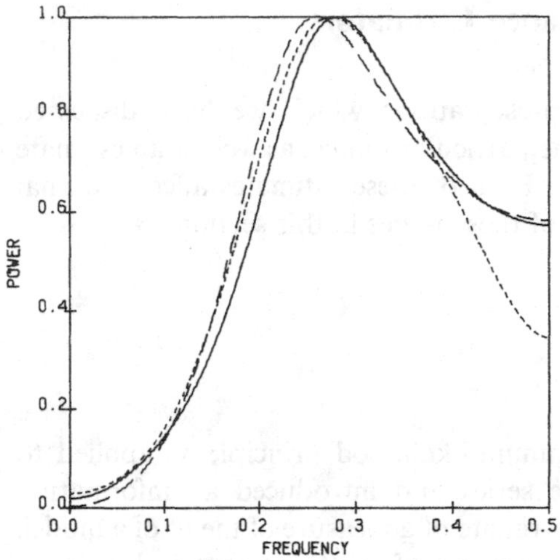

Fig. 3.11. Illustrating the computation of ARMA power spectra. Theoretical ARMA (3, 2) spectrum (——); *Akaike* [3.62] method (– – – –); *Treitel* et al. [3.6] method (· · · · · ·)

The procedure outlined above becomes computationally considerably simpler if the input data are sufficiently long that the matrices C_{00} and C_{jk} can be assumed to be Toeplitz with elements

$$c_{j,k} = {}_1c_\tau = \frac{1}{n} \sum_{i=1}^{n-|\tau|} x_i x_{i+\tau} . \tag{3.164}$$

In this case, the expressions for g' and H are consistent with those presented by *Akaike* [3.63] and *Gersch* et al. [3.65].

We have dealt, in this section, with the scalar ARMA case. The extension of these techniques to the complex case is of importance in many problems of interest and has recently been accomplished by *Ooe* [3.11]. A numerical example using a complex ARMA model is presented in Sect. 3.7.

Remarks Concerning the Computation of ARMA Models

We will now briefly compare the results of the *Akaike* [3.63] and *Treitel* et al. [3.6] methods of computing ARMA spectra. We have chosen as a model the ARMA (3, 2) process used by *Ulrych* and *Bishop* [3.14], and *Treitel* et al. [3.6] the theoretical spectrum of which is shown as the full line in Fig. 3.11. The dashed curve shows the ARMA spectrum computed using the Akaike method. The order of the model was chosen on the basis of the minimum AIC to be (2, 1). The dotted curve is the power spectrum corresponding to the method of *Treitel* et al. for an ARMA (3, 2) model and polynomial lengths chosen by inspection. Both methods yield close approximations to the theoretical model. The Treitel et al. technique is much simpler to implement than the Akaike approach but suffers from the difficulty of requiring an a priori estimate of the ARMA order. The AIC criterion is of considerable importance in the computation of power spectra of ARMA processes.

3.6 AIC, FPE, and Autocovariance Estimators

The fitting of AR and ARMA representations which we have discussed requires an estimate of the order of the particular model, as well as an estimate of the autocovariance of the process. Both of these estimates affect the final result, and we consider some details of these issues in this section.

3.6.1 The AIC Criterion

Akaike [3.29, 75] extended the maximum-likelihood principle as applied to statistical hypothesis testing in time series and introduced an information theoretic criterion, called AIC, as an estimate of a measure of the fit of a model. We look here in some detail at the derivation of this important and versatile statistical identification criterion.

Consider a situation where $x_1, x_2, \ldots, x_n$ are obtained as a result of independent observations of a random variable with probability density function $g(x)$. We define $f(x|\hat{\boldsymbol{\theta}})$ to be a parametric family of density functions, where $\hat{\boldsymbol{\theta}}^{\mathrm{T}} = (\hat{\theta}_1, \hat{\theta}_2, \ldots, \hat{\theta}_m)$ is the estimated vector of parameters which model the process. The true vector is $\boldsymbol{\theta}$.

With the assumption that $f(x|\hat{\boldsymbol{\theta}})$ is fairly close to $g(x)$, a criterion sensitive to the deviation between the two densities is the Kullback-Leibler [3.76] mean information criterion, $I(g, f)$, defined by

$$I(g, f) = \int_{-\infty}^{\infty} g(x) \log g(x) dx - \int_{-\infty}^{\infty} g(x) \log f(x|\hat{\boldsymbol{\theta}}) dx, \tag{3.165}$$

where log indicates the natural logarithm. The best model is that model for which $I(g, f)$ is a minimum. Since $g(x)$ is not known, the problem becomes one of maximizing the second term on the right hand side of (3.165) which we call $I_1(g, f)$, i.e.,

$$I_1(g, f) = \int_{-\infty}^{\infty} g(x) \log f(x|\hat{\boldsymbol{\theta}}) dx. \tag{3.166}$$

The maximization of $I_1(g, f)$ involves the maximization of the log likelihood function $L(\hat{\boldsymbol{\theta}})$ where

$$L(\hat{\boldsymbol{\theta}}) = \frac{1}{n} \sum_{i=1}^{n} \log f(x_i|\hat{\boldsymbol{\theta}}) = \max \frac{1}{n} \sum_{i=1}^{n} \log f(x_i|\boldsymbol{\theta}). \tag{3.167}$$

Expanding $\frac{\partial L(\hat{\boldsymbol{\theta}})}{\partial \boldsymbol{\theta}}$ in terms of Taylor's series, we have

$$\frac{\partial L(\hat{\boldsymbol{\theta}})}{\partial \boldsymbol{\theta}} \simeq \frac{1}{n}\sum_{i=1}^{n} \frac{\partial}{\partial \boldsymbol{\theta}} \log f(x_i|\boldsymbol{\theta}) + \frac{1}{n}\sum_{i=1}^{n} \frac{\partial^2}{\partial \boldsymbol{\theta} \partial \boldsymbol{\theta}^{\mathrm{T}}} \log f(x_i|\boldsymbol{\theta})(\hat{\boldsymbol{\theta}}-\boldsymbol{\theta})=0. \tag{3.168}$$

We define

$$\boldsymbol{D}(\boldsymbol{\theta}) = -\mathrm{E}\left[\frac{\partial^2}{\partial \boldsymbol{\theta} \partial \boldsymbol{\theta}^{\mathrm{T}}} \log f(x|\boldsymbol{\theta})\right] \tag{3.169}$$

and

$$\boldsymbol{C}(\boldsymbol{\theta}) = \mathrm{E}\left[\frac{\partial}{\partial \boldsymbol{\theta}} \log f(x|\boldsymbol{\theta}) \frac{\partial}{\partial \boldsymbol{\theta}^{\mathrm{T}}} \log f(x|\boldsymbol{\theta})\right]. \tag{3.170}$$

With the knowledge, from the central limit theorem, that the probability distribution of

$$\frac{1}{\sqrt{n}}\sum_{i=1}^{n} \frac{\partial}{\partial \boldsymbol{\theta}} \log f(x|\boldsymbol{\theta})$$

approaches a normal distribution $N[0, \boldsymbol{C}(\boldsymbol{\theta})]$ as $n \to \infty$, we can deduce the following from (3.168):

$$\mathrm{E}[\sqrt{n}(\hat{\boldsymbol{\theta}}-\boldsymbol{\theta})]=0, \tag{3.171}$$

and, since $\boldsymbol{D}(\boldsymbol{\theta})$ is a diagonal matrix,

$$\begin{aligned} \mathrm{Var}[\sqrt{n}(\hat{\boldsymbol{\theta}}-\boldsymbol{\theta})] &= \mathrm{E}[\sqrt{n}(\hat{\boldsymbol{\theta}}-\boldsymbol{\theta})\sqrt{n}(\hat{\boldsymbol{\theta}}-\boldsymbol{\theta})^{\mathrm{T}}] \\ &= \boldsymbol{D}^{-1}(\boldsymbol{\theta})\boldsymbol{C}(\boldsymbol{\theta})\boldsymbol{D}^{-1^{\mathrm{T}}}(\boldsymbol{\theta}). \end{aligned} \tag{3.172}$$

Expanding $I_1(g,f)$ and $L(\hat{\boldsymbol{\theta}})$ in a Taylor's series and using (3.171, 172), we get the expression

$$I_1(g,f) \simeq L(\hat{\boldsymbol{\theta}}) - (\hat{\boldsymbol{\theta}}-\boldsymbol{\theta})^{\mathrm{T}} D(\boldsymbol{\theta})(\hat{\boldsymbol{\theta}}-\boldsymbol{\theta}) - Q, \tag{3.173}$$

where

$$Q = \frac{1}{n}\sum_{i=1}^{n} \log f(x_i|\boldsymbol{\theta}) - \int_{-\infty}^{\infty} g(x) \log f(x|\boldsymbol{\theta}) dx. \tag{3.174}$$

Since

$$\mathrm{E}[(\hat{\boldsymbol{\theta}}-\boldsymbol{\theta})^{\mathrm{T}}\boldsymbol{D}(\boldsymbol{\theta})(\hat{\boldsymbol{\theta}}-\boldsymbol{\theta})]=\frac{1}{n}\operatorname{tr}\{\boldsymbol{D}^{-1}(\boldsymbol{\theta})\boldsymbol{C}(\boldsymbol{\theta})\}, \tag{3.175}$$

where $\operatorname{tr}\{\boldsymbol{A}\}$ is the trace of $\boldsymbol{A}$, we can expect (3.173) to approximate

$$I_1(g,f)\simeq L(\hat{\boldsymbol{\theta}})-\frac{1}{n}\operatorname{tr}\{\boldsymbol{D}^{-1}(\boldsymbol{\theta})\boldsymbol{C}(\boldsymbol{\theta})\}-Q. \tag{3.176}$$

Suppose that the true vector of parameters is $\boldsymbol{\theta}_m^{\mathrm{T}}=(\theta_{1m},\theta_{2m},\ldots,\theta_{mm})$ so that $g(x)=f(x|\boldsymbol{\theta}_m)$. Let k be the order of $\hat{\boldsymbol{\theta}}_k$ so that $\hat{\boldsymbol{\theta}}_k^{\mathrm{T}}=(\hat{\theta}_{1k},\hat{\theta}_{2k},\ldots,\hat{\theta}_{kk})$.

We now choose a value of k so as to minimize

$$\begin{aligned}I_2(f_m,f_k)=&\,I_1[f(x|\boldsymbol{\theta}_m),f(x|\hat{\boldsymbol{\theta}}_m)]\\&-I_1[f(x|\boldsymbol{\theta}_m),f(x|\hat{\boldsymbol{\theta}}_k)].\end{aligned} \tag{3.177}$$

When the distance between $\boldsymbol{\theta}_m$ and $\hat{\boldsymbol{\theta}}_k$ is small, we have, using (3.176)

$$\begin{aligned}I_2(f_m,f_k)\simeq&\,L(\hat{\boldsymbol{\theta}}_m)-L(\hat{\boldsymbol{\theta}}_k)\\&+\frac{1}{n}\operatorname{tr}\{\boldsymbol{D}^{-1}(\boldsymbol{\theta}_k)\boldsymbol{C}(\boldsymbol{\theta}_k)\}-\frac{m}{n}.\end{aligned} \tag{3.178}$$

Since $L(\hat{\boldsymbol{\theta}}_m)$ and m/n are independent of k, the information criterion to be minimized can be simplified to

$$I_3(\cdot,f_k)=-L(\hat{\boldsymbol{\theta}}_k)+\frac{1}{n}\operatorname{tr}\{\boldsymbol{D}^{-1}(\boldsymbol{\theta}_k)\boldsymbol{C}(\boldsymbol{\theta}_k)\}. \tag{3.179}$$

In the case that $\boldsymbol{D}(\boldsymbol{\theta}_k)=\boldsymbol{C}(\boldsymbol{\theta}_k)$, a condition which requires that $f(x|\boldsymbol{\theta}_k)$ be close to $f(x|\boldsymbol{\theta}_m)$, $\operatorname{tr}\{\boldsymbol{D}^{-1}(\boldsymbol{\theta}_k)\boldsymbol{C}(\boldsymbol{\theta}_k)\}$ is equal to the number of free parameters, k, and we define AIC as

$$\mathrm{AIC}=2nI_3(\cdot,f_k)=-2nL(\hat{\boldsymbol{\theta}}_k)+2k, \tag{3.180}$$

i.e., AIC is the second-order approximation to the Kullback–Leibler mean information criterion.

3.6.2 AIC and the Normal Distribution

In the case of a normal probability distribution, $f(x|\boldsymbol{\theta})=N(\mu,\sigma)$, the condition $\boldsymbol{D}(\boldsymbol{\theta})=\boldsymbol{C}(\boldsymbol{\theta})$ is fulfilled, since

$$\boldsymbol{D}(\boldsymbol{\theta})=\boldsymbol{C}(\boldsymbol{\theta})=\begin{pmatrix}1/\sigma^2 & 0\\ 0 & 2/\sigma^2\end{pmatrix}. \tag{3.181}$$

In this case $\mathrm{tr}\{\boldsymbol{D}^{-1}(\boldsymbol{\theta})\boldsymbol{C}(\boldsymbol{\theta})\}=2$ which is the number of free parameters.

Consider a linear transformation defined by

$$x_t = T(\boldsymbol{\theta})[y_t], \tag{3.182}$$

where x_t is white noise and the y_t are assumed to be distributed as $N(0,\sigma)$. Hence

$$f(y|\boldsymbol{\theta},\sigma^2) = \frac{|T(\boldsymbol{\theta})|}{(2\pi\sigma^2)^{1/2}}\exp[-\zeta^2(\boldsymbol{\theta})/2\sigma^2], \tag{3.183}$$

where

$$\zeta^2(\boldsymbol{\theta}) = \mathrm{E}[|T(\boldsymbol{\theta})[y_t]|^2]. \tag{3.184}$$

The corresponding log likelihood divided by n is

$$\begin{aligned} L(\boldsymbol{\theta},\sigma) &= \frac{1}{n}\sum^{n}\log f(y|\boldsymbol{\theta},\sigma^2) \\ &\approx -\tfrac{1}{2}\log 2\pi - \log\sigma - \frac{\zeta^2(\boldsymbol{\theta})}{2\sigma^2}. \end{aligned} \tag{3.185}$$

The maximum-likelihood estimates, $\hat{\boldsymbol{\theta}}$ and $\hat{\sigma}$, are determined from

$$\frac{\partial L}{\partial \boldsymbol{\theta}}(\boldsymbol{\theta},\sigma) = \frac{\partial L}{\partial \sigma}(\boldsymbol{\theta},\sigma) = 0$$

which leads to

$$\zeta^2(\hat{\boldsymbol{\theta}}) = \hat{\sigma}^2. \tag{3.186}$$

Hence upon substitution into (3.185) we get

$$L(\hat{\boldsymbol{\theta}},\hat{\sigma}) = -\tfrac{1}{2}\log 2\pi - \tfrac{1}{2}\log\zeta^2(\hat{\boldsymbol{\theta}}) - \tfrac{1}{2}.$$

Omitting constant terms and substituting into (3.180) we obtain the final expression

$$\mathrm{AIC} = n\log\zeta^2(\hat{\boldsymbol{\theta}}) + 2k. \tag{3.187}$$

For an ARMA (p,q) process, $k=p+q$ excluding σ^2 because $\hat{\sigma}^2$ is automatically given by $\zeta^2(\hat{\boldsymbol{\theta}})$ after the maximum-likelihood estimates have been determined.

3.6.3 The FPE Criterion

The final prediction-error criterion was developed by *Akaike* [3.24, 27] for the determination of the order of AR processes. The FPE, given by (3.32), chooses that order which minimizes the prediction error when the error is considered as the sum of the innovation and the error in estimating the AR parameters.

Let us consider an mth order AR process with Gaussian errors. The asymptotic relationship between FPE and AIC may be derived as follows. Since, for $n \gg m$,

$$\log\left|\frac{1+(m+1)/n}{1-(m+1)/n}\right| \simeq \frac{2(m+1)}{n} \tag{3.188}$$

we may write, using (3.32),

$$\begin{aligned}\log(\mathrm{FPE})_m &= \log\zeta_m^2\left|\frac{1+(m+1)/n}{1-(m+1)/n}\right| \\ &\simeq \log\zeta_m^2 + \frac{2(m+1)}{n} \quad \text{for} \quad n \gg m.\end{aligned} \tag{3.189}$$

If we omit the constant term in (3.189), which reflects the subtraction of the mean, and multiply through by n, we get

$$n\log(\mathrm{FPE})_m \simeq n\log\zeta_m^2 + 2m. \tag{3.190}$$

Substituting (3.187) into (3.190) we obtain

$$n\log(\mathrm{FPE})_m \simeq (\mathrm{AIC})_m \quad \text{for} \quad n \gg m. \tag{3.191}$$

It has generally been found that the AR order given by the FPE and AIC criteria is the same for processes of reasonable length. For short records we recommend the use of the AIC.

3.6.4 Autocovariance Estimators

There are a number of ways of estimating the autocovariance of a stationary stochastic process, x_t, which has the property that $\mathrm{E}[x_t]=0$. Estimators with well-known properties are [Ref. 3.34, pp. 171–208]

$${}_1c_l = \frac{1}{n}\sum_{t=1}^{n-|l|} x_t x_{t+l} \tag{3.192}$$

and

$${}_2c_l = \frac{1}{n-|l|} \sum_{t=1}^{n-|l|} x_t x_{t+l} . \tag{3.193}$$

The ${}_1c_l$ estimator is biased but gives rise to a semipositive definite covariance matrix, whereas ${}_2c_l$ is an unbiased estimator but is not necessarily semipositive definite.

Here, we introduce another estimator which arises in the LS estimate of the AR parameters discussed in Sect. 3.3.2. By minimizing the forward and backward errors for a pth order process, *Ulrych* and *Clayton* [3.16] derived the semipositive definite estimator

$$\begin{aligned} {}_3c_{jk} &= \frac{1}{2(n-p)} \left\{ \sum_{t=p+1}^{n} x_{t-j} x_{t-k} + \sum_{t=1}^{n-p} x_{t+j} x_{t+k} \right\} \\ & j,k = 0,1,\ldots,p . \end{aligned} \tag{3.194}$$

In the stationary case ${}_3c_{jk} \approx {}_4c_l$ where

$$\begin{aligned} {}_4c_l &= \frac{1}{2(n-p)} \left\{ \sum_{t=p+1}^{n} x_t x_{t-l} + \sum_{t=1}^{n-p} x_t x_{t+l} \right\} \\ & l = 0,1,\ldots,p , \end{aligned} \tag{3.195}$$

this estimator may not be semipositive definite but is expected to be more stable than ${}_2c_l$.

3.7 Numerical Example: Application of AR and ARMA Models to Polar Motion

Several important questions concerned with the polar motion of the earth can be resolved only by means of the computation of reliable spectral estimates. The possibility of sideband excitation of the Chandler component and the existence of a split resonance peak have been raised in the literature [3.77, 78]. The question of the quality factor, Q, of the Chandler wobble is of considerable interest, and since this factor is estimated from the width of the Chandler resonance in the power spectrum, length of record considerations are of central importance.

In Sect. 3.7.2 we will show some numerical examples which are the result of applying complex AR and ARMA models to synthetic and actual polar motion data. The extension of the scalar case to the complex case has been presented by *Smylie* et al. [3.21] for AR processes, and recently by *Ooe* [3.11] for ARMA processes. The former reference should be consulted for a review of the subject of the earth's rotation.

3.7.1 The ARMA Model of the Chandler Wobble

Let $\tilde{m}_t$ denote the angular displacement of the instantaneous axis of rotation with respect to the right-handed axes fixed in the earth. Then, if ω_0 is the angular frequency of the Chandler wobble and α is the damping constant we may write a difference equation for the motion which is sampled at intervals separated by Δt as

$$\tilde{m}_t - \exp(\mathrm{j}\tilde{\omega}_0 \Delta t)\tilde{m}_{t-\Delta t} = \tilde{F}_t, \tag{3.196}$$

where

$$\tilde{\omega}_0 = \omega_0 + j\alpha .$$

$\tilde{}$ indicates complex quantities

$$\tilde{F}_t = -\mathrm{j}\tilde{\omega}_0 \int_0^{\Delta t} \exp(\mathrm{j}\tilde{\omega}_0 u)\tilde{\psi}(t-u)du \tag{3.197}$$

and $\tilde{\psi}(t)$ represents the nonseasonal part of the excitation function of the Chandler wobble [3.11]. Since the excitation source (for example meteorological phenomena) may not be white, we can express $\tilde{F}_t$ as an MA process of some order q:

$$\tilde{F}_t = \tilde{u}_t + \sum_{l=1}^{q} \tilde{b}_l \tilde{u}_{t-l\Delta t}, \tag{3.198}$$

where $\tilde{u}_t$ is a white noise series.

Equation (3.196) now becomes

$$\tilde{m}_t = \tilde{a}_1 \tilde{m}_{t-\Delta t} + \tilde{u}_t + \sum_{l=1}^{q} \tilde{b}_l \tilde{u}_{t-l\Delta t}, \tag{3.199}$$

where $\tilde{a}_1 = \exp(\mathrm{j}\tilde{\omega}_0 \Delta t)$.

Since the observed polar motion data $\tilde{x}_t$ contain observational errors, $\tilde{e}_t$, which we can assume to be white, we can write

$$\tilde{x}_t = \tilde{m}_t + \tilde{e}_t \tag{3.200}$$

and inserting (3.200) into (3.199) we obtain

$$\tilde{x}_t = \tilde{a}_1 \tilde{x}_{t-\Delta t} + \tilde{e}_t - \tilde{a}_1 \tilde{e}_{t-\Delta t} + \tilde{u}_t + \sum_{l=1}^{q} \tilde{b}_l \tilde{u}_{t-l\Delta t}. \tag{3.201}$$

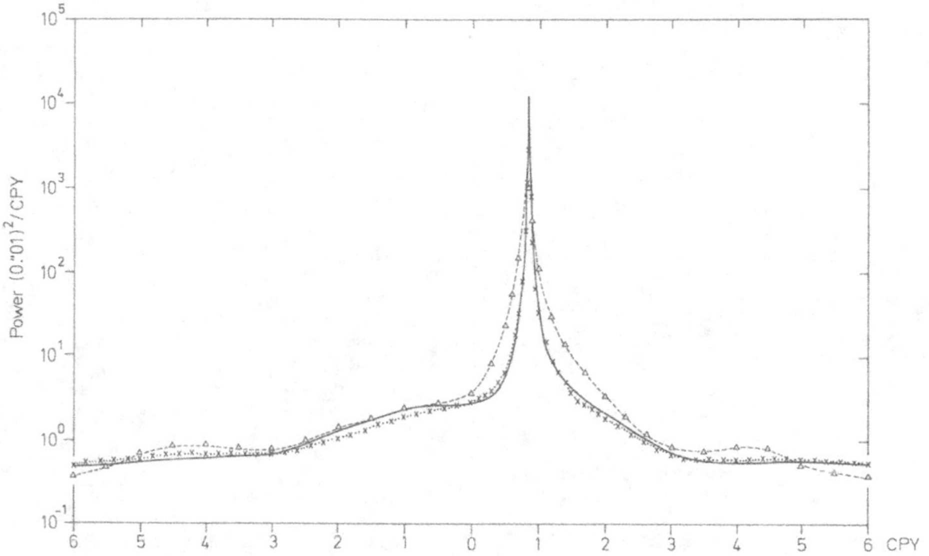

Fig. 3.12. Optimal ARMA (1, 4) spectrum of the Chandler wobble using 66 years of ILS data (———); ARMA (1, 4) spectrum of a synthesized Chandler process using 66 years of data (×); ARMA (1, 4) spectrum of a synthesized Chandler process using 8 years of data (△)

Equation (3.201) is recognized as a modified ARMA equation. We are thus motivated to estimate the power spectrum of the Chandler wobble by fitting an ARMA model to the data using the AIC to obtain the best fit model.

3.7.2 Numerical Computations

We have used ILS polar motion data published by *Vincente* and *Yumi* [3.79, 80] and *Yumi* [3.81]. The seasonal and secular terms were removed from the data prior to spectral analysis.

The first approach in fitting ARMA models to polar motion data was presented by *Ooe* et al. [3.82] who used the scalar TIMSAC package of *Akaike* et al. [3.64] with ${}_1\boldsymbol{c}$ [Equation (3.192)] as the autocovariance estimate. The best fit model chosen by the minimum AIC gave an ARMA order (6, 5) with a Chandler period of 1.19 years and a Q value of 82. The second approach, that of *Ooe* [3.11], extended the modeling to complex ARMA models. Here, we use this second approach to compare the resulting power spectra using various autocovariance estimators.

The application of the AIC to ILS data consisting of 791 points with $\Delta t = 1$ month indicated an ARMA (1, 4) model. The autocovariance estimator used was ${}_1\boldsymbol{c}$. The spectrum for this model, shown as the solid line in Fig. 3.12, gave a Chandler period of 1.19 years and a Q of 39 to 148 with a most probable value of 62. Using the parameters obtained from the ILS data we generated a

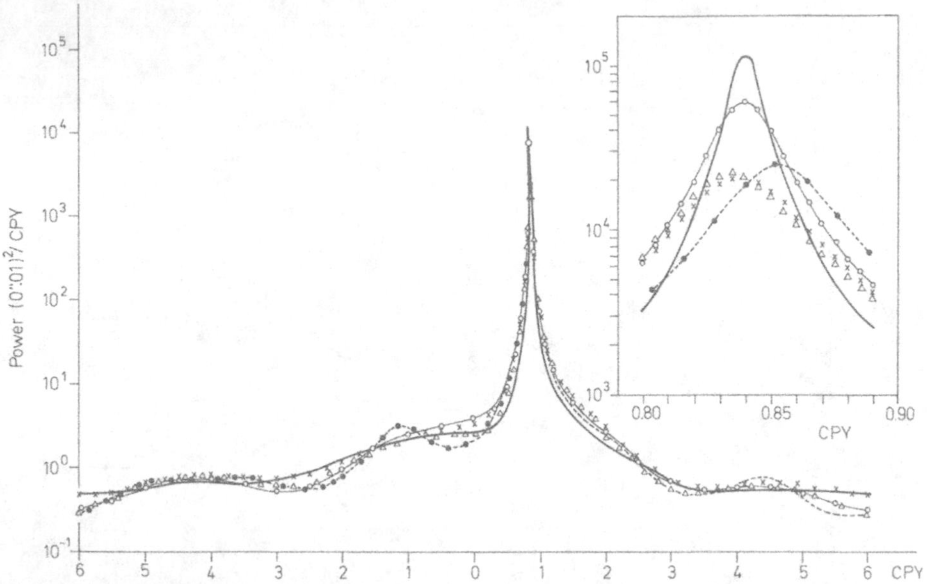

Fig. 3.13. A comparison of autocovariance estimators on ARMA spectral estimates. Optimal ARMA (1, 4) spectrum of 66 years of ILS data (———); ARMA (1, 4) spectrum of 8 years of synthetic data using $_3c$ (○); ARMA (1, 4) spectrum of 8 years of synthetic data using $_4c$ (△); ARMA (1, 4) spectrum of 8 years of synthetic data using Burg covariance estimate (×), and Burg AR/ME spectrum of the same data (●) as for the spectrum of (×). Inset shows a blow up of spectral peak

synthetic ARMA (1, 4) process and repeated the above analysis. The minimum AIC again indicated a (1, 4) model and the computed spectrum is shown in Fig. 3.12. The estimator $_1c$, as discussed previously, is biased, and since the effect of the bias on the damping constant is approximately inversely proportional to the record length the most probable value of Q should be adjusted to 96. The effect of the bias is more serious for short record lengths as indicated by the triangles in Fig. 3.12. This curve shows the ARMA (1, 4) spectrum computed from a synthetic sample consisting of only 100 points.

The effect of various autocovariance estimators as well as a comparison with an AR spectrum is shown in Fig. 3.13. The sample considered was again a 100 point synthetic wobble generated from the ILS ARMA (1, 4) parameters. The AR/ME estimate was computed for an AR process of order 7 as indicated by the minimum AIC. Clearly the estimate closest to the real spectrum is obtained using the autocovariance estimator $_3c$. Further, the ARMA models provide a superior estimate of the Q of the process as compared to AR models.

3.8 Conclusions

The philosophy of this chapter is based on the premise that spectral analysis and time series modeling are intimately connected. This connection is utilized through the parametric representation of the process in question, the parameters being fitted to the data on the basis of the assumption that the process may be represented by either an AR or ARMA model. The benefits of this approach lie in the parsimony which is often associated with the particular representation, the high resolution and the optimally smooth nature of the corresponding spectral estimate and the extension of the stationary analysis to nonstationary processes.

In this chapter we have presented details which are central in the computation of AR and ARMA models and spectra. We have dealt with stationary and nonstationary AR models and spectra and with two quite different approaches to the computation of ARMA parameters. An important model, first developed by *Ulrych* and *Clayton* [3.16], is the special ARMA representation for sinusoids with additive white noise. The representation turns out to be equivalent to the method of spectral decomposition suggested by *Pisarenko* [3.70]. An interesting and potentially useful relationship in terms of the eigenstructure of the autocovariance matrix exists between the Pisarenko and the AR/ME spectral estimators.

We ended this chapter with a numerical example which compares ARMA and AR/ME spectral estimates of synthetic and real polar motion data. We also considered the effect of different autocovariance estimators on the computed spectra. This study indicates that ARMA estimates of the polar motion spectrum from models with orders chosen by the AIC criterion (Sect. 3.6) are superior to AR/ME estimates, particularly as far as the determination of the Q of the Chandler wobble is concerned.

References

3.1 A. van den Bos: IEEE Trans. IT-**17**, 493–494 (1971)

3.2 J.P. Burg: "Maximum entropy spectral analysis", in 37th Ann. Intern. Meet. Soc. Explor. Geophys., Oklahoma City, Okla., Oct. (1967)

3.3 J.P. Burg: "A new analysis technique for time series data". Adv. Study Inst. on Signal Processing NATO, Enschede, Holland (1968)

3.4 J.P. Burg: "Maximum entropy spectral analysis"; Ph.D. Thesis, Stanford University, Palo Alto, Calif. (1975)

3.5 H. Wold: *A Study in the Analysis of Stationary Time Series* (Almqvist and Wiksell, Uppsala 1938)

3.6 S. Treitel, P. R. Gutowski, E. A. Robinson: "Empirical spectral analysis revisited", in *Topics in Numerical Analysis*, ed. by J. H. Miller (Academic Press, New York 1977) pp. 429–446

3.7 R. H. Jones: Technometrics **7**, 531–542 (1965)

3.8 R. B. Blackman, J. W. Tukey: *The Measurement of Power Spectra from the Point of View of Communications Engineering* (Dover, New York 1959)

3.9 L. J. Griffiths: IEEE Trans. ASSP-**23**, 207–222 (1975)

3.10 A. J. Jurkevics, T. J. Ulrych: "Representing and simulating strong ground motion". B.S.S.A. **68** (3), 781–801 (1978)

3.11 M. Ooe: Geophys. J. R. Astron. Soc. **53**, 445–457 (1978)

3.12 G. T. Wilson: SIAM J. Num. Anal. **6**, 1 (1969)

3.13 E. A. Robinson: "Recursive decomposition of stochastic processes", in *Econometric Model Building*, ed. by H. O. Wold (North-Holland, Amsterdam 1964) pp. 111–168

3.14 T.J.Ulrych, T.N.Bishop: Rev. Geophys. **13**, 183–200 (1975)
3.15 J.Makhoul: Proc. IEEE **63**, 561–580 (1975)
3.16 T.J.Ulrych, R.W.Clayton: Phys. Earth Planet. Inter. **12**, 188–200 (1976)
3.17 J.D.Markel, A.H.Gray: *Linear Prediction of Speech*, Communication and Cybernetics, Vol. 12 (Springer, Berlin, Heidelberg, New York 1976)
3.18 J.Scargle: "Studies in Astronomical Time Series Analysis: Modeling Random Processes in the Time Domain". NASA Tech. Memo. (1976)
3.19 G.E.P.Box, G.M.Jenkins: *Time Series Analysis: Forecasting and Control*, rev. ed. (Holden-Day 1976)
3.20 G.U.Yule: Philos. Trans. R. Soc. London, Ser. A **226**, 267–298 (1927)
3.21 D.E.Smylie, G.K.C.Clarke, T.J.Ulrych: Methods Comput. Phys. **13**, 391–430 (1973)
3.22 J.A.Edward, M.M.Fitelson: IEEE Trans. IT-**19**, 232 (1973)
3.23 R.E.Kromer: "Asymptotic properties of the autoregressive spectral estimator"; Ph.D. dissertation, Tech. Rpt. 13, Dept. of Statistics, Stanford University, Stanford, Calif. (1969)
3.24 H.Akaike: Ann. Inst. Statist. Math. **21**, 407–419 (1969)
3.25 K.N.Berk: Ann. Statistics **2**, 489–502 (1974)
3.26 A.B.Baggeroer: IEEE Trans. IT-**22**, 534–545 (1976)
3.27 H.Akaike: Ann. Inst. Statist. Math. **22**, 203–217 (1970)
3.28 H.Akaike: "Use of an information theoretic quantity for statistical model indentification", in Proc. 5th Hawaii Intern. Conf. on System Sciences, 99–101 (1972)
3.29 H.Akaike: IEEE Trans. AC-**19**, 716–723 (1974)
3.30 T.E.Landers, R.T.Lacoss: IEEE Trans. GE-**15** (1), 26–33 (1977)
3.31 E.Parzen: "Multiple time series: Determining the order of approximating autoregressive schemes"; Tech. Rpt. 23, Statistical Sci. Div., State Univ. New York at Buffalo (1976)
3.32 S.L.Marple: "Resolution of conventional Fourier, autoregressive, and special ARMA methods of spectrum analysis". IEEE Intern. Conf. on ASSP, Hartford, Conn. (1977)
3.33 O.I.Frost: "Power spectrum estimation", in 1976 NATO Advanced Study Institute on Signal Processing with Emphasis on Underwater Acoustics, Portovenere (La Spezia), Italy (1976)
3.34 G.M.Jenkins, D.G.Watts: *Spectral Analysis and its Applications* (Holden-Day, San Francisco 1969)
3.35 S.Treitel, T.J.Ulrych: IEEE Trans. ASSP-**27**, 99–100 (1979)
3.36 E.A.Robinson: "Structural properties of stationary stochastic processes with applications", in *Proc. of the Symposium on Time Series Analysis* (Wiley and Sons, New York 1963) pp. 170–196
3.37 E.A.Robinson: *Statistical Communication and Detection* (Charles Griffin, London 1967) pp. 172–174
3.38 E.A.Guillemin: *The Mathematics of Circuit Analysis* (Wiley and Sons, New York 1949)
3.39 K.L.Peacock, S.Treitel: Geophys. **34**, 155–169 (1969)
3.40 N.Levinson: J. Math. Phys. **25**, 261–278 (1947)
3.41 J.F.Claerbout: *Fundamentals of Geophysical Data Processing* (McGraw-Hill, New York 1976)
3.42 N.O.Andersen: Geophys. **39**, 69–72 (1974)
3.43 A.H.Nuttal: "Scientific and Engineering Studies: Spectral Estimation"; Naval Underwater Systems Center (1978)
3.44 R.L.Kayshap, A.R.Rao: Automatica **9**, 175–183 (1973)
3.45 R.L.Kayshap: IEEE Trans. AC-**19**, 13–21 (1974)
3.46 A.J.Berkhout, P.R.Zaanen: Geophys. Prospect. **XXIV** (1), 141–197 (1976)
3.47 W.Y.Chen, G.R.Stegen: J. Geophys. Res. **79**, 3019–3022 (1974)
3.48 D.N.Swingler: "Fundamental limitations of Burg's maximum entropy spectral analysis technique" (unpublished manuscript 1977)
3.49 J.Makhoul: IEEE Trans. ASSP **25**, 423–428 (1977)
3.50 E.Widrow, M.E.Hoff: "Adaptive switching circuits", in IRE WESCON Convention Rec., Pt. 4, pp. 96–104 (1960)
3.51 L.J.Griffiths, R.Prieto-Diaz: IEEE Trans. GE-**15** (1), 13–25 (1977)

3.52 H. Robbins, S. Monroe: Ann. Math. Stat. **22**, 400–407 (1951)
3.53 B. Widrow: "Adaptive filters", in *Aspects of Network and System Theory*, ed. by R. E. Kalman, N. DeClaris (Holt, Rinehart, and Winston, New York 1970)
3.54 R. J. Wang, S. Treitel: Geophys. Prospect. **XIX** (4) 718–728 (1971)
3.55 B. Widrow, J. M. McCool, M. G. Larimore, C. R. Johnson, Jr.: Proc. IEEE **64**, 1521–1529 (1976)
3.56 E. A. Robinson, S. Treitel: Geophys. Prospect. **25**, 434–459 (1977)
3.57 M. S. Bartlett: J. R. Statist. Soc. Suppl. B**8**, 27–41 (1946)
3.58 J. Perl, L. L. Scharf: IEEE Trans. ASSP-**25** (2), 143–151 (1977)
3.59 M. Pagano: Ann. Statistics **2**, 99–108 (1974)
3.60 R. K. Mehra: IEEE Trans. AC-**16**, 12–21 (1971)
3.61 D. J. Krause, D. Graupe: "Identification of autoregressive moving-average predictor models", in Proc. 2nd Symp. on Nonlinear Estimation Theory and Its Applications, Harold W. Sorenson, chairman, 174–182 (1971)
3.62 W. Gersch: IEEE Trans. AC-**15**, 583–588 (1970)
3.63 H. Akaike: Biometrika **60**, 2, 255–265 (1973)
3.64 H. Akaike, E. Arahata, T. Ozaki: TIMSAC-74, A time series analysis and control program package (1). Computer Science Monographs, No. 5, A publication of the Inst. Statist. Math. (1975)
3.65 W. Gersch, N. N. Nielson, H. Akaike: J. Sound Vib. **31** (3), 295–308 (1973)
3.66 H. Akaike: Ann. Inst. Statist. Math. **26**, 363–387 (1974)
3.67 H. Tong: IEEE Trans. IT-**21**, 476–480 (1975)
3.68 H. Tong: IEEE Trans. IT-**22**, 493–496 (1976)
3.69 H. Tong: IEEE Trans. IT-**23**, 409–410 (1977)
3.70 V. F. Pisarenko: Geophys. J. R. Astron. Soc. **33**, 347–366 (1973)
3.71 W. Gersch, D. R. Sharpe: IEEE Trans. AC-**18**, 367–369 (1973)
3.72 E. H. Satorius, S. T. Alexander: "High resolution spectral analysis of sinusoids in correlated noise", in IEEE Intern. Conf. on Acoust., Speech, and Signal Processing, Tulsa, Okla. (1978)
3.73 W. C. Devidon: Comput. J. **10**, 406–410 (1968)
3.74 W. C. Devidon: "Variance algorithms for minimization", in *Optimization*, ed. by R. Fletcher (Academic Press, London 1969) pp. 13–20
3.75 H. Akaike: "Information theory and extension of the maximum likelihood principle", in *Proc. 2nd Intern. Symp. on Information Theory*, ed. by B. N. Petrov, F. Csaki (Akademiai Kiado, Budapest 1973) pp. 267–281
3.76 S. Kullback, R. A. Leibler: Ann. Math. Statistics **22**, 79–86 (1951)
3.77 H. Jeffreys: Mon. Not. Roy. Astron. Soc. **100**, 139–155 (1940)
3.78 G. Colombo, I. I. Shapiro: Nature **217**, 156–157 (1968)
3.79 R. O. Vincente, S. Yumi: Publ. Int. Latit. Obs. Mizusawa **7**, 1, 41–50 (1969)
3.80 R. O. Vincente, S. Yumi: Publ. Int. Latit. Obs. Mizusawa **7**, 2, 109–112 (1970)
3.81 S. Yumi: 1971–1976, Ann. Rpt. of IPMS for the years 1969–1974
3.82 M. Ooe, Y. Kaneko, H. Akaike: Publ. Int. Latit. Obs. Mizusawa **11**, 1, 1–9 (1977)

4. Iterative Least-Squares Procedure for ARMA Spectral Estimation

E. A. Robinson

With 12 Figures

In recent years an important new approach to signal processing has come into importance in science and engineering. It has become practical to represent information-bearing waveforms digitally and to do signal processing on the digital representation of the waveform. We can now manipulate physical signals by use of digital computers in ways that would have been totally impractical with continuous representations of the signals. Once the digital representation of a waveform is in computer memory, the waveform acquires a permanence which permits a wide range of processing operations, and at any point the waveform can be restored to a continuous form with a new time base. As a result, the fitting of finite-parameter models of digital signals is no longer only of theoretical importance but is a processing technique that can be used in day-to-day operations. In the previous chapter a detailed treatment of these finite-parameter models is given. In this chapter we wish to elaborate on the noninvertible autoregressive moving-average (ARMA) model.

The impulse response function of an ARMA model has a z transform given by the ratio of two polynomials. For stability, the denominator polynomial must necessarily be minimum delay. However, the numerator polynomial may or may not be minimum delay. If the numerator polynomial is minimum delay, then the ARMA model is said to be invertible. Otherwise, the ARMA model is said to be noninvertible. This chapter deals with the noninvertible model, and provides an iterative means of estimating both the numerator and denominator polynomials from the impulse response function. From these polynomials, we can then obtain an estimate of the spectral density function.

4.1 Basic Time Series Models

In recent years there has been increased emphasis on methods to construct finite-parameter models of a time series. Such models are based upon the idea that a time series $\{x_n\}$ (where the integer n is the time index) is generated from a series of random uncorrelated innovations $\{\varepsilon_n\}$. Specifically we assume that 1) the innovations have zero expectation, 2) the expectation of the product $\varepsilon_n \varepsilon_k$ of two different innovations (i.e. $n \neq k$) is zero, and 3) the expectation of the square ε_n^2 of any innovation a constant σ^2. Such a series of uncorrelated random variables with zero mean and constant variance is called a white-noise series.

The time series $\{x_n\}$ is then modeled as the output of a linear filter with a white-noise series as input. A finite-parameter model is defined as a model in which the linear filter can be represented by a finite number of coefficients.

Three basic finite-parameter models are the autoregressive model, the moving average model, and the autoregressive-moving average model. The autoregressive model is also called the autorecursive model, or AR model. The moving average is also called the move-ahead or MA model. The autoregressive-moving average model is a hybrid or mixed model with both AR and MA components, and is called the ARMA model.

The AR model is extremely useful in the representation of many physical time series. In this model the current value x_n of the time series is expressed as a finite linear combination of previous values of the time series and the current innovation ε_n. An autoregressive time series of order p may thus be represented as

$$x_n = -a_1 x_{n-1} - a_2 x_{n-2} - \ldots - a_p x_{n-p} + \varepsilon_n . \tag{4.1}$$

The reason for the name "autoregressive" is that a linear model relating a "dependent" variable x_n to a set of "independent" variables $x_{n-1}, x_{n-2}, \ldots, x_{n-p}$ plus an error term ε_n is referred to as a regression model. In this case, of course, the variable x_n is regressed on previous values of itself, and so the model is called autoregressive.

The reason for the minus signs appearing before the coefficients of the autoregressive representation above is that we want to transfer those terms to the left-hand side, and thus obtain

$$x_n + a_1 x_{n-1} + a_2 x_{n-2} + \ldots + a_p x_{n-p} = \varepsilon_n . \tag{4.2}$$

This is the standard representation of an AR model of a stationary time series. We recognize the left-hand side of this equation as the convolution of the operator $\{1, a_1, a_2, \ldots, a_p\}$ with the time series $\{x_n\}$. This operator is called the autoregressive operator. The z transform of this operator is the polynomial

$$A(z) = 1 + a_1 z^{-1} + a_2 z^{-2} + \ldots + a_p z^{-p} . \tag{4.3}$$

The symbol z^{-1} can either be interpreted as a complex variable or as a unit-delay operator. The unit-delay operator z^{-1} is defined as

$$z^{-1}[x_n] = x_{n-1} , \tag{4.4}$$

thus we see that

$$z^{-m}[x_n] = x_{n-m} . \tag{4.5}$$

With the interpretation of z^{-1} as the unit-delay operator, the fundamental autoregressive representation may be written as

$$A(z)[x_n]=\varepsilon_n, \tag{4.6}$$

if we divide each side of this equation by $A(z)$ we obtain

$$x_n=\frac{1}{A(z)}[\varepsilon_n]. \tag{4.7}$$

This equation shows that the autoregressive process x_n is the output of a finite feedback system with input ε_n. The polynomial $A(z)$ is called the feedback polynomial. If we make the expansion

$$H(z)=\frac{1}{A(z)}=1+h_1z^{-1}+h_2z^{-2}+h_3z^{-3}+\dots \tag{4.8}$$

then the coefficients 1, h_1, h_2, ... represent the impulse response of the AR system. In order for the AR system to be stable, and thus for the process x_n to be stationary, the feedback polynomial $A(z)$ must be strictly minimum delay, that is, $A(z)$ must not have any zeros outside or on the unit circle. We can write the autoregressive process as

$$x_n=H(z)[\varepsilon_n] \tag{4.9}$$

which is

$$x_n=\varepsilon_n+h_1\varepsilon_{n-1}+h_2\varepsilon_{n-2}+\dots. \tag{4.10}$$

This equation states that x_n is the convolution of the infinitely long operator $\{1,h_1,h_2,\dots\}$ with the innovation series $\{\varepsilon_n\}$. The operator $\{1,h_1,h_2,\dots\}$ is the inverse of the autoregressive operator $\{1,a_1,a_2,\dots,a_p\}$.

Another model of great physical importance is the moving average model. In this model, the time series $\{x_n\}$ is linearly dependent on a finite number of previous innovations; that is

$$x_n=\varepsilon_n+b_1\varepsilon_{n-1}+b_2\varepsilon_{n-2}+\dots+b_q\varepsilon_{n-q}. \tag{4.11}$$

The name moving average is somewhat misleading because the weights 1, b_1, $b_2,\dots,b_q$ which multiply the innovations need not be positive and need not sum to one, as would ordinarily be implied by the term moving average. However the nomenclature moving average is in common use for this model. The moving average operator has z transform

$$B(z)=1+b_1z^{-1}+b_2z^{-2}+\dots+b_qz^{-q}. \tag{4.12}$$

Interpreting z^{-1} as the unit-delay operator, the moving average model may be written economically as

$$x_n = B(z)[\varepsilon_n]. \tag{4.13}$$

Equation (4.13) shows that the moving average process $\{x_n\}$ is the output of a finite feedforward (i.e., feedfront) system with input $\{\varepsilon_n\}$. The polynomial $B(z)$ is called the feedforward polynomial, or the move-ahead polynomial.

Finally we come to the ARMA model. To achieve greater flexibility in the fitting of physical time series, it is often advantageous to include both autoregressive and moving average components in the model. This leads to the ARMA model

$$A(z)[x_n] = B(z)[\varepsilon_n] \tag{4.14}$$

or

$$x_n + a_1 x_{n-1} + \ldots + a_p x_{n-p} = \varepsilon_n + b_1 \varepsilon_{n-1} + \ldots + b_q \varepsilon_{n-q}.$$

We can write the ARMA model as

$$x_n = \frac{B(z)}{A(z)}[\varepsilon_n]. \tag{4.15}$$

This equation shows that the ARMA process $\{x_n\}$ is the output of a finite feedback-feedfront system with input $\{\varepsilon_n\}$. The polynomial $A(z)$ is the feedback polynomial and the polynomial $B(z)$ is the feedfront polynomial. If we make the expansion

$$H(z) = \frac{B(z)}{A(z)} = h_0 + h_1 z^{-1} + h_2 z^{-2} + \ldots \tag{4.16}$$

then coefficients $h_0, h_1, h_2, \ldots$ represent the impulse response of the ARMA system. In order for the ARMA system to be stable, the feedback polynomial $A(z)$ must be strictly minimum delay, that is, $A(z)$ must have no zeros outside or on the unit circle. A stable ARMA system with white noise input ε_n yields a stationary output x_n.

4.2 Parsimony and the Physical Model

The three models which we discussed in the previous section may be described as models with both statistical and dynamic components. In each case the model expressed the time series as the output of a linear filter with a white noise input. The white noise input represents the statistical aspect of the model and the linear filter represents the dynamic aspect of the model. This approach is in keeping with the concept of a typical time series as being neither completely random nor completely predictable. The white noise series is completely random and thus has zero predictability. On the other hand, the linear filter is characterized by a set of coefficients which are fixed; these coefficients are parameters and not random variables. The resulting time series has both elements of randomness and non randomness. Such a time series can be predicted with some degree of success but not perfectly. In this respect, we think of the innovation series as being the random component of the time series, and the impulse response function of the filter as the dynamic component. The time series itself is thus the convolution of the random and dynamic components. Given a stationary time series $\{x_n\}$ there is, of course, no way to unravel its random and dynamic components without further information. This unraveling process, which takes apart the convolution of the random with the dynamic, is called deconvolution.

In the above discussion we have mentioned the need for further information. This remark is a key point in time series analysis. As a result we would like to link time series methods closely with the empirical and theoretical evidence provided by the subject matter under investigation. This approach is used in conjunction with the parsimony approach to time series analysis [4.1]. The parsimony approach is so named because the general principle of parsimony is used. As we have seen, the mathematical models which we employ contain certain coefficients or parameters whose values must be estimated from the data. The parsimony approach is the approach based upon the concept of employing in practice the smallest possible number of parameters for adequate representation. The central role played by this principle of parsimony in the estimation of parameters is evident in almost all scientific research.

With this in mind let us now return to the subject of deconvolution. Ideally in our time series model, we want to identify the random component and the dynamic component with actual physical entities or processes. The success of the deconvolution is not measured solely by the use of the least number of parameters but also by how well the estimated quantities agree with the physical world. A few coefficients that fit the time series but have nothing to do with the physical phenomenon do not represent a physical deconvolution. However, a physical deconvolution with a small number of parameters is to be preferred over one with a large number of parameters. The key point is that the number of parameters should be minimized under the condition that results can be verified in practice or in theory by direct physical measurements in the real world. In summary, the principle of parsimony must be used in conjunction with the physics of the problem.

Much is known about the physics of seismic wave propagation through the earth, and the results of a seismic time series analysis made on the basis of surface observations can be directly verified by drilling a hole into the earth (that is, an oil well). In order to illustrate these points, we will now discuss the process of seismic deconvolution and the type of spectral estimate obtained by such an approach.

4.3 Seismic Deconvolution and the Resulting Spectral Estimate

In order to understand seismic deconvolution, we must simplify. It is the process of simplification which brings out the essential features of the sedimentary layers when they are subjected to a seismic source of energy. A mathematical equation has little value unless we can link its meaning with the physical structure of the earth. Of course such linkage cannot usually be made in all generality, so we resort to simplification in order to gain as much understanding as we can. It is better to understand a few simple things well than to have a whole book of mathematical equations which are not connected to the physical action of the earth. The science of geophysics is indeed the study of the physics of the earth.

With this approach in mind let us look at a single horizontal interface between two sedimentary layers. Throughout we measure wave motion in terms of units which represent square root of energy. A unit spike by definition has amplitude one at time zero and amplitude zero elsewhere; thus the energy content of a unit spike is one-squared, which is one unit of energy. Similarly, a spike of amplitude 0.5 would contain 0.5 squared, or 0.25 units of energy. A wavelet with amplitudes

$$a_0=1\,, \quad a_1=0.5\,, \quad a_2=-0.2\,, \quad a_t=0 \quad \text{for} \quad t>2$$

has total energy

$$1+(0.5)^2+(-0.2)^2=1.29\,.$$

Suppose now that a downgoing unit spike strikes an interface between two layers. As we know from classical physics, some of the energy is transmitted through the interface and some is reflected back from the interface. The law of conservation of energy says that the sum of the reflected and transmitted energies is equal to the incident energy. By physical reasoning we also know that we can define a reflection coefficient ε and a transmission coefficient τ. For the given downgoing unit spike the upgoing reflected spike has amplitude ε and the transmitted downgoing spike has amplitude τ. This relationship is illustrated in Fig. 4.1. In all the figures we are representing plane waves propagating at normal incidence, but ray paths are given a horizontal displacement to simulate the passage of time in the horizontal direction.

The law of conservation of energy states that

$$1^2=\varepsilon^2+\tau^2$$

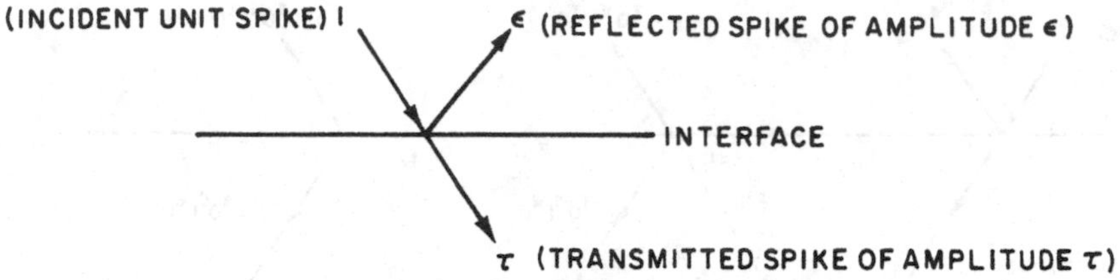

Fig. 4.1. Reflected and transmitted spikes due to a downgoing incident spike

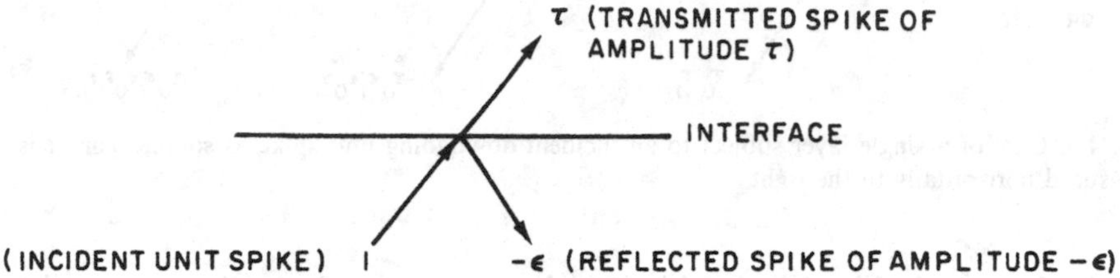

Fig. 4.2. Reflected and transmitted spikes due to an upgoing incident spike

so ε and τ are related by

$$\tau=\sqrt{1-\varepsilon^2}\,. \tag{4.17}$$

From physical reasoning it also follows that a given upgoing unit spike striking the interface from below gives rise to a downgoing reflected spike of amplitude $-\varepsilon$ and an upgoing transmitted spike of amplitude τ, see Fig. 4.2.

The law of conservation of energy is

$$1^2=(-\varepsilon)^2+\tau^2$$

which is the same as before.

The transfer function is defined as the ratio of the z transforms of the output to the input. If we let $T(z)$ represent the transfer function of the transmitted wave as output with respect to a downgoing incident wave we have

$$T(z)=\frac{\tau}{1}$$

which in this simple case is the constant τ.

Let us now look at a single sedimentary layer with air above and granite below. We have two interfaces with reflection coefficients ε_0 and ε_1 and the corresponding transmission coefficients

$$\tau_0=\sqrt{1-\varepsilon_0^2} \quad\text{and}\quad \tau_1=\sqrt{1-\varepsilon_1^2}\,, \tag{4.18}$$

see Fig. 4.3.

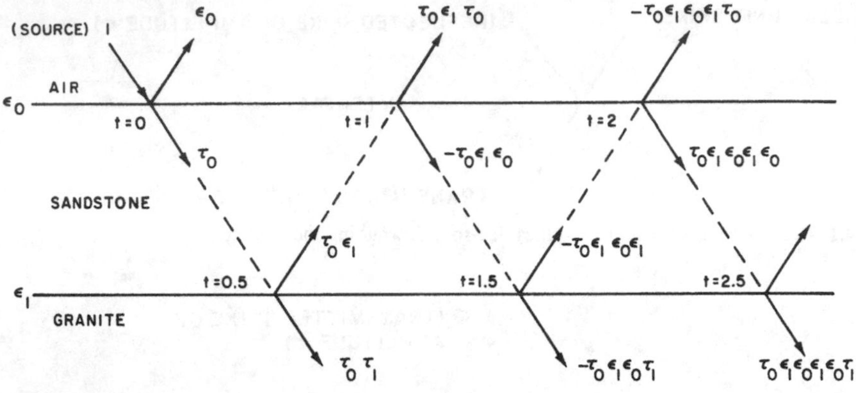

Fig. 4.3. Case of a single layer subject to an incident downgoing unit spike as source. Time t is measured horizontally to the right

As depicted in Fig. 4.3 we subject the layer to an incident downgoing spike from above. This source energy results in a sequence of reflected and transmitted waves. Let us trace this action. We suppose that it takes 0.5 time units for a spike to travel through the sedimentary layer. Thus the two-way travel time in the sandstone is one time unit. At time 0, the incident spike gives rise to a reflected spike ε_0 and a transmitted spike τ_0. The reflected spike escapes but the transmitted spike travels down to the lower interface, arrives at time 0.5, and gives rise to a transmitted spike $\tau_0\tau_1$ and a reflected spike $\tau_0\varepsilon_1$. This transmitted spike escapes but the reflected spike travels up to the upper interface, arriving at time 1, and gives rise to a transmitted spike $\tau_0\varepsilon_1\tau_0$ and a reflected spike $\tau_0\varepsilon_1(-\varepsilon_0)$. This transmitted spike escapes but the reflected spike travels down to the lower interface, arriving at time $t=1.5$, and the process is repeated again and again as the energy reverberates within the sandstone layer. At each whole unit of time (i.e., $0, 1, 2, 3, \ldots$), some energy escapes into the air, and each unit plus one-half (i.e., $0.5, 1.5, 2.5, \ldots$) of time some energy escapes into the granite, so after an infinite amount of time there is no more energy left in the sandstone.

At all times the law of conservation of energy holds. The total amount of energy is the energy of the source, namely one. This total energy must be equal at any time to the sum of the energy in the layer plus the escaped energy. For example, just after $t=0.5$ the wave motion in the layer is a spike of amplitude $\tau_0\varepsilon_1$ and the escaped wave motion is the spike of amplitude ε_0 in the air and the spike of amplitude $\tau_0\tau_1$ in the granite. We have

$$
\begin{aligned}
(\tau_0\varepsilon_1)^2+\varepsilon_0^2+(\tau_0\tau_1)^2 &= \tau_0^2(\varepsilon_1^2+\tau_1^2)+\varepsilon_0^2 \\
&= \tau_0^2+\varepsilon_0^2 = 1
\end{aligned}
$$

as it should.

The transmitted wavelet is made up of all the spikes that escape into the granite. Thus the transmitted wavelet is

$$(\tau_0\tau_1, -\tau_0\varepsilon_1\varepsilon_0\tau_1, \tau_0(\varepsilon_1\varepsilon_0)^2\tau_1, -\tau_0(\varepsilon_1\varepsilon_0)^3\tau_1, \ldots),$$

where the initial coefficient is at $t=0.5$, the next at $t=1.5$, and so on. However, for convenience we can choose a new time scale, so we may regard the initial coefficient in the transmitted wavelet as occurring at time 0, the next at time 1, and so on. In other words, the new time scale has its origin at the time break of the transmitted wavelet. The z transform of the transmitted wavelet is then

$$\begin{aligned} T(z) &= \tau_0\tau_1 - \tau_0\varepsilon_1\varepsilon_0\tau_1 z^{-1} + \tau_0(\varepsilon_1\varepsilon_0)^2\tau_1 z^{-2} - \tau_0(\varepsilon_1\varepsilon_0)^3\tau_1 z^{-3} + \ldots \\ &= \tau_0\tau_1[1-(\varepsilon_1\varepsilon_0 z^{-1}) + (\varepsilon_1\varepsilon_0 z^{-1})^2 - (\varepsilon_1\varepsilon_0 z^{-1})^3 + \ldots]. \end{aligned} \tag{4.19}$$

The expression within brackets is a geometric series which may be summed. As a result we have

$$T(z) = \frac{\tau_0\tau_1}{1+\varepsilon_0\varepsilon_1 z^{-1}}. \tag{4.20}$$

This expression is the transfer function of a single layer.

Instead of the complicated way in which we derived this transfer function, we can derive it directly as follows. The layer acts as a feedback loop. Each round trip the wave makes in the layer represents one passage through the feedback loop. Let x_t be an arbitrary source wavelet incident on the top interface and let y_t be the resulting transmitted wavelet. Let us choose the point that represents the mixer of the feedback system as a point in space just above the bottom layer, see Fig. 4.4. The input wavelet x_t must go through the top interface and then travel to the bottom layer to arrive at the mixer. Thus at the mixer the input wavelet appears as $x_t\tau_0 z^{-1/2}$, where τ_0 represents the transmission through the top interface and $z^{-1/2}$ represents the time delay of 0.5 required for the wavelet to travel through the layer. The output of the mixer is y_t/τ_1. That is, we must divide out the transmission coefficient τ_1 as the mixer is just above the bottom interface. The basic feedback loop then delays the output of the mixer by one time unit (representing the round trip passage time in the layer) and multiplies this delayed output by ε_1 representing the bounce from the bottom interface and also by $-\varepsilon_0$ representing the bounce from the top interface. That is, the basic feedback loop forms the quantity

$$(y_t/\tau_1)z^{-1}\varepsilon_1(-\varepsilon_0).$$

The mixer then adds this quantity to the input to give the output; that is

$$(y_t/\tau_1)z^{-1}\varepsilon_1(-\varepsilon_0) + x_t\tau_0 z^{-1/2} = y_t/\tau_1.$$

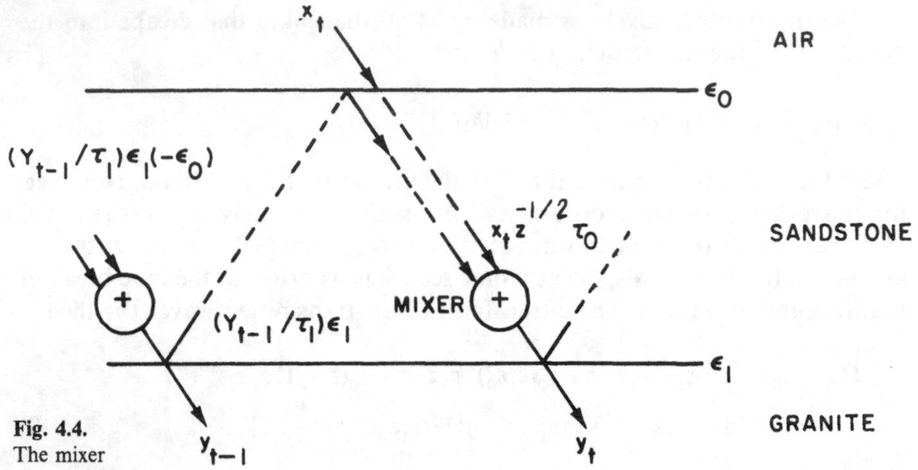

Fig. 4.4.
The mixer

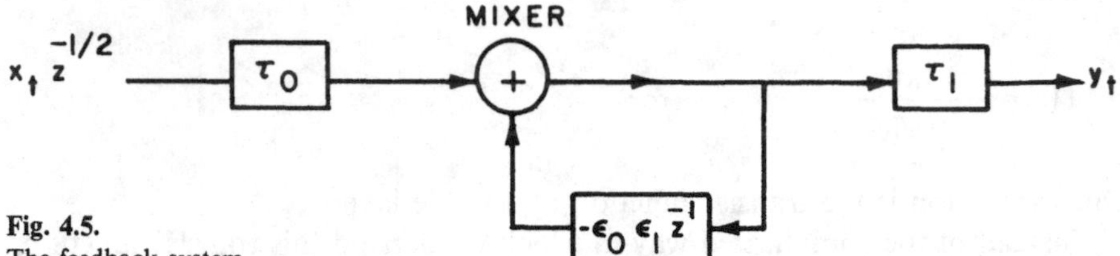

Fig. 4.5.
The feedback system

Solving this equation we have

$$(y_t/\tau_1)(1+\varepsilon_0\varepsilon_1 z^{-1})=x_t\tau_0 z^{-1/2}$$

or

$$\frac{y_t z^{1/2}}{x_t}=\frac{\tau_0\tau_1}{1+\varepsilon_0\varepsilon_1 z^{-1}}. \tag{4.21}$$

Note that the $z^{1/2}$ on the left corresponds to the time advance of 0.5 which we introduced when we measured the transmitted wavelet at its first break. Thus we obtain the same transfer function as before (see Fig. 4.5), namely,(20).

The principle of reciprocity states that if we interchange source and receiver we still obtain the same transfer function. Figure 4.6 (left side) depicts such an interchange, where $T'(z)$ denotes the new transfer function. However, we can flip the entire picture upside down, as shown in Fig. 4.6 (right side) to obtain the equivalent situation. Note that the order and sign of the reflection coefficients have been reversed. Thus we see that $T'(z)$ can be obtained from $T(z)$ by using $-\varepsilon_1$, $-\varepsilon_0$ instead of ε_0, ε_1. We have

$$T'(z)=\frac{\tau_1\tau_0}{1+(-\varepsilon_1)(-\varepsilon_0)z^{-1}}, \tag{4.22}$$

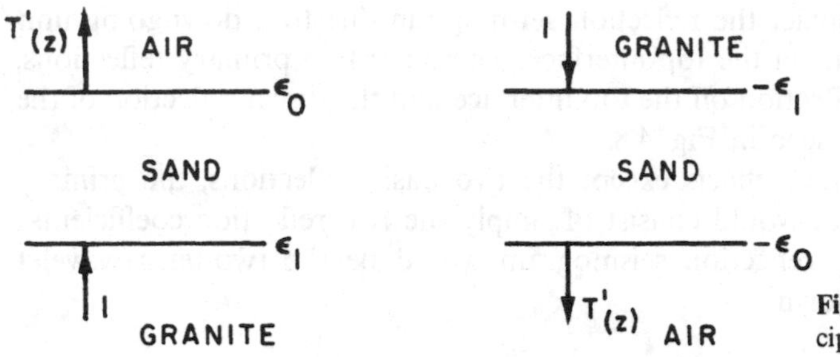

Fig. 4.6. Reciprocity principle illustrated

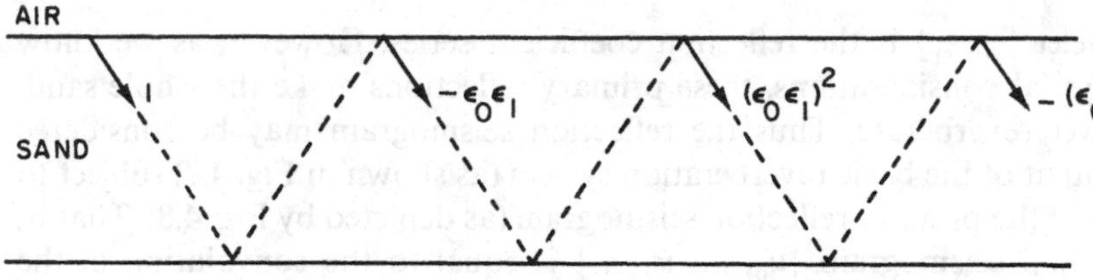

Fig. 4.7. The basic reverberating system

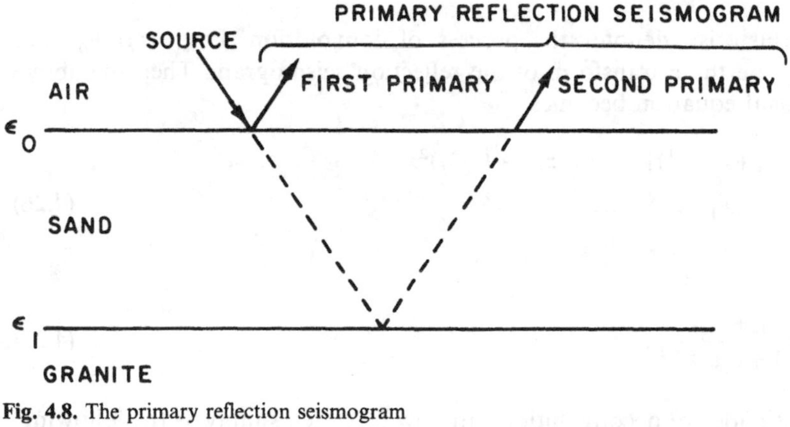

Fig. 4.8. The primary reflection seismogram

and we see that $T'(z)$ is identical to $T(z)$, as predicted by the principle of reciprocity.

Let us now look at the transfer function $T(z)$. In the expression for $T(z)$ the quantity

$$\frac{1}{1+\varepsilon_0\varepsilon_1 z^{-1}} = 1-\varepsilon_0\varepsilon_1 z^{-1}+(\varepsilon_0\varepsilon_1)^2 z^{-2}-(\varepsilon_0\varepsilon_1)^3 z^{-3}+\ldots \tag{4.23}$$

represents the basic reverberation within the sedimentary layer, as depicted in Fig. 4.7.

Now let us construct the reflection seismogram due to a downgoing unit spike source incident on the top interface. There are two primary reflections, namely, the direct reflection off the top interface and the direct reflection of the bottom interface, as seen in Fig. 4.8.

Neglecting all other effects except the two basic reflections, the primary reflection seismogram would consist of simply the two reflection coefficients; that is, the primary reflection seismogram would be the two-term wavelet $\{\varepsilon_0, \varepsilon_1\}$ with z transform

$$E(z)=\varepsilon_0+\varepsilon_1 z^{-1}. \tag{4.24}$$

The wavelet $\{\varepsilon_0, \varepsilon_1\}$ is the reflection coefficient series. However, as we know from physical considerations, these primary reflections make the whole sandstone layer reverberate. Thus the reflection seismogram may be considered as the output of the basic reverberation system (as shown in Fig. 4.7) subject to the input of the primary reflection seismogram (as depicted by Fig. 4.8). That is, the reflection seismogram $\{u_0, u_1, u_2, \ldots\}$ is equal to the convolution of the reflection coefficient series with the basic reverberation; i.e.,

$$\{u_0, u_1, u_2, \ldots\}=\{\varepsilon_0, \varepsilon_1\}*\{1, -\varepsilon_0\varepsilon_1, (\varepsilon_0\varepsilon_1)^2, -(\varepsilon_0\varepsilon_1)^3, \ldots\} \tag{4.25}$$

where the asterisk denotes the process of convolution. Let $U(z)=u_0+u_1 z+u_2 z^2+\ldots$ be the z transform of the reflection seismogram. Then the above convolutional equation becomes

$$U(z)=(\varepsilon_0+\varepsilon_1 z^{-1})[1-\varepsilon_0\varepsilon_1 z^{-1}+(\varepsilon_0\varepsilon_1)^2 z^{-2} -(\varepsilon_0\varepsilon_1)^3 z^{-3}+\ldots] \tag{4.26}$$

which is

$$U(z)=\frac{\varepsilon_0+\varepsilon_1 z^{-1}}{1+\varepsilon_0\varepsilon_1 z^{-1}}. \tag{4.27}$$

The basic idea of deconvolution can now be stated simply as the following. Given the reflection seismogram, find the reflection coefficient series. From the above equation we can write

$$(1+\varepsilon_0\varepsilon_1 z^{-1})(u_0+u_1 z^{-1}+u_2 z^{-2}+\ldots)=\varepsilon_0+\varepsilon_1 z^{-1} \tag{4.28}$$

which in the time domain is

$$\{1, \varepsilon_0\varepsilon_1\}*\{u_0, u_1, u_2, \ldots\}=\{\varepsilon_0, \varepsilon_1\}. \tag{4.29}$$

This equation states that the reflection coefficient series is the convolution of the operator $\{1, \varepsilon_0\varepsilon_1\}$ with the reflection seismogram. We say that the

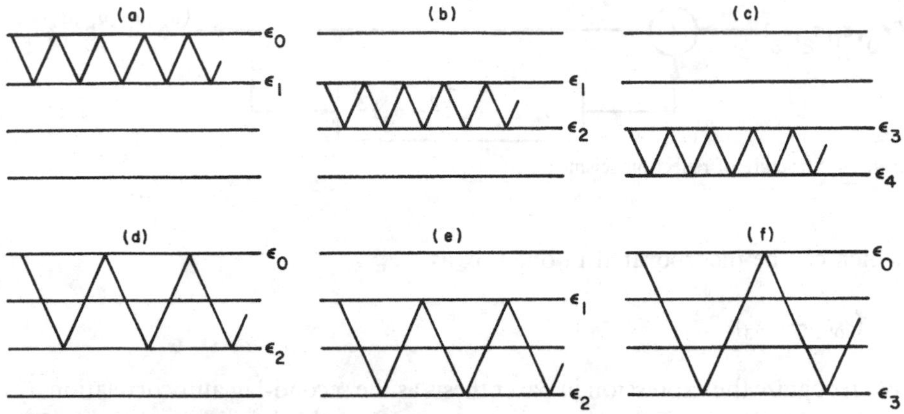

Fig. 4.9a–f. Reverberations. **(a–c)** First-order; **(d, e)** second-order linked; **(f)** third-order

operator $\{1, \varepsilon_0\varepsilon_1\}$ deconvolves the reflection seismogram to yield the reflection coefficient series. Thus the operator $\{1, \varepsilon_0\varepsilon_1\}$ is the required deconvolution operator.

Because the above discussion contains all the basic ideas, we can now write down the results for the multilayered sedimentary system which occurs in actual seismic prospecting. First we must examine the types of reverberations which can occur. We add enough hypothetical layers as necessary in our mathematical model so that the round trip travel time in each layer is one. A first-order reverberation is defined as one in which the closed loop travel time is one; a second-order reverberation is defined as one with time 2 and so on. Furthermore, we define a linked reverberation as one which involves only physically adjacent (i.e., connected) layers. These ideas will become clear if we examine a three-layer sedimentary system, see Fig. 4.9. Of course, all first-order reverberations are connected because they all involve just one layer. The feedback boxes for the three first-order reverberations shown in Fig. 4.9 are respectively

$$-\varepsilon_0\varepsilon_1 z^{-1}, \qquad -\varepsilon_1\varepsilon_2 z^{-1}, \qquad -\varepsilon_2\varepsilon_3 z^{-1},$$

which can be incorporated in one box as

$$-(\varepsilon_0\varepsilon_1 + \varepsilon_1\varepsilon_2 + \varepsilon_2\varepsilon_3)z^{-1}.$$

We recognize the expression in parenthesis as the first lag autocorrelation a_1 of the reflection coefficient series $\{\varepsilon_0, \varepsilon_1, \varepsilon_2, \varepsilon_3\}$, so this box is simply $-a_1 z^{-1}$. The feedback boxes of the two second-order linked reverberations are

$$-\varepsilon_0\varepsilon_2 z^{-2}, \qquad -\varepsilon_1\varepsilon_3 z^{-2}$$

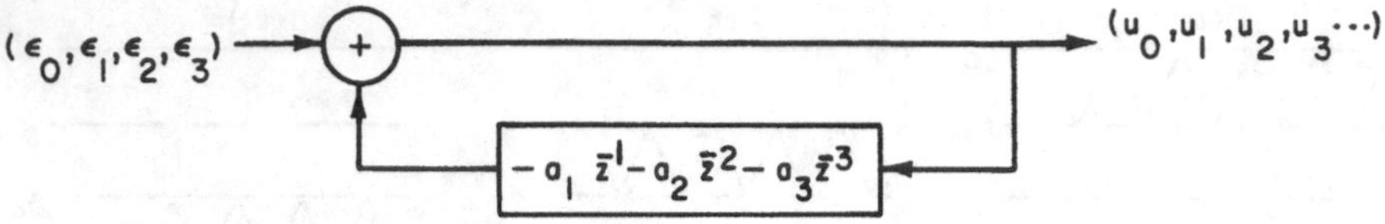

Fig. 10. Structure of reflection seismogram

which can be incorporated in one box as

$$(\varepsilon_0\varepsilon_2+\varepsilon_2\varepsilon_3)z^{-2}.$$

We recognize the expression in parenthesis as the second-lag autocorrelation a_2 of the reflection coefficient series $\{\varepsilon_0, \varepsilon_1, \varepsilon_2, \varepsilon_3\}$ so this box is simply $-a_2z^{-2}$. The third-order reverberation for a three-layer system must necessarily be linked. Its feedback box is

$$-\varepsilon_0\varepsilon_3z^{-3}$$

which may be written as $-a_3z^{-3}$. The entire feedback box for the linked reverberations is thus

$$-a_1z^{-1}-a_2z^{-2}-a_3z^{-3}.$$

The reflection coefficient series has z transform

$$\varepsilon_0+\varepsilon_1z^{-1}+\varepsilon_2z^{-2}+\varepsilon_3z^{-3}.$$

The reflection seismogram is (approximately) obtained by passing the reflection coefficient series through the feedback filter, as seen in Fig. 4.10.

The z transform of the reflection seismogram is therefore (approximately) given by [4.2]

$$U(z)=\frac{\varepsilon_0+\varepsilon_1z+\varepsilon_2z^2+\varepsilon_3z^3}{1+a_1z+a_2z^2+a_3z^3} \tag{4.30}$$

which we write as

$$U(z)=\frac{E(z)}{A(z)}. \tag{4.31}$$

The deconvolution operator is thus $\{1, a_1, a_2, a_3\}$, as given by the denominator coefficients in the above expression.

The above discussion of course is an ideal one, as it involves the entire reflection seismogram. In the case of N layers the above expression for $U(z)$

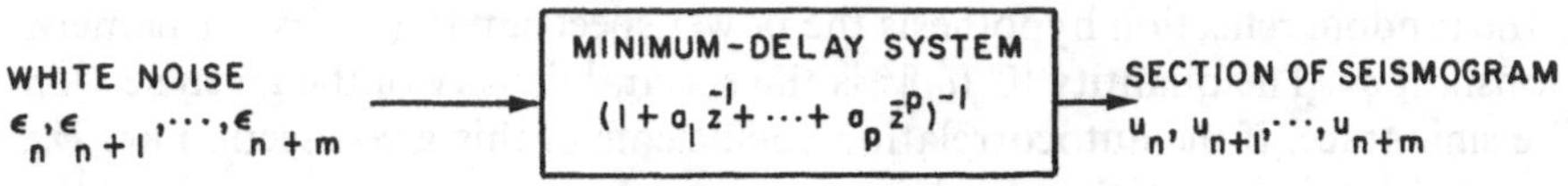

Fig. 4.11. Model of section of seismogram

becomes

$$U(z)=\frac{\varepsilon_0+\varepsilon_1 z^{-1}+\ldots+\varepsilon_N z^{-N}}{1+a_1 z^{-1}+\ldots+a_N z^{-N}}. \tag{4.32}$$

In actual practice we assume that the autocorrelation $a_1, a_2, \ldots, a_N$ damps out so that we need only consider the first p coefficients $a_1, a_2, \ldots, a_p$ where $p<N$. Also we only analyze a section of the seismogram at a time, say the section u_n, $u_{n+1}, u_{n+2}, \ldots, u_{n+m}$ from $t=n$ to $t=n+m$. Then the z transform of this section is approximately

$$u_n+u_{n+1}z^{-1}+\ldots+u_{n+m}z^{-m}=\frac{\varepsilon_n+\varepsilon_{n+1}z^{-1}+\ldots+\varepsilon_{n+m}z^{-m}}{1+a_1 z^{-1}+\ldots+a_p z^{-p}} \tag{4.33}$$

which we write as

$$U_s(z)=\frac{E_s(z)}{A(z)}. \tag{4.34}$$

Now we appeal to the random reflection hypothesis, which states that the reflection coefficients $\varepsilon_n, \varepsilon_{n+1}, \ldots, \varepsilon_{n+m}$ within the given section are mutually uncorrelated (i.e., white). The random reflection hypothesis is generally true for many actual sections of sedimentary layers in the earth's crust. Also, as a physical fact, we know that a reflection seismogram damps out, so that the feedback system that generates the seismogram must be stable. This stability property means that the denominator polynomial $1+a_1 z^{-1}+\ldots+a_p z^{-p}$ is minimum delay. Therefore, its reciprocal $(1+a_1 z^{-1}+\ldots+a_p z^{-p})^{-1}$ is also minimum delay. Thus, we see that we have the following model for the given section shown in Fig. 4.11.

Using the above expression for $U_s(z)$ we have

$$U_s(z)U_s(z^{-1})=\frac{E_s(z)E_s(z^{-1})}{A(z)A(z^{-1})} \tag{4.35}$$

which, if we let $z=e^{j\omega}$, becomes

$$|U_s(\omega)|^2=\frac{|E_s(\omega)|^2}{|A(\omega)|^2}. \tag{4.36}$$

By the random reflection hypothesis the power spectrum $|E_s(\omega)|^2$ is flat, namely, a constant σ^2. The quantity $|U_s(\omega)|^2$ is the spectral density of the given section of seismic trace. If the autocorrelation coefficients of this given section are ϕ_0, $\phi_1, \ldots$ (where $\phi_{-t}=\phi_t$), then we have

$$|U_s(\omega)|^2 = \sum_{t=-\infty}^{\infty} \phi_t e^{-i\omega t}. \tag{4.37}$$

Returning now to z, we have

$$\sum_{t=-\infty}^{\infty} \phi_t z^{-t} = \frac{\sigma^2}{A(z)A(z^{-1})} \tag{4.38}$$

which is

$$A(z)\sum \phi_t z^{-t} = \frac{\sigma^2}{A(z^{-1})}. \tag{4.39}$$

Because the right-hand side of (4.39) has an expansion in only nonpositive powers of z^{-1}, the left-hand side also must have an expansion in only nonpositive powers of z^{-1}. Hence, in particular the coefficients of z^{-k} for $k=1,2,\ldots,p$ must be zero. These coefficients obtained from the left-hand side are

$$\begin{gathered} \phi_1 + a_1\phi_0 + \ldots + a_p\phi_{p-1} = 0 \\ \ldots \\ \phi_p + a_1\phi_{p-1} + \ldots + a_p\phi_0 = 0. \end{gathered} \tag{4.40}$$

This set of equations is the set of well-known normal equations, and they can be solved for the unknown $a_1, a_2, \ldots, a_p$. The constant σ^2 can then be found from the equation

$$\phi_0 + a_1\phi_1 + \ldots + a_p\phi_p = \sigma^2. \tag{4.41}$$

We can now summarize the procedure. Compute the autocorrelation coefficients $\phi_0, \ldots, \phi_p$ of the given section of the seismogram. Then solve the normal equations for the deconvolution operator $a_1, a_2, \ldots, a_p$. Finally, deconvolve the seismogram (i.e., convolve the given section of the seismogram by the deconvolution operator $\{1, a_1, a_2, \ldots, a_p\}$) to obtain the white reflection coefficient series $\{\varepsilon_n, \varepsilon_{n+1}, \ldots, \varepsilon_{n+m}\}$. This series is called the deconvolved seismic trace.

This method of deconvolution is called statistical deconvolution because it is based on the random reflection hypothesis, namely, the hypothesis that the reflection coefficients within a given earth section form a white noise series. Statistical deconvolution provides an estimate of the spectral density of the given section of seismic trace. This spectral estimate is the maximum-entropy spectral estimate of *Burg* [4.3]; namely

$$|U_s(\omega)|^2 = \frac{\sigma^2}{|A(\omega)|^2}, \tag{4.42}$$

where $A(\omega)$ is the discrete Fourier transform of the operator; that is,

$$A(\omega) = 1 + a_1 e^{-j\omega} + a_2 e^{-j\omega 2} + \ldots + a_p e^{-j\omega p}. \tag{4.43}$$

4.4 Design of Shaping and Spiking Filters

In this section we want to consider the case where the forms of the input signal and the desired output signal are known.

The design problem to be considered may be formulated as follows. Given a signal $\{x_n\}$ we want to operate on $\{x_n\}$ in some manner so as to obtain the best approximation to the signal $\{s_n\}$. That is, let $T[x_n]$ be the best approximation to $\{s_n\}$. More generally, we may let $T[x_n]$ be the best approximation to the shifted signal $\{s_{n+\alpha}\}$, where α is a time shift. At this point, the criterion of best approximation and the form of the operator are unspecified.

Let us now consider an ideal system $\{f_n\}$. An ideal system would transform the input into the desired output. If the ideal system is a linear shift-invariant system, then

$$\{s_{n+\alpha}\} = \{f_n\} * \{x_n\}. \tag{4.44}$$

In terms of Fourier transforms, this equation becomes

$$S(\omega)e^{j\omega\alpha} = F(\omega)X(\omega). \tag{4.45}$$

The solution of the above equation gives the filter characteristics of the *ideal filter* as

$$F(\omega) = \frac{S(\omega)e^{j\omega\alpha}}{X(\omega)}. \tag{4.46}$$

This result represents the formal solution to the problem. However, for actual computations this equation is of no help, because the impulse response f_n of the ideal system $F(\omega)$ will generally be an infinitely long two-sided operator. For practical purposes, a digital filter with a finite number of coefficients is required, so let us try to approximate the ideal filter f_n by a finite-length moving average

(MA) filter h_n where

$$h_n = 0 \quad \text{for} \quad n<0 \quad \text{and for} \quad n>M. \tag{4.47}$$

(We recall that an MA filter is defined as a causal linear shift-invariant filter with a finite-length impulse response. Alternatively, a MA filter can be defined as a linear shift-invariant filter that has a transfer function equal to a polynomial in z^{-1}). Thus, we want to approximate the frequency response $F(\omega)$ of the ideal filter by the frequency response

$$H(\omega) = \sum_{n=0}^{M} h_n e^{-j\omega n} \tag{4.48}$$

of the required MA filter.

Such an approximation procedure means that a certain amount of information contained in the infinite-length impulse response $\{f_n\}$ must be lost in order to obtain the approximate finite-length impulse response $\{h_n\}$. Consequently, we shall need some sort of averaging process to carry out this approximation. We want the difference

$$F(\omega) - H(\omega)$$

to be small in some sense. For example, we might choose the coefficients $\{h_n\}$ so that the mean-squared difference

$$\frac{1}{2\pi}\int_{-\pi}^{\pi} |F(\omega) - H(\omega)|^2 d\omega$$

is a minimum. However, in this expression we see that the squared difference $|F(\omega) - H(\omega)|^2$ is given a uniform weighting for all frequencies. It would make more sense to weight the squared difference according to the energy spectrum $|X(\omega)|^2$ of the input signal. Thus, let us choose the MA coefficients $\{h_n\}$ so that the mean-weighted-squared difference, or mean-square error,

$$I = \frac{1}{2\pi}\int_{-\pi}^{\pi} |F(\omega) - H(\omega)|^2 |X(\omega)|^2 d\omega \tag{4.49}$$

is a minimum. This equation can be written as

$$I = \frac{1}{2\pi}\int_{-\pi}^{\pi} |F(\omega)X(\omega) - H(\omega)X(\omega)|^2 d\omega. \tag{4.50}$$

Substituting (4.46) in (4.50), we obtain

$$I = \frac{1}{2\pi}\int_{-\pi}^{\pi} |S(\omega)e^{j\omega\alpha} - H(\omega)X(\omega)|^2 d\omega. \tag{4.51}$$

Now $S(\omega)$ is the spectrum of the desired output signal $\{s_n\}$ and $H(\omega)X(\omega)$ is the spectrum of the actual output signal $\{h_n * x_n\}$. If we apply Parseval's theorem to this equation, we obtain

$$I = \sum_{n=-\infty}^{\infty} |s_{n+\alpha} - h_n * x_n|^2 . \tag{4.52}$$

Because we are dealing with real signals, the above equation becomes

$$I = \sum_{n=-\infty}^{\infty} \left(s_{n+\alpha} - \sum_{k=0}^{M} h_k x_{n-k} \right)^2 . \tag{4.53}$$

In order to minimize I, we set its partial derivatives with respect to each of h_0, $h_1, \ldots, h_M$ equal to zero. We obtain

$$\frac{\partial I}{\partial h_j} = -2 \sum_{n=-\infty}^{\infty} \left(s_{n+\alpha} - \sum_{k=0}^{M} h_k x_{n-k} \right) x_{n-j} = 0$$

for $j = 0, 1, 2, \ldots, M$. We may therefore write

$$\sum_{k=0}^{M} h_k \left(\sum_{n=-\infty}^{\infty} x_{n-k} x_{n-j} \right) = \sum_{n=-\infty}^{\infty} s_{n+\alpha} x_{n-j} \tag{4.54}$$

for $j = 0, 1, 2, \ldots, M$. We recognize the autocorrelation coefficient

$$r_{j-k} = \sum_{n=-\infty}^{\infty} x_{n-k} x_{n-j} . \tag{4.55}$$

Also, we define the cross-correlation coefficient

$$g_{j+\alpha} = \sum_{n=-\infty}^{\infty} s_{n+\alpha} x_{n-j} . \tag{4.56}$$

Then, (4.54) becomes

$$\sum_{k=0}^{M} h_k r_{j-k} = g_{j+\alpha} \tag{4.57}$$

for $j = 0, 1, 2, \ldots, M$. This set of equations are called the *Toeplitz normal equations.* The autocorrelation coefficients and the cross-correlation coefficients can be computed from the known input signal x_n and desired output signal s_n. Then the Toeplitz normal equations can be solved to find the required filter coefficients h_n. Because the output $\{h_n\} * \{x_n\}$ of this filter approximates the desired output signal $\{s_n\}$ in a least-squares sense, this filter is called the least-squares shaping filter.

Because of the special form of the simultaneous equations called the Toeplitz form, the equations may be solved by an efficient recursive procedure called the Toeplitz recursion [4.4, 5].

A special case of the least-squares shaping filter is the least-squares *spiking filter*. A spiking filter is a shaping filter for which the desired output signal is a spike. Thus, the signal $\{s_n\}$ is the unit impulse with coefficients

$$s_n = \delta_n = \begin{cases} 1, & n=0 \\ 0, & n \neq 0 . \end{cases} \tag{4.58}$$

A spiking filter for time-shift $\alpha = 0$ is called a zero-delay spiking filter. A well-known theorem [4.6] states that the least-squares zero-delay spiking filter is minimum delay.

Because the concepts of minimum phase and minimum delay are identical, the least-squares zero-delay spiking filter has a minimum-phase-lag spectrum.

Let us now examine the normal equations given above. We notice that the left-hand side of the normal equations depends only upon the autocorrelation of the input $\{x_n\}$ and not on $\{x_n\}$ itself. Because the desired output for a zero-delay spiking filter is the spike $\{1, 0, 0, \ldots, 0\}$, the right-hand side of the normal equations are

$$g_0 = x_0, g_1 = 0, g_2 = 0, \ldots, g_m = 0 . \tag{4.59}$$

Thus, in the case of a zero-delay spiking filter, the wave shape of the input sequence $\{x_n\}$ itself does not enter into the normal equations, except for the initial value x_0, which affects the filter as only a scale factor. Therefore, the normal equations in the case of a zero-delay spiking filter depend upon the input sequence only through its autocorrelation. Any two input sequences with the same autocorrelation would have the same zero-delay spiking filter (except for a constant factor). Because autocorrelations contain no phase information, it is an interesting consequence that the phase spectrum of the filter is the minimum-phase-lag spectrum.

For a zero-delay spiking filter the minimum value of the mean-square spiking error is

$$I = 1 - h_0 x_0 . \tag{4.60}$$

In order for this expression for I to be small when we consider all input sequences with the same autocorrelation, the leading coefficient x_0 should be large in magnitude. In other words, the smallest mean-square spiking error occurs when the input sequence is minimum delay. In summary, all zero-delay spiking filters are necessarily minimum delay, and the best case of spiking occurs when the input sequence is also minimum delay. In fact, for a minimum delay input sequence the mean-square spiking error tends to zero as the number of coefficients in the zero-delay spiking filter $\{h_0, h_1, \ldots, h_M\}$ tend to infinity ($M \to \infty$). On the other hand, for a nonminimum-delay input sequence, the mean-square spiking error does not tend to zero as the number of coefficients in the zero-delay spiking filter tend to infinity. We, therefore, conclude that perfect spiking can be achieved by a stable infinitely long causal spiking filter if and only if the input sequence is minimum delay. In other words, given the causal input x_n we can find a stable infinitely long causal filter $\{h_n\}$ such that

$$\{h_n\} * \{x_n\} = \{1, 0, 0, \ldots\} \tag{4.61}$$

if and only if $\{x_n\}$ is minimum delay. Any stable causal sequence $\{x_n\}$ that can be converted into a zero-delay spike by a stable causal filter is said to be invertible. Thus a stable causal sequence is invertible if and only if it is minimum delay. For that reason, we can use the terms "invertible sequence" and "minimum-delay sequence" interchangeably, even as we use the terms "minimum-phase sequence" and "minimum-delay sequence" interchangeably.

In summary, the least-squares finite-length zero-delay spiking filter for a stable causal sequence $\{x_n\}$ as input may be described as the least-squares inverse of $\{x_n\}$. In case $\{x_n\}$ is minimum delay, then if the number of terms in the least-squares inverse is allowed to increase, the least-squares inverse tends to the exact stable causal inverse of $\{x_n\}$.

4.5 Invertibility

In many applications we know from the physics of the situation that a certain finite-length causal sequence must be invertible. However, suppose that the estimation method used does not necessarily give an invertible sequence. Thus we must have a method for converting a sequence which is not necessarily invertible into an invertible sequence. In order to obtain a meaningful procedure, we must impose some sort of restriction on this conversion process. Let the restriction be the condition that the given sequence and the resulting invertible sequence have (approximately) the same autocorrelation.

Let us designate the given finite causal sequence $\{a_0, a_1, \ldots, a_M\}$ simply by the voctor $\boldsymbol{a}$. Let us denote the least-squares inverse (i.e., zero-delay spiking filter) of $\boldsymbol{a}$ by $\boldsymbol{a}^{-1}$. In turn let us denote the least-squares inverse (i.e., zero-delay spiking filter) of $\boldsymbol{a}^{-1}$ by $(\boldsymbol{a}^{-1})^{-1}$.

We can now describe a method of converting a finite-length causal sequence to an invertible finite-length causal sequence with approximately the same autocorrelation. This method consists of the following two steps:

1) Compute the zero-delay spiking filter $\boldsymbol{a}^{-1}$ of the given filter $\boldsymbol{a}$. As we have seen in the preceding section, the filter $\boldsymbol{a}^{-1}$ is necessarily invertible.

2) Now take this approach one step further; namely, compute the zero-delay spiking filter $(\boldsymbol{a}^{-1})^{-1}$ of the filter $\boldsymbol{a}^{-1}$. This second zero-delay spiking filter is necessarily invertible, and in fact it is the required invertible counterpart of the given filter $\boldsymbol{a}$.

In summary, the invertible counterpart of the finite-length causal sequence $\boldsymbol{a}$ is the least-squares inverse of the least-squares inverse of $\boldsymbol{a}$.

4.6 Spectral Estimates of the Components of a Noninvertible ARMA System

As we have seen in Sect. 4.3, a reflection seismogram $\{u_0, u_1, u_2, \ldots\}$ has the z transform

$$U(z) = u_0 + u_1 z^{-1} + u_2 z^{-2} + \ldots \tag{4.62}$$

which can be approximately modeled as the ratio

$$U(z) = \frac{E(z)}{A(z)}. \tag{4.63}$$

In this equation the feedfront polynomial $E(z)$ is the z transform

$$E(z) = \varepsilon_0 + \varepsilon_1 z^{-1} + \varepsilon_2 z^{-2} + \ldots + \varepsilon_N z^{-N} \tag{4.64}$$

of the reflection coefficient series $\{\varepsilon_0, \varepsilon_1, \ldots, \varepsilon_N\}$ of the earth. There is no physical reason why $E(z)$ should be minimum delay. However, because the layered earth is a stable filter, the feedback polynomial

$$A(z) = 1 + a_1 z^{-1} + a_2 z^{-2} + \ldots + a_N z^{-N} \tag{4.65}$$

must be minimum delay. The coefficients $a_1, a_2, \ldots, a_N$, respectively, are approximately equal to the autocorrelation coefficients $\gamma_1, \gamma_2, \ldots, \gamma_N$ of the reflection coefficient series provided that each reflection coefficient is much smaller than one in magnitude [4.2].

The z transform of an ARMA system is a rational function, where the denominator polynomial (which represents the autoregressive component) must be invertible and where the numerator polynomial (which represents the moving average component) may or may not be invertible. If the numerator polynomial is invertible, then the ARMA process is invertible. If the numerator is not invertible, then the ARMA process is not invertible.

The equation $U(z) = E(z)/A(z)$ represents a model of a seismogram as a noninvertible ARMA system. We find it sometimes convenient to write the seismogram in symbolic form as

$$\boldsymbol{u}=\frac{\boldsymbol{\varepsilon}}{\boldsymbol{a}}, \tag{4.66}$$

where $\boldsymbol{u}$ is the seismogram sequence, $\boldsymbol{\varepsilon}$ is the reflection coefficient sequence, and $\boldsymbol{a}$ is the inverse reverberation filter sequence.

Most treatments of ARMA systems are concerned with systems where the MA part is invertible, so the entire ARMA system is invertible. However, as we have seen, the ARMA model for a seismogram in general is not invertible. We can observe the seismogram $\boldsymbol{u}$. We would like to determine the sequence $\boldsymbol{\varepsilon}$ and $\boldsymbol{a}$ from the sequence $\boldsymbol{u}$. The purpose of this chapter is to give an algorithm which carries out this identification.

The algorithm is based on the least-squares wave-form shaping filter. We recall that, in order to derive the shaping filter $\boldsymbol{h}$, we must specify the input $\boldsymbol{x}$ and the desired output $\boldsymbol{s}$. The shaping filter $\boldsymbol{h}$ is determined in such a way that the actual output $\boldsymbol{h}*\boldsymbol{x}$ approximates the desired output $\boldsymbol{s}$ in a least-squares sense. (In this discussion we let the time-shift parameter α be zero.)

Let us now introduce a short-hand notation to describe the operation of a shaping filter. This short-hand notation has the general form (here special use is made of the approximately equal symbol $\simeq$)

$$\boldsymbol{x}*\boldsymbol{h}\simeq\boldsymbol{s}, \tag{4.67}$$

where we adhere to the convention that $\boldsymbol{x}$ and $\boldsymbol{s}$ are given, and $\boldsymbol{h}$ is determined in such a way that the convolution $\boldsymbol{x}*\boldsymbol{h}$ approximates $\boldsymbol{s}$ in the least-squares sense. We remember the sequences on each end (i.e., $\boldsymbol{x}$ and $\boldsymbol{s}$) are given, and the sequence in the middle (i.e., $\boldsymbol{h}$) is determined. Subroutine SHAPE [4.7] computes the least-squares shaping filter; in subroutine SHAPE the sequences $\boldsymbol{x}$ and $\boldsymbol{s}$ are subroutine inputs and the filter sequence $\boldsymbol{h}$ is the subroutine output.

Let us now describe the algorithm. At the beginning of step k we have the seismogram $\boldsymbol{u}$ and the approximate reflection coefficient sequence $\boldsymbol{\varepsilon}^{(k-1)}$. We apply subroutine SHAPE as

$$\boldsymbol{u}*\boldsymbol{a}^{(k)}\simeq\boldsymbol{\varepsilon}^{(k-1)} \tag{4.68}$$

to yield the preliminary estimate of $\boldsymbol{a}^{(k)}$. However, we want to guarantee that $\boldsymbol{a}^{(k)}$ is an invertible sequence, and in order to do so, we apply subroutine SHAPE as

$$\boldsymbol{a}^{(k)}*[\boldsymbol{a}^{-1}]^{(k)}\simeq\{1,0,0,\ldots,0\} \tag{4.69}$$

to yield the least-squares inverse $[\boldsymbol{a}^{-1}]^{(k)}$. We then apply subroutine SHAPE as

$$[\boldsymbol{a}^{-1}]^{(k)}*\boldsymbol{a}^{(k)}\simeq\{1,0,0,\ldots,0\} \tag{4.70}$$

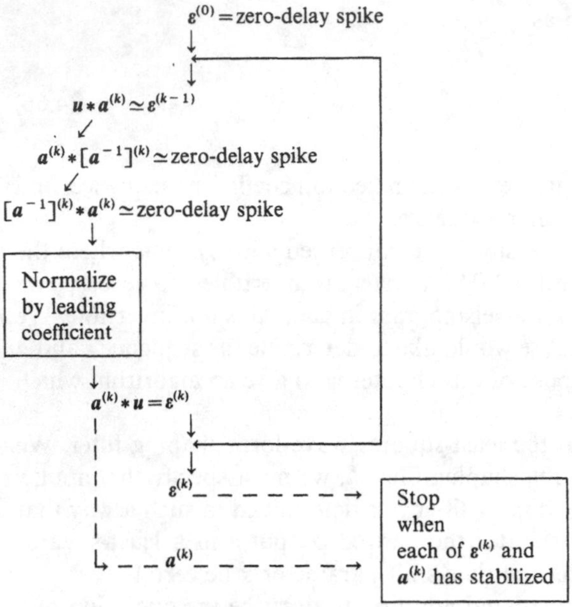

Fig. 4.12. Flow graph of the algorithm

to yield the second estimate of $\boldsymbol{a}^{(k)}$. Because of the minimum-delay property of least-squares inverses, this second estimate $\boldsymbol{a}^{(k)}$ is necessarily invertible. Next we normalize this $\boldsymbol{a}^{(k)}$ by dividing each element of $\boldsymbol{a}^{(k)}$ by its leading element; the result is the final estimate of $\boldsymbol{a}^{(k)}$. The final estimate of $\boldsymbol{\varepsilon}^{(k)}$ is given as the convolution of $\boldsymbol{a}^{(k)}$ with $\boldsymbol{u}$, that is,

$$\boldsymbol{\varepsilon}^{(k)} = \boldsymbol{a}^{(k)} * \boldsymbol{u}. \tag{4.71}$$

We thus have found $\boldsymbol{a}^{(k)}$ and $\boldsymbol{\varepsilon}^{(k)}$, and the description of the algorithm is complete. Initially, we start with $\boldsymbol{\varepsilon}^{(0)} = (1, 0, 0, \ldots, 0)$ and we end when each of $\boldsymbol{\varepsilon}^{(k)}$ and $\boldsymbol{a}^{(k)}$ has stabilized.

We can diagram the algorithm as shown in Fig. 4.12.

4.6.1 Example

In order to construct a simple numerical example, let us consider the case of a marine seismogram. Let the two-way travel time in the water layer represent one time unit, and let the water bottom reflection coefficient be c. Let the reflection coefficients of the deep interfaces be $\{\varepsilon_0, \varepsilon_1, \ldots, \varepsilon_N\}$. Then the z transform of the seismogram is approximately

$$U(z) = \frac{\varepsilon_0 + \varepsilon_1 z^{-1} + \varepsilon_2 z^{-2} + \ldots + \varepsilon_M z^{-M}}{(1 + cz^{-1})^2} \tag{4.72}$$

which we can write as

$$U(z)=\frac{E(z)}{A(z)} \tag{4.73}$$

where

$$E(z)=\varepsilon_0+\varepsilon_1 z^{-1}+\varepsilon_2 z^{-2}+\ldots+\varepsilon_M z^{-M} \tag{4.74}$$

and

$$A(z)=1+2cz^{-1}+c^2z^{-2}. \tag{4.75}$$

Let us now give a numerical example. Suppose there are $M+1=10$ deep interfaces, and suppose that the first 19 observations of the resulting marine seismogram are (reading consecutively as words on a page)

$$\begin{aligned} U=\{&1.0000, && 0.0000, && 0.1875, && -0.9063, && -0.7148,\\ &0.1992, && -0.6057, && -0.8153, && -0.1198, && 0.9911,\\ &0.5030, && 0.1896, && 0.0633, && 0.0198, && 0.0060,\\ &0.0017, && 0.0005, && 0.0001, && 0.0000\}. \end{aligned}$$

Given these observations, the problem is to find the water bottom reflection coefficient c and the deep reflection coefficient sequence $\boldsymbol{\varepsilon}=\{\varepsilon_0, \varepsilon_1, \varepsilon_2, \ldots, \varepsilon_9\}$. Because the observed seismogram is not known in absolute terms but only to within an arbitrary scale factor, the resulting reflection coefficient sequence $\boldsymbol{\varepsilon}$ will be equal to the actual physical reflection coefficient sequence only to within a scale factor.

The method of solution is the iterative scheme based on the algorithm which, for the kth step, has inputs $\boldsymbol{u}$ and $\boldsymbol{\varepsilon}^{(k-1)}$ and outputs $\boldsymbol{a}^{(k)}$ and $\boldsymbol{\varepsilon}^{(k)}$. The seismogram sequence $\boldsymbol{u}$ is the numerical sequence of 19 observations given above, and remains fixed for each step. The outputs $\boldsymbol{a}^{(k)}$ and $\boldsymbol{\varepsilon}^{(k)}$ of the kth step are supposed to converge to the sequence $\boldsymbol{a}$ and $\boldsymbol{\varepsilon}$ respectively.

In order to start the iteration, we set $\boldsymbol{\varepsilon}^{(0)}$ equal to the zero-delay spike;

$$\boldsymbol{\varepsilon}^{(0)}=\{1,0,0,\ldots,0\}.$$

The first iteration gives:

$$\boldsymbol{a}^{(1)}=\{1,-0.32,0.18\}$$

$$\boldsymbol{\varepsilon}^{(1)}=\{1,-0.32,0.37,-0.97,-0.39,0.27,-0.80,-0.58,0.04,0.88\}.$$

The second iteration gives:

$$a^{(2)}=\{1,-0.37,0.14\}$$
$$\varepsilon^{(2)}=\{1,-0.37,0.33,-0.98,-0.35,0.34,-0.78,-0.56,0.10,0.92\}.$$

The third iteration gives:

$$a^{(3)}=\{1,-0.41,0.12\}$$
$$\varepsilon^{(3)}=\{1,-0.41,0.30,-0.98,-0.32,0.39,-0.77,-0.54,0.14,0.95\}.$$

Continue iterating. The twentieth iteration gives:

$$a^{(20)}=\{1,-0.50,0.624\}$$
$$\varepsilon^{(20)}=\{1,-0.50,0.25,-1.00,-0.25,0.50,-0.75,-0.50,0.25,1.00\}.$$

At this point the values of the sequences have stabilized. As a result, we conclude that $\varepsilon^{(20)}$ gives the deep reflection coefficients (to within a scale factor) and

$$a^{(20)}=\{1,-0.50,0.624\}=\{1,2c,c^2\}$$

gives the water bottom reflection coefficient as $c=-0.25$.

In conclusion, this method allows one to decompose the impulse response of an ARMA system into estimates of its feedback and feedfront components. In this example, the spectral estimate of the feedfront component is

$$E(\omega)=1-0.5e^{-j\omega}+0.25e^{-j\omega 2}+\ldots+0.25e^{-j\omega 8}+1.0e^{-j\omega 9}$$

and the spectral estimate of the feedback component is

$$A(\omega)=1-0.5e^{-j\omega}+0.624e^{-j\omega 2}.$$

This feedback spectral estimate necessarily has minimum phase lag.

4.7 Conclusions

Three fundamental types of finite-parameter time series models are the autoregressive (AR) process which is a feedback model, the moving average (MA) process which is a feedfront model, and the hybrid (ARMA) process which is a feedback-feedfront model. The fitting of time series models involves the use of the principle of parsimony together with the linking of the model with physically observable quantities in the real world. These principles are illustrated with the model for seismic deconvolution where the parameters of the model can be identified with feedback loops within the sedimentary layers of the earth. Shaping and spiking filters are interpreted as filters which are data dependent, that is, the mean-squared error is weighted according to the energy spectrum of the input signal. The zero-delay spiking filter corresponding to a given input signal is called the least-squares inverse of that input signal. Any signal may be converted to minimum delay by simply computing the least-squares inverse of its least-squares inverse. Given the impulse response of an ARMA system, its feedback and feedfront components can be computed by an iterative scheme. The feedback component of a stable ARMA system must necessarily be minimum delay, but the feedfront component may or may not be minimum delay. If the feedfront component is minimum delay the ARMA system is invertible; if the feedfront component is not minimum delay then the ARMA system is not invertible. Each step of the iterative scheme involves first computing the feedback component and converting it to minimum delay and then estimating the feedfront component. This conversion at each step guarantees that the ARMA system is stable at each step of the iteration. The results of the iterative scheme allow us to compute the spectrum of the feedfront component which may or may not be minimum phase lag, and the spectrum of the feedback component which is necessarily minimum phase lag.

We have given an algorithm to compute the parameters of the moving average (MA) and the autoregressive (AR) components of a noninvertible ARMA system from its impulse response function. We have applied this algorithm to a reflection seismogram, which can be regarded as the impulse response of a noninvertible ARMA system, in order to determine the reflection coefficients of the deep interfaces. A diagram of the algorithm is given. From the general principles illustrated in this chapter, other algorithms along similar lines may be readily devised.

References

4.1 G.E.P.Box, G.M.Jenkins: *Time Series Analysis, Forecasting, and Control* (Holden-Day, San Francisco 1970)

4.2 E.A.Robinson, M.T.Silvia: *Digital Signal Processing and Time Series Analysis* (Holden-Day, San Francisco 1978)

4.3 J.P.Burg: "Maximum Entropy Spectral Analysis"; Ph.D. Thesis, Stanford University, Stanford, Calif. (1975)

4.4 N.Levinson: J. Math. Phys. **25**, 261–278 (1946)

4.5 R.A.Wiggins, E.A.Robinson: J. Geophys. Res. **70**, 1885–1891 (1965)

4.6 E.A.Robinson, H.Wold: "Minimum-Delay Structure of Least-Squares and eo-ipso Predicting Systems", in *Brown University Symposium on Time Series Analysis* (Wiley and Sons, New York 1963) pp. 192–196

4.7 E.A.Robinson: *Multichannel Time Series Analysis with Digital Computer Programs* (Holden-Day, San Francisco 1978)

5. Maximum-Likelihood Spectral Estimation

J. Capon

With 1 Figure

The use of an array of sensors for determining the properties of propagating waves is of considerable importance in many areas. As an example, such phased arrays find application in radar, where an array of receiving antennas is used to determine the spatial coordinates of radar targets. In seismic applications, the requirement is to use an array of sensors to facilitate the discrimination between earthquakes and underground nuclear explosions. The present chapter is concerned with the use of an array of sensors to determine the vector velocity of propagating waves.

5.1 Explanatory Remarks

The spectral density of a stationary stochastic process provides information concerning the power as a function of frequency for the process. In a similar manner, propagating waves, or a homogeneous random field, can be characterized by a frequency-wavenumber spectral density function. Loosely speaking, this function provides the information concerning the power as a function of frequency and the vector velocities of the propagating waves. The definition and properties of the frequency wave number spectrum will be given subsequently. However, the main purpose of the present chapter is to discuss the measurement, or estimation, of the frequency wave number spectrum. Previous methods of estimation were based on the use, at a given frequency, of a fixed wave number window. These conventional methods were limited to a wave number resolution which was determined primarily by the natural beam pattern of the array.

A high-resolution estimation method, known as the maximum-likelihood method (MLM)[1], will be introduced which is based on the use of a wave number window which is not fixed, but is variable at each wave number considered. As a consequence, the wave number resolution achievable by this method is considerably greater than that of the conventional method and is limited primarily by signal-to-noise ratio considerations.

---

1 The development of the maximum-likelihood method as presented in this chapter is based on material presented by *Capon* [5.2–4]. These references also present a detailed account of the application of the method to data obtained from a large aperture seismic array located in Eastern Montana. A description of this array is given in Chap. 6.

It is noteworthy that although the development of the maximum-likelihood method will be presented in the context of wave number analysis with arrays, nevertheless, it is a straightforward matter to adapt the method to the spectral analysis of a single time series [5.1].

5.2 Seismic Signals and Noise

Seismic signals propagating as body waves can be described as nondispersive, linearly polarized group arrivals with power in the 0.1 to 3 Hz range for natural and artificial events of small to moderate magnitude. The body-wave phases are separable into P (compressional) and S (shear) phases on the basis of both velocity of propagation and polarization. The P wave is the only short-period (SP) phase which is considered. The maximum observed power of the P wave is usually in the 0.6 to 2.0 Hz band, and its horizontal phase velocity is between 15 and 25 km s^{-1}.

Surface waves of Rayleigh type from small to moderate magnitude events can be described as dispersive group arrivals, generally elliptically polarized with maximum observed power in the period range from 2 to 100 s. The higher modes of these surface waves are normally observed in the shorter period range from 2 to 20 s and can have either prograde or retrograde elliptically polarized particle motion, while fundamental-mode Rayleigh waves have retrograde elliptical particle motion. Surface waves of the Love type from these sources are dispersive and rectilinearly polarized in a horizontal plane orthogonal to the direction of propagation. The period range of observation of both fundamental- and higher-mode Love waves corresponds roughly to that for Rayleigh waves. The only long-period (LP) wave considered is the fundamental-mode Rayleigh wave. The phase velocity for this wave along a continental path is between 3.2 and 4.0 km s^{-1} in the period range of interest. In addition, the maximum observed power for this phase is usually in the 0.025 to 0.05 Hz band.

The background microseismic noise seems primarily to consist of fundamental- and higher-mode Rayleigh waves, fundamental-mode Love waves, and compressional body waves, from many sources distributed in azimuth and distance around the point of observation. Some of the microseismic noise is nonpropagating and appears to be incoherent between sensors which are separated by small spatial lags. The spectral distribution of the background noise is predominantly in the range from 0.01 to 1.0 Hz. The ambient microseismic noise spectrum is usually peaked at about 0.125 Hz, with a secondary peak at about 0.067 Hz and a minimum near 1.0 Hz. It is for this reason that the SP and LP phases are observed separately with SP and LP seismometers, respectively. The SP and LP seismometers have a peak response at about 1.0 and 0.04 Hz, respectively.

5.2.1 Statistical Models for Seismic Noise and Signal

We assume that an array of K sensors is available and that the ith sensor is located at the vector position $\boldsymbol{x}_i$. The signal is assumed to be known and to propagate across the array as a uniform plane wave. In addition, the output of each sensor is assumed to be available in sampled form

$$x_{im} = s_{m+\boldsymbol{\alpha}\cdot\boldsymbol{x}_i} + n_{im}, \quad \begin{array}{l} i=1,\ldots,K \\ m=0, \pm 1, \pm 2, \ldots \end{array} \tag{5.1}$$

where $-\boldsymbol{\alpha}$ is the vector of time delays per unit distance for the signal, measured in multiples of the sampling period, T, along each coordinate axis, and points in the direction of propagation and n_{im} is the noise present in the ith sensor. It is assumed that $\boldsymbol{\alpha}$ is known; however, the modifications in the theory that are required when it is unknown are discussed in Sect. 5.4.3. In addition, the assumption is made that $\{n_{im}\}$ is a weakly stationary discrete stochastic process with zero mean and correlation matrix

$$R_{il}(m-n) = \mathrm{E}[n_{im} n_{ln}] \tag{5.2}$$

where E denotes ensemble expectation, and $(m-n)$ denotes the time difference between the noise samples n_{im} and n_{ln} at the ith and l^{th} sensors, respectively. The cross-power spectral density is

$$S_{il}(\lambda) = \sum_{m=-\infty}^{\infty} R_{il}(m) \exp(jm\lambda), \tag{5.3}$$

and

$$R_{il}(m) = \int_{-\pi}^{\pi} S_{il}(\lambda) \exp(-jm\lambda) \frac{d\lambda}{2\pi}, \tag{5.4}$$

where $\lambda = 2\pi f T$ is the normalized frequency, f is the frequency in Hertz, and T is the sampling period of the data in seconds.

5.3 Optimum Detector for Seismic Arrays

A derivation is now given to determine the optimum detector to be used in a seismic array for the purpose of detecting a known seismic signal in a background of microseismic noise. It is well known that the optimum detector is based on a likelihood ratio, as discussed by *Helstrom* [5.5]. Using the statistical models proposed above for noise and signal, this likelihood ratio is evaluated to determine optimum detection receivers. It is shown that a central role in the design of optimum detectors is played by the maximum-likelihood filter.

5.3.1 Evaluation of the Likelihood Ratio

In order to find the likelihood ratio, the probability densities of the signal plus noise, and of the noise alone, must be evaluated. This step requires the assumption that $\{n_{im}\}$ has a multidimensional Gaussian distribution. In addition, the likelihood ratio is evaluated for the case where the observation time is long, i.e., the memory of the filter is infinite. Thus, a finite-dimensional approximation for this likelihood ratio is

$$L_N = \frac{p_1}{p_0} \tag{5.5}$$

where p_1, and p_0 are the probability densities of the signal plus noise and of the noise alone, respectively, and may be evaluated as

$$p_1 = (2\pi)^{-K(2N+1)/2} |\boldsymbol{R}|^{-1/2} \cdot \exp\left\{ -\frac{1}{2} \sum_{i,k=1}^{K} \sum_{m,n=-N}^{N} R_{ik}^{-1}(m,n)(x_{im} - s_{m+\boldsymbol{\alpha}\cdot\boldsymbol{x}_i})(x_{kn} - s_{n+\boldsymbol{\alpha}\cdot\boldsymbol{x}_k}) \right\}, \tag{5.6}$$

$$p_0 = (2\pi)^{-K(2N+1)/2} |\boldsymbol{R}|^{-1/2} \cdot \exp\left\{ -\frac{1}{2} \sum_{i,k=1}^{K} \sum_{m,n=-N}^{N} R_{ik}^{-1}(m,n) x_{im} x_{kn} \right\}, \tag{5.7}$$

where $|\boldsymbol{R}|$ denotes the determinant of the block Toeplitz matrix $\boldsymbol{R}$ which is a matrix of K by K submatrices, the mn^{th} submatrix has the elements $R_{ik}(m-n)$, $i,k=1,\ldots,K$, $m,n=-N,\ldots,N$, with a corresponding notation for the inverse matrix $\boldsymbol{R}^{-1}$ whose elements are $R_{ik}^{-1}(m,n)$. We assume throughout that the matrix $\boldsymbol{R}$ is positive-definite.

If we use (5.5–7) to evaluate the logarithm of the likelihood ratio, and retain only those terms which depend on the x_{im} then we obtain

$$\log L'_N = \sum_{i=1}^{K} \sum_{m=-N}^{N} h_{im} x_{im}, \tag{5.8}$$

where

$$h_{im} = \sum_{k=1}^{K} \sum_{n=-N}^{N} s_{n+\boldsymbol{\alpha}\cdot\boldsymbol{x}_k} R_{ik}^{-1}(m,n), \qquad \begin{array}{l} i=1,\ldots,K \\ m=-N,\ldots,N. \end{array} \tag{5.9}$$

We see from (5.9) that the h_{im} satisfy the following system of equations

$$\sum_{i=1}^{K} \sum_{m=-N}^{N} h_{im} R_{ik}(m-n) = s_{n+\alpha \cdot x_k}, \quad \begin{matrix} k=1,\dots,K \\ n=-N,\dots,N. \end{matrix} \tag{5.10}$$

The sequence of likelihood ratios $\{L_N\}$ is a martingale[2], so that it is known that the likelihood ratio for the case of infinite memory can be evaluated as

$$\log L' = \sum_{i=1}^{K} \sum_{m=-\infty}^{\infty} h'_{im} x'_{im}, \tag{5.11}$$

where $x'_{im} = x_{i,m-\alpha \cdot x_i}$ represents the data after appropriate time delays have been introduced to align the signal in each channel, and the h'_{im} now satisfy the following systems of equations:

$$\sum_{i=1}^{K} \sum_{m=-\infty}^{\infty} h'_{im} R'_{ik}(m-n) = s_n, \quad \begin{matrix} k=1,\dots,K \\ n=0, \pm 1, \pm 2, \dots \end{matrix} \tag{5.12}$$

where $R'_{ik}(m) = R_{ik}[m - \alpha \cdot (x_i - x_k)]$ is the correlation of the noise $\{n'_{im}\}$ after the appropriate time delays have been introduced.

If we transform both sides of (5.12), we obtain the following system of equations:

$$\sum_{k=1}^{K} G_k(\lambda) S'_{ik}(\lambda) = H_s^*(\lambda), \quad \begin{matrix} i=1,\dots,K \\ -\pi \leqq \lambda \leqq \pi, \end{matrix} \tag{5.13}$$

$$H_s(\lambda) = \sum_{m=-\infty}^{\infty} s_m \exp(-jm\lambda), \tag{5.14}$$

$$G_i(\lambda) = \sum_{m=-\infty}^{\infty} h'_{im} \exp(jm\lambda), \tag{5.15}$$

$$\begin{aligned} S'_{ik}(\lambda) &= \sum_{m=-\infty}^{\infty} R'_{ik}(m) \exp(jm\lambda) \\ &= \exp[j\lambda\alpha \cdot (x_i - x_k)] S_{ik}(\lambda). \end{aligned} \tag{5.16}$$

Thus, $S'_{ik}(\lambda)$ is the cross-power spectral density of the noise after the appropriate time delays have been applied. The function $H_s(\lambda)$ is the frequency response of a convolutional digital filter which is matched to the signal. We can solve for the

2 The definition of a martingale has been given by [Ref. 5.6, Chap. VII.1] while the important convergence theorems concerning martingales have been presented in [Ref. 5.6, Chap. VII.4]. The application of these results to likelihood ratios in statistics has been given by [Ref. 5.6, Chap. VII.9]. These results have also been used by *Capon* [5.7] to determine the properties of the likelihood ratio in certain detection problems.

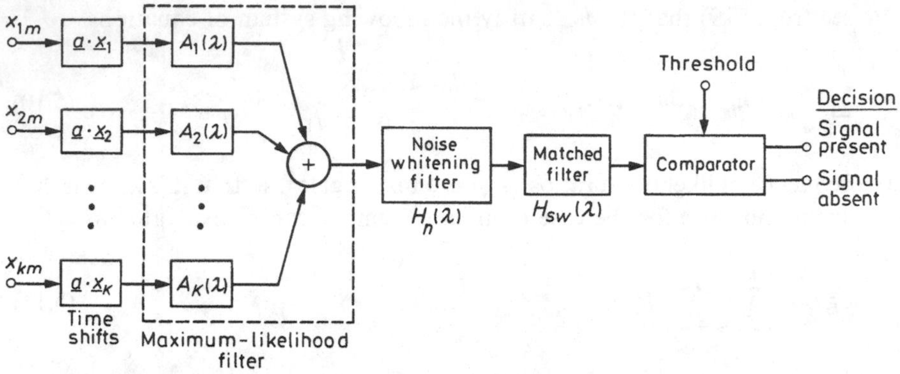

Fig. 5.1. Optimum detector for a known signal in additive Gaussian noise when the observation interval is large

filter frequency response $G_i(\lambda)$ from (5.13) as

$$G_i(\lambda) = H_n(\lambda) H_{sw}(\lambda) A_i(\lambda), \tag{5.17}$$

where

$$H_n(\lambda) = \left[\sum_{i,k=1}^{K} q'_{ik}(\lambda) \right]^{1/2}, \tag{5.18}$$

$$H_{sw}(\lambda) = H_s^*(\lambda) H_n(\lambda), \tag{5.19}$$

$$A_i(\lambda) = \frac{\sum_{k=1}^{K} q'_{ik}(\lambda)}{\sum_{i,k=1}^{K} q'_{ik}(\lambda)}, \qquad i = 1, \ldots, K \tag{5.20}$$

and $\{q'_{ik}\}$ is the inverse of the matrix $\{S'_{ik}(\lambda)\}$. It should be noted that if $\{q_{ik}(\lambda)\}$ is the inverse of $\{S_{ik}(\lambda)\}$ then $q'_{ik}(\lambda) = \exp[j\lambda \boldsymbol{\alpha} \cdot (\boldsymbol{x}_i - \boldsymbol{x}_k)] q_{ik}(\lambda)$.

The implementation of the optimum detector, for the case of a known signal, is shown in Fig. 5.1. The output of the ith sensor is delayed, or advanced, by the appropriate time and then applied to a convolutional digital filter whose frequency function is $A_i(\lambda)$. All such outputs are then summed and applied to the convolutional digital filters whose frequency responses are $H_n(\lambda)$, $H_{sw}(\lambda)$. The filter $H_n(\lambda)$ is a whitening filter since the spectrum of the noise at its input is $[H_n(\lambda)]^{-2}$, as will be shown in Sect. 5.5.2. The filter $H_{sw}(\lambda)$ is matched to the whitened signal. The output of the matched filter is applied to the comparator where it is compared with a threshold. This threshold is determined in the usual way, as discussed by *Helstrom* [5.5]. If the threshold is exceeded, then a decision is made that a signal is present; otherwise, the decision is made that signal is absent.

The array processing filter shown inside the dotted lines in Fig. 5.1 is important and is known as a maximum-likelihood filter. The reason for this is presented in Sect. 5.4, where it is shown that this filter provides a minimum-variance unbiased estimate for the input signal, when it is not known, which is the same as the maximum-likelihood estimate of the signal if the noise is a multidimensional Gaussian process. It should also be noted that the maximum-likelihood filter is a distortionless filter in the sense that it reproduces at its output a replica of any input signal which is propagating across the array with a velocity corresponding to the vector of time delays $\boldsymbol{\alpha}$.

5.4 Signal Estimation Procedures

In Sect. 5.3.1 it was shown that the maximum-likelihood filter plays an important role in the implementation of the optimum detector for a seismic array. It is now shown that this filter is also important in the case where it is desired to estimate the signal waveform in the presence of noise. The assumption is made once again, as in Sect. 5.3.1, that the observation time is long.

5.4.1 Maximum-Likelihood Estimation

A derivation is now given for the maximum-likelihood estimate of the signal in noise. In order to do this it is necessary to find the probability density function of the signal plus noise. This step requires the assumption that $\{n_{im}\}$ has a multidimensional Gaussian distribution, as was done in Sect. 5.3.1. In addition, it is assumed that appropriate time delays have been introduced to align the signal in each channel. A discussion is given in Sect. 5.4.3 of the modifications required when the slowness vector of the signal is not known so that these time delays cannot be computed.

The probability density function of the signal plus noise is

$$p'_1 = (2\pi)^{-K(2N+1)/2} |R'|^{-1/2} \cdot \exp\left\{-\frac{1}{2} \sum_{i,k=1}^{K} \sum_{m,n=-N}^{N} R'^{-1}_{ik}(m,n)(x'_{im}-s_m)(x'_{kn}-s_n)\right\}. \tag{5.21}$$

As usual, the maximum-likelihood estimator of the signal is found by differentiating the logarithm of this likelihood function with respect to s_n, $n=-N,\ldots,N$, and equating the results to zero, that is,

$$\frac{\partial \log p'_1}{\partial s_n} = 0, \quad n = -N, \ldots, N. \tag{5.22}$$

Thus, we obtain from (5.21)

$$\sum_{i,k=1}^{K} \sum_{m=-N}^{N} R_{ik}'^{-1}(m,n)(\hat{s}_m - x_{im}') = 0, \qquad n = -N, \ldots, N \tag{5.23}$$

where $\hat{s}_m$ is the maximum-likelihood estimator for s_m, $m = -N, \ldots, N$.

It is possible to rewrite (5.23) as

$$\sum_{m=-N}^{N} \hat{s}_m \sum_{i,k=1}^{K} R_{ik}'^{-1}(m,n) = \sum_{i,k=1}^{K} \sum_{m=-N}^{N} x_{im}' R_{ik}'^{-1}(m,n),$$
$$n = -N, \ldots, N \tag{5.24}$$

We now define the $(2N+1)$ by $(2N+1)$ matrix $\{\alpha_{mn}\}$ in terms of its inverse $\{\alpha_{mn}^{-1}\}$ whose elements are defined as

$$\alpha_{mn}^{-1} = \sum_{i,k=1}^{K} R_{ik}'^{-1}(m,n), \qquad m, n = -N, \ldots, N \tag{5.25}$$

The only estimator of interest is $\hat{s}_0$, which can be obtained from (5.24) as

$$\hat{s}_0 = \sum_{m=-N}^{N} \sum_{i=1}^{K} a_{im} x_{im}', \tag{5.26}$$

where

$$a_{im} = \sum_{n=-N}^{N} \sum_{k=1}^{K} \alpha_{n0} R_{ik}'^{-1}(m,n). \qquad \begin{matrix} i = 1, \ldots, K \\ m = -N, \ldots, N \end{matrix} \tag{5.27}$$

We see from (5.27) and the definition for α_{mn} in (5.25) that

$$\sum_{i=1}^{K} a_{im} = \begin{cases} 1, & m = 0 \\ 0, & m \neq 0. \end{cases} \tag{5.28}$$

We can rewrite (5.27) as

$$\sum_{n=-N}^{N} \sum_{k=1}^{K} a_{kn} R_{ik}'(m-n) - \alpha_{m0} = 0. \qquad \begin{matrix} i = 1, \ldots, K \\ m = -N, \ldots, N \end{matrix} \tag{5.29}$$

Now using the martingale property for the likelihood function, as indicated previously, it is possible to use (5.29) to obtain the filter functions for the case of infinite memory as

$$\int_{-\pi}^{\pi} \sum_{k=1}^{K} S_{ik}'(\lambda) A_k(\lambda) \exp(-jm\lambda) \frac{d\lambda}{2\pi} - \alpha_{m0} = 0, \qquad m = 0, \pm 1, \pm 2, \ldots \tag{5.30}$$

where

$$A_k(\lambda)=\sum_{m=-\infty}^{\infty} a_{km}\exp(jm\lambda), \qquad k=1,\dots,K \tag{5.31}$$

where the filter weighting coefficients for the case of infinite memory are denoted by a'_{im}. If we let

$$\alpha_0(\lambda)=\sum_{m=-\infty}^{\infty} \alpha_{m0}\exp(jm\lambda) \tag{5.32}$$

then

$$\alpha_{m0}=\int_{-\pi}^{\pi} \alpha_0(\lambda)\exp(-jm\lambda)\frac{d\lambda}{2\pi}, \tag{5.33}$$

so that using (5.33) in (5.30) we obtain

$$\int_{-\pi}^{\pi} \exp(-jm\lambda)\left[\sum_{k=1}^{K} S'_{ik}(\lambda)A_k(\lambda)-\alpha_0(\lambda)\right]\frac{d\lambda}{2\pi}=0\,.$$
$$m=0,\pm 1,\pm 2,\dots \tag{5.34}$$

According to (5.34), all of the Fourier coefficients of the quantity in the brackets in (5.34) must be zero. It follows that the quantity itself must be zero so that

$$\sum_{k=1}^{K} S'_{ik}(\lambda)A_k(\lambda)=\alpha_0(\lambda)\,. \qquad \begin{matrix}-\pi\leqq\lambda\leqq\pi\\ i=1,\dots,K\end{matrix} \tag{5.35}$$

The set of constraints in (5.28) imply that

$$\sum_{k=1}^{K} A_k(\lambda)=1\,. \qquad -\pi\leqq\lambda\leqq\pi \tag{5.36}$$

However, we have from (5.35)

$$A_i(\lambda)=\alpha_0(\lambda)\sum_{k=1}^{K} q'_{ik}(\lambda)\,. \qquad i=1,\dots,K \tag{5.37}$$

Using (5.36, 37) we obtain

$$\alpha_0(\lambda)=\left[\sum_{i,k=1}^{K} q'_{ik}(\lambda)\right]^{-1}, \tag{5.38}$$

and

$$A_i(\lambda)=\frac{\sum_{k=1}^{K} q'_{ik}(\lambda)}{\sum_{i,k=1}^{K} q'_{ik}(\lambda)}, \qquad i=1,\ldots,K \tag{5.39}$$

which agrees with (5.20). Thus, it has been shown that the convolutional digital filter, whose frequency function is $A_i(\lambda)$, which is used in the implementation of the optimum detector is also the filter that provides a maximum-likelihood estimate of the signal. It is for this reason that the filter is known as a maximum-likelihood filter. We stress once again that the maximum-likelihood filter is a distortionless filter in the sense that it reproduces at its output a replica of any input signal which is propagating across the array with a velocity corresponding to the vector of time delays $\boldsymbol{\alpha}$. The reason for this can be seen from (5.36) which states that the sum of all the frequency functions $A_i(\lambda)$ is unity for all frequencies.

5.4.2 Minimum-Variance Unbiased Estimation

We now wish to derive the minimum-variance unbiased estimate of the signal in noise. It is important to point out that to do this it is not necessary to assume that the noise has a multidimensional Gaussian distribution as was done previously in Sects. 5.3.1, 4.1. However, we do assume once again that appropriate time delays have been introduced to align the signal in each sensor.

The minimum-variance unbiased estimator of s_0, denoted by $\hat{s}_0$, can be expressed as

$$\hat{s}_0=\sum_{m=-N}^{N}\sum_{i=1}^{K} a_{im}x'_{im}, \tag{5.40}$$

with the constraints

$$\sum_{i=1}^{K} a_{im}=\begin{cases}1, & m=0\\ 0, & m\neq 0.\end{cases} \tag{5.41}$$

The variance of $(\hat{s}_0-s_0)$ is

$$\begin{aligned}\mathrm{E}[(\hat{s}_0-s_0)^2]&=\mathrm{E}\left[\left(\sum_{m=-N}^{N}\sum_{i=1}^{K} a_{im}(x'_{im}-s_m)\right)^2\right]\\&=\mathrm{E}\left[\left(\sum_{m=-N}^{N}\sum_{i=1}^{K} a_{im}n'_{im}\right)^2\right]\\&=\sum_{m,n=-N}^{N}\sum_{i,k=1}^{K} a_{im}a_{kn}R'_{ik}(m-n).\end{aligned} \tag{5.42}$$

Using the calculus of variations, we obtain that the minimum-variance unbiased estimator has weights which satisfy the system of equations

$$\sum_{n=-N}^{N}\sum_{k=1}^{K} a_{kn}R'_{ik}(m-n)-\lambda_m=0, \quad \begin{array}{l} i=1,\dots,K \\ m=-N,\dots,N \end{array} \tag{5.43}$$

where the λ_m are $(2N+1)$ Lagrangian multipliers chosen to satisfy the constraints in (5.41). However, (5.41, 43) are identical to (5.28, 29), respectively. Hence, it has been shown that the maximum-likelihood filter also provides a minimum-variance unbiased estimate of the signal.

5.4.3 Case When Slowness Vector of Signal is Unknown

There are many practical situations in which the slowness vector of the signal is not known. In these cases it is not possible to apply the appropriate time delays to align the signal in each channel, as was done previously. Thus, it is necessary to modify the theory to take these situations into account. We begin by considering the modifications required in the optimum detector that was discussed previously and is shown in Fig. 5.1.

One way in which this modification can be done is to make use of the maximum-likelihood principle as applied to the likelihood ratio, as suggested by *Davenport* and *Root* [Ref. 5.8, pp. 327–332]. In this case the structure shown in Fig. 5.1, up through the comparator, is duplicated many times and these sections are used in parallel to form a new detector. Each section uses a trial set of time shifts corresponding to some trial slowness vector of the signal. The sections are identical except for the set of time shifts employed. Thus, the output of the matched filter in each section is observed and the section in which the maximum output occurs is noted. The slowness vector used in this section is taken as that for the signal. In addition, the output of this section can be compared to a threshold to determine if a signal is present. If it is determined that a signal is present, then the slowness vector of the signal can be assumed to be known as that which was found in the section having the maximum output. This slowness can then be used in the maximum-likelihood, or minimum-variance unbiased, estimation procedures discussed previously.

5.5 Frequency Wave Number Power Spectrum

A stationary stochastic process can be characterized by means of a power spectral density function. This function provides the information concerning the power as a function of frequency for the stationary stochastic process. In a similar manner, propagating waves, or a homogeneous random field, can be characterized by a frequency wave number power spectral density function.

This function provides the information concerning the power as a function of frequency and the vector velocities of the propagating waves. In addition, it can be shown that the performance of the optimum detector can be described in terms of the frequency wave number power spectrum of the background noise. Thus, this spectrum plays an important part in the analysis and design of seismic arrays. The definition and properties of the frequency wave number power spectrum are now given for random noise. The case of deterministic signals is relatively well known and is not presented.

5.5.1 Definition

If the sensor output noise field is space stationary, then for fixed λ, $S_{il}(\lambda)$ depends only on the vector difference $\boldsymbol{x}_i - \boldsymbol{x}_l$. In this case the sensor outputs are said to comprise a homogeneous random field, as discussed by *Yaglom* [Ref. 5.9, pp. 81–84], and it is convenient to introduce a cross-power spectral density $S(\lambda, \boldsymbol{r})$ and correlation $R(m, \boldsymbol{r})$ as

$$S(\lambda, \boldsymbol{r}) = S_{il}(\lambda), \tag{5.44}$$

$$R(m, \boldsymbol{r}) = R_{il}(m) \tag{5.45}$$

whenever $\boldsymbol{x}_i - \boldsymbol{x}_l = \boldsymbol{r}$.

Following *Yaglom* [5.9] any homogeneous random field has a spectral representation

$$N_{im} = \int_{-\pi}^{\pi} \int_{-\infty}^{\infty} \int_{-\infty}^{\infty} \exp[-\mathrm{j}(m\lambda + \boldsymbol{k} \cdot \boldsymbol{x}_i)] Z(d\lambda, d\boldsymbol{k}) \tag{5.46}$$

where $\boldsymbol{k}$ is the vector wave number. We have that $Z(\Delta\lambda, \Delta\boldsymbol{k})$ is a random function of the frequency interval $\Delta\lambda$ and the elemental wave number area, or interval $\Delta\boldsymbol{k}$, with the following properties:

1) $\mathrm{E}[Z(\Delta\lambda, \Delta\boldsymbol{k})] = 0$, for all $\Delta\lambda$, $\Delta\boldsymbol{k}$

2) $Z(\Delta\lambda, \Delta_1\boldsymbol{k} + \Delta_2\boldsymbol{k}) = Z(\Delta\lambda, \Delta_1\boldsymbol{k}) + Z(\Delta\lambda, \Delta_2\boldsymbol{k})$, if $\Delta_1\boldsymbol{k}$ and $\Delta_2\boldsymbol{k}$ are disjoint intervals, and $Z(\Delta_1\lambda + \Delta_2\lambda, \Delta\boldsymbol{k}) = Z(\Delta_1\lambda, \Delta\boldsymbol{k}) + Z(\Delta_2\lambda, \Delta\boldsymbol{k})$ if $\Delta_1\lambda$ and $\Delta_2\lambda$ are disjoint intervals

3) $\mathrm{E}[Z(\Delta_1\lambda, \Delta_1\boldsymbol{k})\, Z^*(\Delta_2\lambda, \Delta_2\boldsymbol{k})] = 0$, if $\Delta_1\boldsymbol{k}$ and $\Delta_2\boldsymbol{k}$ are disjoint intervals, or if $\Delta_1\lambda$ and $\Delta_2\lambda$ are disjoint intervals

4) $\mathrm{E}[|Z(\Delta\lambda, \Delta\boldsymbol{k})|^2] = P(\lambda, \boldsymbol{k})\, (\Delta\lambda/2\pi)\, \Delta k_x \Delta k_y$,

where $P(\lambda, \boldsymbol{k})$ is the frequency wave number power spectral density function and $2\pi k_x$, $2\pi k_y$ are the x, y components, respectively, of the vector $\boldsymbol{k}$ in rad km^{-1}. It should be noted that $Z(\Delta\lambda, \Delta\boldsymbol{k})$ is a random function with uncorrelated increments, where the increments can be taken either in frequency λ or in vector wave number $\boldsymbol{k}$. The covariance and cross-power spectral density can be written as

$$R(m,\boldsymbol{r})=\int_{-\pi}^{\pi}\int_{-\infty}^{\infty}\int_{-\infty}^{\infty}P(\lambda,\boldsymbol{k})\cdot\exp[-\mathrm{j}(m\lambda+\boldsymbol{k}\cdot\boldsymbol{r})]\frac{d\lambda}{2\pi}dk_x dk_y, \tag{5.47}$$

$$S(\lambda,\boldsymbol{r})=\int_{-\infty}^{\infty}\int_{-\infty}^{\infty}P(\lambda,\boldsymbol{k})\exp(-\mathrm{j}\boldsymbol{k}\cdot\boldsymbol{r})dk_x dk_y. \tag{5.48}$$

It is possible to write the frequency wave number spectrum as

$$\begin{aligned}P(\lambda,\boldsymbol{k})&=\sum_{m=-\infty}^{\infty}\int_{-\infty}^{\infty}\int_{-\infty}^{\infty}R(m,\boldsymbol{r})\exp[\mathrm{j}(m\lambda+\boldsymbol{k}\cdot\boldsymbol{r})]dr_x dr_y\\&=\int_{-\infty}^{\infty}\int_{-\infty}^{\infty}S(\lambda,\boldsymbol{r})\exp(\mathrm{j}\boldsymbol{k}\cdot\boldsymbol{r})dr_x dr_y\end{aligned} \tag{5.49}$$

where r_x, r_y are the x, y components, respectively, of the vector $\boldsymbol{r}$, in kilometers.

If there are M independent and random plane waves, each of which propagates across the array with a phase velocity $v_l\,\mathrm{km\,s^{-1}}$, $l=1,\ldots,M$, then

$$n_{im}=\sum_{l=1}^{M}a_l\cos(2\pi f_l mT+\boldsymbol{k}_l\cdot\boldsymbol{x}_i+\theta_l),\qquad \begin{array}{l}i=1,\ldots,K\\ m=0,\pm1,\pm2,\ldots\end{array} \tag{5.50}$$

where f_l is the frequency of the lth wave, T is the sampling period, $k_l=2\pi f_l\alpha_l$, $-\boldsymbol{\alpha}_l$ is a slowness vector which points in the direction of propagation of the lth wave, $|\boldsymbol{\alpha}_l|=1/v_l$ and θ_l, $l=1,\ldots,M$ are mutually independent random phase angles uniformly distributed on $[0,2\pi]$. Thus we have

$$R(m,\boldsymbol{r})=\frac{1}{2}\sum_{l=1}^{M}a_l^2\cos(m\lambda_l+\boldsymbol{k}_l\cdot\boldsymbol{r}), \tag{5.51}$$

and

$$\begin{aligned}(2\pi)^{-1}S(\lambda,\boldsymbol{r})=\frac{1}{4}\sum_{l=1}^{M}a_l^2[&\exp(-\mathrm{j}\boldsymbol{k}_l\cdot\boldsymbol{r})\delta(\lambda-\lambda_l)\\&+\exp(\mathrm{j}\boldsymbol{k}_l\cdot\boldsymbol{r})\delta(\lambda+\lambda_l)]\end{aligned} \tag{5.52}$$

so that

$$\begin{aligned}(2\pi)^{-1}P(\lambda,\boldsymbol{k})=\frac{1}{4}\sum_{l=1}^{M}a_l^2[&\delta(\lambda-\lambda_l,\boldsymbol{k}-\boldsymbol{k}_l)\\&+\delta(\lambda+\lambda_l,\boldsymbol{k}+\boldsymbol{k}_l)]\end{aligned} \tag{5.53}$$

which is a set of delta functions located at the frequency wave number points $(\lambda_l, \boldsymbol{k}_l)$, $(-\lambda_l, -\boldsymbol{k}_l)$, $\lambda_l = 2\pi f_l T$, $l = 1, \ldots, M$. It should now be apparent how $P(\lambda, \boldsymbol{k})$ provides the information concerning the speed and azimuth, or vector velocity, of propagating waves.

5.5.2 Properties

We now wish to show how to describe the performance of the maximum-likelihood filter in terms of the structure of the frequency wave number spectrum of the noise. This description applies equally well to any array processing filter, such as shown in Fig. 5.1. The power output of the maximum-likelihood filter due to noise alone, at a particular frequency λ, is obtained by first finding the power spectral density of the noise at the output of this filter, denoted by $p_{\mathrm{F}}(\lambda, \boldsymbol{k}_0)$, where $\boldsymbol{k}_0 = \lambda \boldsymbol{\alpha}_0$, $-\boldsymbol{\alpha}_0$ is the vector of time delays per unit distance for the signal, discussed previously in Sect. 5.3.1 and is measured in multiples of the sampling period T along each coordinate axis and points in the direction of propagation. We see from Fig. 5.1 that

$$\begin{aligned} p_{\mathrm{F}}(\lambda, \boldsymbol{k}_0) &= \sum_{i,l=1}^{K} S'_{il}(\lambda) A_i(\lambda) A_l(\lambda) \\ &= \left[\sum_{i,l=1}^{K} q'_{il}(\lambda)\right]^{-1} \sum_{i=1}^{K} A_i(\lambda) \\ &= \left\{\sum_{i,l=1}^{K} q_{il}(\lambda) \exp[\mathrm{j}\boldsymbol{k}_0 \cdot (\boldsymbol{x}_i - \boldsymbol{x}_l)]\right\}^{-1}, \end{aligned} \tag{5.54}$$

where we have made use of (5.35, 36, 38). The power output of the maximum-likelihood filter due to noise alone, in the frequency band $(\lambda, \lambda + \Delta\lambda)$ is obtained from (5.54) as

$$\begin{aligned} P_{\mathrm{F}}(\lambda, \boldsymbol{k}_0) &= \left\{\sum_{i,l=1}^{K} q_{il}(\lambda) \exp[\mathrm{j}\boldsymbol{k}_0 \cdot (\boldsymbol{x}_i - \boldsymbol{x}_l)]\right\}^{-1} \Delta\lambda \\ &= \sum_{i,l=1}^{K} A_i^*(\lambda, \boldsymbol{k}_0) A_l(\lambda, \boldsymbol{k}_0) S_{il}(\lambda) \exp[\mathrm{j}\boldsymbol{k}_0 \cdot (\boldsymbol{x}_i - \boldsymbol{x}_l)] \Delta\lambda \\ &= \Delta\lambda \int_{-\infty}^{\infty} \int_{-\infty}^{\infty} P(\lambda, \boldsymbol{k}) |B'(\lambda, \boldsymbol{k}, \boldsymbol{k}_0)|^2 dk_x dk_y, \end{aligned} \tag{5.55}$$

where $A_i(\lambda, \boldsymbol{k}_0)$ is obtained from (5.20), and now the fact that $A_i(\lambda)$ depends on the vector wave number of the signal $\boldsymbol{k}_0$ is brought out by the notation and

$$B'(\lambda, \boldsymbol{k}, \boldsymbol{k}_0) = \sum_{i=1}^{K} A_i(\lambda, \boldsymbol{k}_0) \exp[\mathrm{j}(\boldsymbol{k} - \boldsymbol{k}_0) \cdot \boldsymbol{x}_i]. \tag{5.56}$$

Thus, the output power of the maximum-likelihood filter is obtained by means of a frequency wave number filter $|B'(\lambda, \boldsymbol{k}, \boldsymbol{k}_0)|^2$ which is designed to suppress the output due to noise subject to the constraint that a signal propagating with a velocity corresponding to the vector wave number $\boldsymbol{k}_0$ be passed with no distortion, i.e., we see from (2.20, 56) that $B'(\lambda, \boldsymbol{k}_0, \boldsymbol{k}_0) = 1$, $-\pi \leqq \lambda \leqq \pi$. The maximum-likelihood filter will be particularly effective if $P(\lambda, \boldsymbol{k})$ is highly concentrated in the vicinity of a few points in wave number space, sufficiently different from the wave number corresponding to the signal $\boldsymbol{k}_0$. In this case, the frequency wave number filter $|B'|^2$ can be designed to pass the signal and steer nulls in its wave number response to lie at those points where $P(\lambda, \boldsymbol{k})$ is highly concentrated. If $P(\lambda, \boldsymbol{k})$ is not highly concentrated, but has a structure which is diffuse, so to speak, in wave number space, then the ability to steer nulls in wave number space, by means of the function $|B'|^2$, loses much of its importance. Since the use of the maximum-likelihood filter entails the insertion of a filter $A_i(\lambda, \boldsymbol{k}_0)$ in each channel of the array of sensors, followed by a summing operation, this type of processing is called filter-and-sum (FS) processing.

An important simple special case occurs when simple delay-and-sum (DS) processing is done. In this case we do not use the filter functions given by (5.20) but merely use an amplitude weight of $1/K$, so that the output power of the array processing filter due to noise alone is

$$
\begin{aligned}
P_{\mathrm{D}}(\lambda, \boldsymbol{k}_0) &= \frac{1}{K^2} \sum_{i,l=1}^{K} S_{il}(\lambda) \exp[\mathrm{j}\boldsymbol{k}_0 \cdot (\boldsymbol{x}_i - \boldsymbol{x}_l)] \Delta\lambda \\
&= \Delta\lambda \int_{-\infty}^{\infty} \int_{-\infty}^{\infty} P(\lambda, \boldsymbol{k}) |B(\boldsymbol{k} - \boldsymbol{k}_0)|^2 \, dk_x dk_y
\end{aligned}
\tag{5.57}
$$

where $|B(\boldsymbol{k})|^2$ is the beam-forming array response pattern

$$
B(\boldsymbol{k}) = \frac{1}{K} \sum_{i=1}^{K} \exp(\mathrm{j}\boldsymbol{k} \cdot \boldsymbol{k}_i) \, . \tag{5.58}
$$

It should be noted that $B(0) = 1$, and that the functional form, or sidelobe level, of $|B|^2$ remains the same for a fixed array geometry, i.e., sensor locations. This is to be contrasted with the functional form of $|B'|^2$ which depended on the spectral matrix of the noise as well as the array geometry. Thus, using DS processing it is not possible to steer nulls at those coordinates in wave number space where $P(\lambda, \boldsymbol{k})$ may be highly concentrated, subject to the constraint that the signal be passed with no distortion. In spite of this disadvantage, DS processing has been found to be very important. The reason for this is that the frequency wave number spectrum of the microseismic noise has been found to have a diffuse structure in wave number space so that the signal-to-noise ratio gain obtained with DS processing is almost the same as that for FS processing.

5.5.3 Relation to Coherence

There are some situations in which it is desired to know the coherence of the noise, as well as the frequency wave number power spectrum. These problems involve primarily the determination of the relative amounts of propagating and nonpropagating noise, as is now discussed. The coherence of the noise between the ith and l^{th} sensors is defined as

$$\varrho_{il}(\lambda)=\frac{|S_{il}(\lambda)|}{[S_{ii}(\lambda)S_{ll}(\lambda)]^{1/2}}. \tag{5.59}$$

It is easily seen from the definition that $\varrho_{il}(\lambda)\geqq 0$. The spectral matrix is nonnegative-definite and Hermitian so that

$$S_{ii}(\lambda)S_{ll}(\lambda)-|S_{il}(\lambda)|^2\geqq 0 \tag{5.60}$$

and thus $\varrho_{il}(\lambda)\leqq 1$. Hence we have

$$0\leqq\varrho_{il}(\lambda)\leqq 1. \tag{5.61}$$

The physical significance of the fact $\varrho_{il}(\lambda)=0$ is that the spectral components of the noise, at frequency λ, observed on the ith and l^{th} sensors are uncorrelated. In addition, if $\varrho_{il}(\lambda)=1$, then the spectral components of the noise, at frequency λ, observed on the ith and l^{th} sensors are linearly related to each other. It is also obvious that as the position of the ith sensor approaches that of the lth sensor, $\varrho_{il}(\lambda)$ approaches unity.

The coherence may, in general, be obtained from the frequency wave number power spectrum by noting that (5.48) may be written as

$$S_{il}(\lambda)=\int_{-\infty}^{\infty}\int_{-\infty}^{\infty}P(\lambda,\boldsymbol{k})\exp[-\mathrm{j}\boldsymbol{k}\cdot(\boldsymbol{x}_i-\boldsymbol{x}_l)]dk_x dk_y. \tag{5.62}$$

We now give three examples of noise fields for which the coherence can be computed:

1) The first example is one in which the noise consists of a single uniform plane wave propagating with a wave number k_0, so that

$$P(\lambda_0,\boldsymbol{k})=g(\lambda_0)\delta(\boldsymbol{k}-\boldsymbol{k}_0) \tag{5.63}$$

where $g(\lambda)$ is the power spectral density of the noise. There is no loss of generality in assuming $g(\lambda_0)=1$. Thus, using (5.63) in (5.62) we have

$$S_{il}(\lambda_0)=\exp[-\mathrm{j}\boldsymbol{k}_0\cdot(\boldsymbol{x}_i-\boldsymbol{x}_l)] \tag{5.64}$$

so hat $S_{ii}(\lambda_0)=1$ and

$$\varrho_{il}(\lambda_0)=1. \tag{5.65}$$

2) The next example is one in which the noise consists of a large number of independent and random plane waves each propagating with the same velocity v_0, in $\mathrm{km\,s^{-1}}$, and frequency f_0, in hertz, and also propagating uniformly from all directions or azimuths, simultaneously. Thus, its frequency wave number power spectrum is located on a ring in wave number space, i.e.,

$$P(\lambda_0,\varrho,\theta)=g(\lambda_0)\delta(\varrho-\boldsymbol{k}_0/2\pi), \quad 0\leqq\theta\leqq 2\pi \tag{5.66}$$

where $k_0=2\pi f_0/v_0$ is the wave number of the noise, in $\mathrm{rad\,km^{-1}}$, and a transformation to polar coordinates has been made in which $k_x=\varrho\cos\theta$, and $k_y=\varrho\sin\theta$. Thus, using (5.66) in (5.62) and assuming once again that $g(\lambda_0)=1$, we obtain

$$\begin{aligned}S_{il}(\lambda_0)&=\int_0^{\infty}\int_0^{2\pi}P(\lambda_0,\varrho,\theta)\exp(-\mathrm{j}2\pi\varrho\Delta x_{il}\cos\theta)\varrho d f d\theta\\&=\frac{k_0}{2\pi}\int_0^{2\pi}\exp(-\mathrm{j}k_0\Delta x_{il}\cos\theta)d\theta\\&=k_0\mathrm{J}_0(k_0\Delta x_{il})\end{aligned} \tag{5.67}$$

where $\mathrm{J}_0(u)$ is the Bessel function of zero order, $\Delta x_{il}=|x_i-x_l|$, and it has been assumed, with no loss of generality, that the coordinates of the ith and l^{th} sensors are $(\Delta x_{il},0)$, $(0,0)$, respectively. Using (5.67), we have $S_{ii}(\lambda_0)=k_0$ and

$$\varrho_{il}(\lambda_0)=|\mathrm{J}_0(k_0\Delta x_{il})|. \tag{5.68}$$

An illustration of this type of noise is provided by the situation in which the noise consists of either fundamental-mode, or higher mode, Rayleigh waves propagating uniformly from all azimuths simultaneously.

3) The final example is that of a disk noise of frequency f_0, in hertz, defined as consisting of a large number of plane waves propagating uniformly from all directions and with velocities between v_0 and infinity, in $\mathrm{km\,s^{-1}}$. Thus, its frequency wave number power spectrum is located on the disk in the wave number plane located within the circle $|k|=k_0=2\pi f_0 v_0$, where k_0 is in $\mathrm{rad\,km^{-1}}$. We can assume that disk noise is composed of the superposition of infinitely many ring-type noise fields with wave numbers ranging from zero to k_0. It is apparent from (5.62) that $S_{il}(\lambda)$ can be obtained by means of a

superposition principle from (5.67) as

$$S_{il}(\lambda_0) = \int_0^{k_0/2\pi} uJ_0(2\pi u \Delta x_{il}) du = \frac{1}{(2\pi\Delta x_{il})^2} \int_0^{k_0\Delta x_{il}} vJ_0(v) dv = \frac{k_0}{(2\pi)^2 \Delta x_{il}} J_1(k_0 \Delta x_{il}) \tag{5.69}$$

where $J_1(u)$ is the Bessel function of first order. We have from (5.69) that $S_{ii}(\lambda_0) = k_0^2/(8\pi^2)$ so that

$$\varrho_{il}(\lambda_0) = \frac{2}{k_0 \Delta x_{il}} |J_1(k_0 \Delta x_{il})| . \tag{5.70}$$

We see from (5.65, 68, 70) that the manner in which the coherence decreases, for increasing spatial lags Δx_{il}, provides an indication of the spatial extent of the frequency wave number spectrum in wave number space. Thus, the measurement of the coherence can provide some information about the structure of the frequency wave number power spectrum. However, this structure is much better defined by a measurement of the frequency wave number spectrum made directly with the high-resolution method to be described in Sect. 5.6.3. In order to use this method, however, an array must be available with a large number of sensors located within a reasonably large aperture. If such an array is not available, then the measurement of coherence, using a small number of sensors, assumes an important role in describing the frequency wave number power spectrum of the noise.

Another important application of the use of the coherence is the determination of the relative amounts of propagating and nonpropagating noise. We assume that the noise consists of the sum of two components. One component is propagating noise whose frequency wave number spectrum is located on an arc or arcs of a circle in wave number space. The nonpropagating or incoherent noise is characterized by a frequency wave number power spectrum $P_1(\lambda, \boldsymbol{k})$ which is determined by a cross-power spectral density function $S_{1il}(\lambda_0$, defined as

$$S_{1il}(\lambda_0) = g_1(\lambda_0)\delta(\varepsilon, \Delta x_{il}) \tag{5.71}$$

where

$$\delta(\varepsilon, x) = \begin{cases} 1, & |x| \leqq \varepsilon \\ 0, & |x| > \varepsilon \end{cases} \tag{5.72}$$

and $g_1(\lambda)$ is the power spectral density of the incoherent noise. We assume $g_1(\lambda) = r$, where r is the ratio of the incoherent to total noise power, and the total noise power is taken to be unity. Hence, $(1-r)$ is the relative amount of propagating noise power. The distance ε, in kilometers, is assumed to be very small and close to zero.

The frequency wave number power spectrum of the total noise is

$$P(\lambda, k) = P_I(\lambda, k) + P_p(\lambda, k) \tag{5.73}$$

where $P_p(\lambda, k)$ is the frequency wave number spectrum of the propagating component of the noise, and is defined as

$$P_p(\lambda_0, \varrho, \theta) = [(1-r)/\varrho P_c]\delta(\varrho - k_0/2\pi), \quad \theta \in C, \tag{5.74}$$

where C is an arc or arcs located on the circle $|k| = k_0$ in wave number space and $P_c = \int_c d\theta$. We have

$$\begin{aligned} S_{il}(\lambda_0) &= r\delta(\varepsilon, \Delta x_{il}) + \int_0^\infty \int_0^{2\pi} P_p(\lambda_0, \varrho, \theta) \\ &\quad \cdot \exp(-j2\pi\varrho\Delta x_{il}\cos\theta)\varrho d\varrho d\theta \\ &= r\delta(\varepsilon, \Delta x_{il}) + \frac{1-r}{P_c}\int_c \exp(-jk_0\Delta x_{il}\cos\theta)d\theta \\ &\simeq r\delta(\varepsilon, \Delta x_{il}) + (1-r)[1 - j(P_c'/P_c)k_0\Delta x_{il}] \end{aligned} \tag{5.75}$$

where $P_c' = \int_c \cos\theta \, d\theta$. We see that $|P_c'/P_c| \leqq 1$, so that the second term in the brackets in (5.75) is negligible if $k_0\Delta x_{il} \ll 1$, and we have

$$\varrho_{il}(\lambda_0) \simeq 1 - r, \quad \varepsilon < \Delta x_{il} \ll 1/k_0. \tag{5.76}$$

Since $k_0 = 2\pi/\lambda_{w0}$, λ_{w0} is the wavelength in kilometers, we see that the coherence provides an indication of the relative amount of propagating noise power, provided the separation between the sensors satisfies the conditions $\varepsilon < \Delta x_{il} \ll \lambda_{w0}/2\pi$.

5.6 Estimation of Frequency Wave Number Power Spectrum

In Sect. 5.5 it was shown that the frequency wave number power spectrum of the microseismic noise, as well as that of the seismic signals, play an important role in the analysis and design of seismic arrays. Thus, the estimation of this spectrum for both signals and noise is very important. A description is given in this section of the conventional and high-resolution methods for the estimation of the frequency wave number power spectrum. In order to do this it is necessary to begin with a discussion of a spectral matrix estimation procedure.

5.6.1 Spectral Matrix Estimation Procedure

The present method of spectral matrix estimation may be termed a direct segment, or block averaging, method. The number of data points in each channel which is to be used in the estimation, namely L, is divided into M nonoverlapping segments of N data points, $L=MN$. The data in each segment and each channel are transformed into the frequency domain, and these transforms are used to obtain an estimate of the cross spectra in the segment. The stability of the estimate is then increased by averaging over the M segments. We now describe the method in some detail.

The Fourier transform of the data in the nth segment, i^{th} channel, and normalized frequency λ, is

$$N_{in}(\lambda)=(N)^{-1/2}\sum_{m=1}^{N} a_m n_{i,m+(n-1)N}\exp(jm\lambda)\,. \qquad \begin{matrix} i=1,\ldots,K \\ n=1,\ldots,M \end{matrix} \tag{5.77}$$

The a_m are weights which are used to control the shape of the frequency window used in estimating $S_{ik}(\lambda)$. For simplicity, we assume $a_m=1, m=1,\ldots,N$. As an estimate for $S_{ik}(\lambda)$, we take

$$\hat{S}_{ik}(\lambda)=\frac{1}{M}\sum_{n=1}^{M} N_{in}(\lambda)N_{kn}^{*}(\lambda)\,. \qquad i,k=1,\ldots,K \tag{5.78}$$

The mean value of this estimate is

$$\begin{aligned} \mathrm{E}[\hat{S}_{ik}(\lambda_0)] &= \mathrm{E}\left[\frac{1}{M}\sum_{n=1}^{M} N_{in}(\lambda_0)N_{kn}^{*}(\lambda_0)\right] \\ &= \mathrm{E}\left[\frac{1}{MN}\sum_{n=1}^{M}\sum_{m,m'=1}^{N} n_{i,m+(n-1)N}n_{k,m'+(n-1)N}\right. \\ &\qquad \left.\cdot\exp[j(m-m')\lambda_0]\right] \\ &= \frac{1}{N}\sum_{m,m'=1}^{N} R_{ik}(m-m')\exp[j(m-m')\lambda_0] \\ &= \int_{-\pi}^{\pi} S_{ik}(\lambda)\,|W_N(\lambda-\lambda_0)|^2\frac{d\lambda}{2\pi} \end{aligned} \tag{5.79}$$

where $|W_N(\lambda)|^2$ is the Bartlett window, discussed by *Blackman* and *Tukey* [5.10], and is defined as

$$|W_N(\lambda)|^2 = \frac{1}{N}\left|\frac{\sin[(N/2)\lambda]}{\sin[(1/2)\lambda]}\right|^2 . \tag{5.80}$$

As usual, the frequency resolution of the measurement is determined by the spectral width of $|W_N(\lambda)|^2$, which will be about $1/NT$ Hz. We see that $W_N(\lambda)$ is a reasonably good frequency window in the sense that in the range $-\pi \leqq \lambda \leqq \pi$, $|W_N(\lambda)|^2$ approaches a delta function in such a manner that

$$\int_{-\pi}^{\pi} |W_N(\lambda)|^2 \frac{d\lambda}{2\pi} = 1 . \tag{5.81}$$

Thus $\hat{S}_{ik}(\lambda)$ is an asymptotically unbiased estimate for $S_{ik}(\lambda)$, and using (5.79) we obtain

$$\mathrm{E}[\hat{S}_{ik}(\lambda_0)] \simeq S_{ik}(\lambda_0) . \tag{5.82}$$

The mean-square value of the estimator is

$$\begin{aligned} &\mathrm{E}[\hat{S}_{ik}^2(\lambda_0)] \\ &= \mathrm{E}\Bigg[\frac{1}{M^2N^2} \sum_{n,n'=1}^{M} \sum_{m,m',m'',m'''=1}^{N} n_{i,m+(n-1)N} n_{k,m'+(n-1)N} \\ &\quad \cdot n_{i,m''+(n'-1)N} n_{k,m'''+(n'-1)N} \exp[\mathrm{j}(m-m'+m''-m''')\lambda]\Bigg] . \end{aligned} \tag{5.83}$$

In order to proceed with the analysis, we must at this point assume, once again, that $\{N_{im}\}$ is a multidimensional Gaussian process, so that

$$\begin{aligned} &\mathrm{E}[n_{i,m+(n-1)N} n_{k,m'+(n-1)N} n_{i,m''+(n'-1)N} n_{k,m'''+(n'-1)N}] \\ &= R_{ik}(m-m') R_{ik}(m''-m''') \\ &\quad + R_{ii}[m-m''+(n-n')N] R_{kk}[m'-m'''+(n-n')N] \\ &\quad + R_{ik}[m-m'''+(n-n')N] R_{ki}[m'-m'''+(n-n')N] . \end{aligned} \tag{5.84}$$

Using (5.84) in (5.83) and employing (5.79), we obtain, after some manipulations, for the variance of the estimator

$$\begin{aligned} &\mathrm{Var}[\hat{S}_{ik}(\lambda_0)] \\ &= \int_{-\pi}^{\pi} \int_{-\pi}^{\pi} \Bigg\{\frac{1}{M} + \frac{2}{M} \sum_{n=1}^{M-1} \left(\frac{M-n}{M}\right) \exp[\mathrm{j}nN(\lambda-\lambda')] \\ &\quad \times [S_{ii}(\lambda) S_{kk}(\lambda') |W_N(\lambda-\lambda_0) W_N(\lambda'-\lambda_0)|^2 \\ &\quad + S_{ik}(\lambda) S_{ik}(\lambda') W_N(\lambda-\lambda_0) W_N^*(\lambda+\lambda_0) W_N(\lambda'+\lambda_0) W_N^*(\lambda-\lambda_0)\Bigg\} \\ &\quad \times \frac{d\lambda}{2\pi} \frac{d\lambda'}{2\pi} , \end{aligned} \tag{5.85}$$

where W_N^* denotes complex conjugate of W_N. If $S_{ik}(\lambda)$ is reasonably constant in the vicinity of the frequency λ_0, then we can simplify (5.85) considerably as

$$\begin{aligned}\mathrm{Var}[\hat{S}_{ik}(\lambda_0)] &\simeq \frac{1}{M} S_{ii}(\lambda_0) S_{kk}(\lambda_0), \quad \lambda_0 \neq 0, \pi \\ &\simeq \frac{1}{M}[S_{ii}(\lambda_0) S_{kk}(\lambda_0) + S_{ik}^2(\lambda_0)], \quad \lambda_0 = 0, \pi. \end{aligned} \tag{5.86}$$

Thus, $\hat{S}_{ik}(\lambda)$ is a consistent estimate for $S_{ik}(\lambda)$, since its variance approaches zero as M approaches infinity.

We consider finally the important special case of estimation of the power spectrum for a single time series where $i=k$. In this case, we follow [5.10] and assume $\hat{S}_{ii}(\lambda_0)$ is a multiple of a chi-square variable so that to establish confidence intervals the chi-square distribution can be used with number of degrees of freedom or stability k_s, given by

$$k_s = 2\{\mathrm{E}[\hat{S}_{ii}(\lambda_0)]\}^2 / \mathrm{Var}[\hat{S}_{ii}(\lambda_0)]. \tag{5.87}$$

If we use (5.82, 86) we get $k_s = 2M$, $\lambda_0 \neq 0, \pi$. The 90% confidence limits in decibels are approximately $\pm 10/\sqrt{k_s} = \pm 10/\sqrt{2M}$. If $M=36$ then these confidence limits are about ± 1.2 dB, provided $\lambda_0 \neq 0, \pi$. If $\lambda_0 = 0, \pi$, then these confidence limits are about ± 1.7 dB.

5.6.2 Conventional Method

We now briefly describe the conventional method for the estimation of the frequency wave number power spectrum $P(\lambda, k)$. A more detailed discussion has been given by *Capon* [5.2]. The conventional estimate is

$$\hat{P}(\lambda, \boldsymbol{k}) = \frac{1}{K^2} \sum_{i,l=1}^{K} w_i w_l^* \hat{S}_{il}(\lambda) \exp[\mathrm{j}\boldsymbol{k} \cdot (\boldsymbol{x}_i - \boldsymbol{x}_l)] \tag{5.88}$$

where the w_i are weights which are used to control the shape of the wave number window used in estimating $P(\lambda, k)$ and $\hat{S}_{il}(\lambda)$ is an estimate for $S_{il}(\lambda)$ obtained by the direct segment, or block averaging, method of estimation. This method was described in Sect. 5.6.1 and it has been shown to be very desirable from the point of view of computational efficiency by *Capon* et al. [5.11]. It is assumed, for simplicity, that $w_i = 1$, $i = 1, \ldots, K$. If we compare (5.88) with (5.75) we see that $\hat{P}$ can be considered as an estimate for P_D.

We now show that $\hat{P}$ is an asymptotically unbiased and consistent estimate for cP, where c is some positive constant. In order to do this, the results of [5.3] may be used which show that if $\{n_{im}\}$ is a multidimensional Gaussian process, then $\hat{P}$ is a multiple of a chi-square variable with $2M$ degrees of freedom and mean, variance, given by

$$E[\hat{P}(\lambda_0, \boldsymbol{k}_0)] = \int_{-\pi}^{\pi} \int_{-\infty}^{\infty} \int_{-\infty}^{\infty} P(\lambda, \boldsymbol{k}) |W_N(\lambda-\lambda_0) B(\boldsymbol{k}-\boldsymbol{k}_0)|^2 \cdot \frac{d\lambda}{2\pi} dk_x dk_y \tag{5.89}$$

$$\begin{aligned} \mathrm{Var}[\hat{P}(\lambda_0, \boldsymbol{k}_0)] &= \frac{1}{M}\{E[\hat{P}(\lambda_0, \boldsymbol{k}_0)]\}^2, \quad \lambda_0 \neq 0, \pi \\ &= \frac{2}{M}\{E[\hat{P}(\lambda_0, \boldsymbol{k}_0)]\}^2, \quad \lambda_0 = 0, \pi \end{aligned} \tag{5.90}$$

where $|W_N(\lambda)|^2$ is the Bartlett window defined in (5.80) and $|B(\boldsymbol{k})|^2$ is the beamforming array response pattern defined previously in (5.58). Thus, $E[\hat{P}(\lambda_0, \boldsymbol{k}_0)]$ is obtained by means of a frequency wave number window $|W_n(\lambda-\lambda_0)B(\boldsymbol{k}-\boldsymbol{k}_0)|^2$. Hence, $\hat{P}$ is an asymptotically unbiased estimate for cP if $|W_N(\lambda)B(\boldsymbol{k})|^2$ approaches a delta function in such a way that

$$\int_{-\pi}^{\pi} \int_{-\infty}^{\infty} \int_{-\infty}^{\infty} |W_N(\lambda)B(\boldsymbol{k})|^2 \frac{d\lambda}{2\pi} dk_x dk_y = c. \tag{5.91}$$

According to (5.90) the variance of $\hat{P}$ approaches zero as M approaches infinity, so that $\hat{P}$ is a consistent estimate for cP. The confidence limits for $\hat{P}$ can be computed in a manner similar to that for $\hat{S}_{ii}(\lambda_0)$ described previously in Sect. 5.6.1.

5.6.3 High-Resolution Method

The high-resolution estimate for $P(\lambda, k)$ is defined as

$$P'(\lambda, \boldsymbol{k}) = \left\{ \sum_{i,l=1}^{K} \hat{q}_{il}(\lambda) \exp[j\boldsymbol{k} \cdot (\boldsymbol{x}_i - \boldsymbol{x}_l)] \right\}^{-1} \tag{5.92}$$

where $\{\hat{q}_{il}(\lambda)\}$ is the inverse of the spectral matrix $\{\hat{S}_{il}(\lambda)\}$. If we compare (5.92) with (5.55) we see that P' is an estimate for P_F. Thus, $P'(\lambda, \boldsymbol{k}_0)$ is an estimate for the power output of the array processor, known as the maximum-likelihood filter. It is recalled that this filter passes undistorted any monochromatic plane wave traveling at a velocity corresponding to the wave number $\boldsymbol{k}_0$ and suppresses in an optimum least-squares sense the power of those waves traveling at velocities corresponding to wave numbers other than $\boldsymbol{k}_0$. This important property of the maximum-likelihood filter, which follows from the minimum-variance unbiased estimation property discussed in Sect. 5.4.2 can be considered as providing the justification for the use of P' as an estimate for P. It should also be noted that the amount of computation required to obtain P' is almost the same as that to get $\hat{P}$, since only an additional Hermitian matrix inversion is required. It is this fact which makes the computation of P' feasible.

It has also been shown by [5.3] that P' is a multiple of a chi-square variable with $2(M-K+1)$ degrees of freedom and mean, variance, given by

$$\mathrm{E}[P'(\lambda_0, \boldsymbol{k}_0)] = \left(\frac{M-K+1}{M}\right) \int_{-\pi}^{\pi} \int_{-\infty}^{\infty} \int_{-\infty}^{\infty} \times P(\lambda, \boldsymbol{k}) \, |W_N(\lambda-\lambda_0) B'(\lambda_0, \boldsymbol{k}, \boldsymbol{k}_0)|^2 \frac{d\lambda}{\partial\pi} dk_x dk_y \tag{5.93}$$

$$\begin{aligned} \mathrm{Var}[P'(\lambda_0, \boldsymbol{k}_0)] &= \left(\frac{1}{M-K+1}\right) \{\mathrm{E}[P'(\lambda_0, \boldsymbol{k}_0]\}^2, \quad \lambda_0 \neq 0, \pi \\ &= \left(\frac{2}{M-K+1}\right) \{\mathrm{E}[P'(\lambda_0, \boldsymbol{k}_0)]\}^2. \quad \lambda_0 = 0, \pi \end{aligned} \tag{5.94}$$

It should be noted that the functional form or shape of B', defined previously in (5.56), changes as a function of the wave number $\boldsymbol{k}_0$. Thus, $\mathrm{E}[P'(\lambda_0, \boldsymbol{k}_0)]$ is obtained by means of a frequency wave number window

$$|W_N(\lambda-\lambda_0) B'(\lambda_0, \boldsymbol{k}, \boldsymbol{k}_0)|^2 .$$

Hence, P' is an asymptotically unbiased estimate for cP if this window approaches a three-dimensional delta function in such a way that

$$\left(\frac{M-K+1}{M}\right) \int_{-\pi}^{\pi} \int_{-\infty}^{\infty} \int_{-\infty}^{\infty} |W_N(\lambda-\lambda_0) B'(\lambda_0, \boldsymbol{k}, \boldsymbol{k}_0)|^2 \frac{d\lambda}{2\pi} dk_x dk_y = c . \tag{5.95}$$

The confidence limits for P' can be obtained in a manner similar to that for $\hat{P}$ discussed in Sect. 5.6.2. If $M=36$, and $K=21$, then there are 32 degrees of freedom and the 90% confidence limits are about ± 1.8 dB, $\lambda_0 \neq 0, \pi$. If $\lambda_0 = 0, \pi$, then these confidence limits are about ± 2.5 dB.

5.6.4 Estimation of Coherence

The estimate for the coherence between the noise in the ith and l^{th} sensors, at frequency λ, is given by

$$\varrho_{il}(\lambda) = \frac{|\hat{S}_{il}(\lambda)|}{[\hat{S}_{ii}(\lambda) \hat{S}_{ll}(\lambda)]^{1/2}} . \tag{5.96}$$

The sampling distribution for $\hat{\varrho}_{il}(\lambda)$ is known and has been tabulated by *Amos* and *Koopmans* [5.12]. These results can be used to establish the 95% confidence limits for $\hat{\varrho}_{il}(\lambda)$. Such a computation has been given by *Capon* [5.13] for the case when there are 36 independent segments used in the estimation of the coherence by means of the direct segment method described in Sect. 5.6.1. We only note here that since $\hat{\varrho}_{il}(\lambda)$ is always positive, the coherence measurement level for uncorrelated noise spectral components is about 0.2.

5.7 Conclusions

A description has been given of some signal processing methods in large array seismology. The optimum detector for a known signal in additive Gaussian noise was shown to consist of the tandem combination of appropriate time delays, maximum-likelihood filter, noise whitening filter, matched filter, and a threshold comparator. The maximum-likelihood filter plays an important rolc in determining the structure of the optimum detector. This filter also provides a minimum-variance unbiased estimate for the input signal when it is not known, which is the same as the maximum-likelihood estimate of the signal if we have Gaussian noise.

If the noise is stationary in both time and space then it can be characterized by a frequency wave number power spectral density function. The performance of array processing filters, such as the maximum-likelihood filter, is relatively simple to explain in terms of the structure of this function.

References

5.1 R.T.Lacoss: Geophys. **36**, 661–675 (1971)
5.2 J.Capon: Proc. IEEE **57**, 1408 (1969)
5.3 J.Capon, N.R.Goodman: Proc. IEEE **58**, 1785 (1970)
5.4 J.Capon: *Methods in Computational Physics*, Vol. 13 (Academic Press, New York 1973)
5.5 C.W.Helstrom: *Statistical Theory of Signal Detection* (Pergamon, New York 1960)
5.6 J.L.Doob: *Stochastic Processes* (Wiley and Sons, New York 1953)
5.7 J.Capon: IEEE Trans. IT-**11**, 247 (1965)
5.8 W.L.Davenport, W.L.Root: *An Introduction to the Theory of Random Signals and Noise* (McGraw-Hill, New York 1958)
5.9 A.M.Yaglom: *An Introduction to the Theory of Stationary Random Functions* (Prentice-Hall, Englewood Cliffs, NJ 1962)
5.10 R.B.Blackman, J.W.Tukey: *The Measurement of Power Spectra from the Point of View of Communications Engineering* (Dover, New York 1959)
5.11 J.Capon, R.J.Greenfield, R.J.Kolker: Proc. IEEE **55**, 192 (1967)
5.12 D.E.Amos, L.H.Koopmans: "Tables of the Distribution of the Coefficient of Coherence for Stationary Bivariate Gaussian Processes"; SCR-483, Sandia Corp. New Mexico (1963)
5.13 J.Capon: J. Geophys. Res. **74**, 3182 (1969)

6. Application of the Maximum-Likelihood Method and the Maximum-Entropy Method to Array Processing

R. N. McDonough

6.1 Overview

"Array Processing" has to do with the processing of data from a spatially distributed array of sensors. The objective is to carry out an analysis of the radiation field in which the sensor array is immersed, in order to determine the radiated power impinging on the array location from various (ideally all) directions in space. In geophysical applications, we may be interested in the geometrical direction from which pressure waves of some frequency and velocity of interest are reaching the array, in which case some distribution of seismometers is used to collect the pressure-wave amplitude data as a function of time at each space point of the array. In ocean sound, an array of hydrophones may provide the data for determining a directional map of background sound power in the deep ocean. In radio astronomy, a uniformly distributed "array" (antenna), or some number of discrete antennas (an interferometer), may be used to determine the intensity of radiation of some specific frequency reaching the earth from various regions of the sky.

In all of these application areas, spectral analysis is used. The radiation field in which the array is immersed is assumed (usually implicitly) to be created by some continuous distribution of far-distant energy sources, each sending out traveling waves which sweep across the array as plane wave fronts. The various elemental wave sources are assumed to be statistical, and independent of one another. Let $x(t, \boldsymbol{z})$ be the signal available at space point $\boldsymbol{z}$ in the array, where there may or may not actually be a sensor located. Then we hypothesize that $x(t, \boldsymbol{z})$ is the superposition of an infinitely large number of plane waves of the type

$$\exp[\mathrm{j}2\pi(ft+\boldsymbol{v}\cdot\boldsymbol{z})], \tag{6.1}$$

where $\boldsymbol{v}$ is the vector wave number. The wave (6.1) propagates in the direction of $-\boldsymbol{v}$, with speed (phase velocity)

$$v=f/|\boldsymbol{v}|, \tag{6.2}$$

and has frequency f when observed at any fixed space point. (Often $2\pi\boldsymbol{v}=\boldsymbol{k}$, the vector propagation constant, is called the vector wave number.) Since we must allow for waves of any frequency and wave number, we write

$$x(t,z)=\int_{-\infty}^{\infty}\int_{V}\tilde{X}(f,\boldsymbol{v})\exp[j2\pi(ft+\boldsymbol{v}\cdot\boldsymbol{z})]d\boldsymbol{v}df, \tag{6.3}$$

where V indicates all 3-space, and $\tilde{X}(f,\boldsymbol{v})$ is the complex phasor representing the amplitude and phase of waves in the differential frequency and wave number ranges $(f, f+df)$, $(\boldsymbol{v}, \boldsymbol{v}+d\boldsymbol{v})$. The phasors $\tilde{X}(f,\boldsymbol{v})$ are taken to be independent random variables at each f, $\boldsymbol{v}$, heuristically, since (see below) such an $\tilde{X}(f,\boldsymbol{v})$ is not well defined.

What we would like to determine from the array data $x(t,z)$ is the average power of each traveling wave component in (6.3), i.e., we want to find

$$S_x(f,\boldsymbol{v})=\mathrm{E}[|\tilde{X}(f,\boldsymbol{v})|^2] \tag{6.4}$$

by processing whatever values of $x(t,z)$ we have available. One approach becomes clear if we examine the time–space covariance function of the data (assuming zero mean, for simplicity)

$$C_x(\tau,\boldsymbol{r})=\mathrm{E}[x(t+\tau,z+\boldsymbol{r})x^*(t,z)], \tag{6.5}$$

which we assume to depend only on the lag variables τ and $\boldsymbol{r}$. This will be the case if the data field $x(t,z)$ is a stationary random process in both space and time (i.e., homogeneous). Using (6.3) in (6.5), and remembering that the elemental sources are assumed independent, we obtain

$$C_x(\tau,\boldsymbol{r})=\int_{-\infty}^{\infty}\int_{V}S_x(f,\boldsymbol{v})\exp[j2\pi(f\tau+\boldsymbol{v}\cdot\boldsymbol{r})]d\boldsymbol{v}df. \tag{6.6}$$

Observing here that we have a multidimensional inverse Fourier transform, we may write at once

$$S_x(f,\boldsymbol{v})=\int_{-\infty}^{\infty}\int_{V}C_x(\tau,\boldsymbol{r})\exp[-j2\pi(f\tau+\boldsymbol{v}\cdot\boldsymbol{r})]d\boldsymbol{r}d\tau, \tag{6.7}$$

where V refers to $\boldsymbol{r}$-space.

Thus a general strategy is clear: estimate the data space-time covariance function, and analyze its spectrum. All the conventional ways of doing this encounter a basic problem. Due to the finite extent of our ability to collect data, the available values of $x(t,z)$ are limited to some region, say $-T/2\leqq t\leqq T/2$ and $-L/2\leqq z_i\leqq L/2$, so that the potentially available values $C_x(\tau,\boldsymbol{r})$ lie within $-T\leqq\tau\leqq T$, $-L\leqq r_i\leqq L$. The classical procedure is then to construct an approximation to the spectrum,

$$\hat{S}_x(f, \boldsymbol{v}) = \int_{-T}^{T} \int_{V_L} C_x(\tau, \boldsymbol{r}) \exp[-\mathrm{j}2\pi(f\tau + \boldsymbol{v} \cdot \boldsymbol{r})] d\boldsymbol{r} d\tau, \tag{6.8}$$

where V_L is the obvious cube in $\boldsymbol{r}$-space. But then the resolving ability of this estimator is limited, of the order of

$$\Delta f = 1/T, \quad \Delta v_i = 1/L, \tag{6.9}$$

i.e., with the linear estimator (6.8), field components spaced closer in frequency or wave number than (6.9) are essentially not resolved.

With linear processing, then, resolution improves only with an increase in data span. In frequency, resolution is easily "bought," simply by leaving the recorder run longer, to increase the data time span (up to the limits imposed by the stationarity assumption, in principle). In wave number, however, improved resolution involves increasing the spatial extent of the measuring apparatus, which may be expensive indeed, or making recourse to some such time-coherent data scanning technique as used in synthetic aperture radar [6.1] or sonar [6.2], or long baseline interferometry in radio astronomy [6.3]. Hence, the trade-off between data processing effort as opposed to increased data collection heavily favors the former, in the spatial domain, and nonlinear spectral estimation methods have been of great interest in array processing for that reason.

In the discussions below, we will restate more carefully what has been sketched in the last few paragraphs. We will discuss the origin of the frequency wave number spectrum, and summarize the usual well-known facts about its estimation by conventional linear techniques. We will then suppress the time domain, and concentrate on wave number aspects of the problem. We will discuss the "maximum-likelihood" estimator, and link it to the wave number spectrum. A difficulty which can arise in practice will then be discussed namely, the fact that the resolution of this nonlinear spectral estimator may deteriorate seriously if the actual data structure does not fit our assumptions about it. Finally, in regard to maximum-likelihood processing, we will discuss a few of the many iterative realizations of the scheme which have been proposed.

In regard to the "maximum-entropy" method, we shall summarize the approach to this second nonlinear processor through information theory, going back to some of the early work of *Jaynes* [6.4]. We will discuss maximum-entropy processing as a way to treat correlation values, and describe the interrelations with autoregressive modeling and linear predictive coding.

Little of what we shall discuss below is new. Our hope is that the exposition will provide in a convenient place some summary of the many aspects of the two nonlinear estimation techniques which are treated in the literature. In order to emphasize the spatial nature of our discussion, we will usually suppress the time variable. However, the isomorphism between time/frequency and space/wave number exists, and whenever convenient to the exposition we will revert to the simple time domain alone to make a point.

6.2 The Wave Number Spectrum and Its Estimation by Linear Processing

By the "spectrum" of a signal, we mean its Fourier transform. Which version of the Fourier theory applies depends on our model of the signal to be analyzed. We may think of the signal as being deterministic, or as being noiselike, a sample function of some probabilistically defined stochastic process. In either case, we may think of the signal $x(t)$ as vanishing outside our data interval $-T/2 \leqq t \leqq T/2$, as repeating periodically (with period T) outside that interval, or as continuing indefinitely in some unknown fashion. We may (and ultimately must) assume we have available only sampled values $x(k\Delta t)$, in which case we must assume $x(t)$ is appropriately band limited. The samples may be uniformly spaced in time, or they may not be. (In the spatial domain, they probably are not.) We shall assume the time-domain signal to be a sample function of a continuous-time stochastic process, which "goes on" indefinitely in both directions away from our data interval. We shall not necessarily assume that the process is band limited, and thus we work in continuous time. This framework leads to the least awkward exposition when the switch to the spatial domain is made.

6.2.1 Spectral Representation of a Stochastic Process

Suppose that a time-domain data waveform $x(t)$ is to be analyzed into constituent harmonic components $\exp(\mathrm{j}2\pi ft)$. The usual exposition of Fourier analysis, found in engineering texts such as [6.5], begins by considering the waveform $x(t)$ to be of interest only on $-T/2 \leqq t \leqq T/2$, and to repeat periodically with period T outside that interval. A series representation then exists, under the usual mild restrictions, mainly that $x(t)$ be absolutely integrable on the interval $(-T/2, T/2)$:

$$x(t) = \sum_{n=-\infty}^{\infty} X'_n \exp(\mathrm{j}n2\pi t/T)\,, \qquad -T/2 \leqq t \leqq T/2\,, \tag{6.10}$$

$$X'_n = \frac{1}{T} \int_{-T/2}^{T/2} x(t) \exp(-\mathrm{j}n2\pi t/T) dt\,. \tag{6.11}$$

In a heuristic passage to the limit as $T \to \infty$, the frequency separation $1/T$ of the constituent harmonics in (6.10) approaches a differential df and the quantity n/T approaches a continuous variable f, since for sufficiently small $1/T$ an integer n can be found such that n/T is near any selected value of f. Then by associating the factor $1/T$ in (6.11) with (6.10), we may write

$$x(t) = \frac{1}{T} \sum_{n=-\infty}^{\infty} X''_n \exp(\mathrm{j}n2\pi t/T)\,, \qquad -T/2 \leqq t \leqq T/2\,, \tag{6.12}$$

$$X_n'' = \int_{-T/2}^{T/2} x(t) \exp(-jn2\pi t/T) dt . \tag{6.13}$$

Passing to the limit as $T \to \infty$ in (6.12, 13), it is argued that we obtain

$$x(t) = \int_{-\infty}^{\infty} X''(f) \exp(j2\pi ft) df , \qquad -\infty < t < \infty , \tag{6.14}$$

$$X''(f) = \int_{-\infty}^{\infty} x(t) \exp(-j2\pi ft) dt . \tag{6.15}$$

In (6.15), the behavior of $x(t)$ in decaying to zero at $t = \pm\infty$ is relied upon to assure convergence of the integral. [The result (6.14, 15) is of course correct, but the heuristic argument suffers from a defect: it cannot easily be argued convincingly that the limit $n \to \infty$ in (6.12) should pass over to $f = n/T \to \infty$ in (6.14), since $1/T \to 0$. We will not pursue the question here.]

If we regard $x(t)$ as a random, noiselike waveform, however, we should not assume decay of the type required for convergence of the integral (6.15), e.g., we do not want to require finite energy, so that we want to allow (6.15) to diverge. On the other hand, finite average power fits the model we have in mind:

$$\lim_{T \to \infty} \frac{1}{T} \int_{-T/2}^{T/2} |x(t)|^2 dt < \infty . \tag{6.16}$$

Use of this requires returning to the expressions (6.10, 11), and defining the spectrum $X(f)$ as a limit based on

$$x(t) = \frac{1}{\sqrt{T}} \sum_{n=-\infty}^{\infty} X_n \exp(jn2\pi t/T) , \tag{6.17}$$

$$X_n = \frac{1}{\sqrt{T}} \int_{-T/2}^{T/2} x(t) \exp(-jn2\pi t/T) dt . \tag{6.18}$$

If (6.16) holds, the $1/\sqrt{T}$ factor in (6.18) is just enough to assure that $X(f) = \lim_{T \to \infty} X_n$ is well defined, but then in (6.17) we have that $\lim_{T \to \infty} X_n/\sqrt{T}$ behaves as $1/\sqrt{T}$, i.e., as $\sqrt{df}$, and we do not obtain an ordinary integral as the heuristic limit of (6.17).

Wiener [6.6] dealt rigorously with this situation, by expanding the time autocorrelation function of $x(t)$,

$$\phi_x(\tau) = \lim_{T \to \infty} \frac{1}{T} \int_{-T/2}^{T/2} x(t+\tau) x^*(t) dt , \tag{6.19}$$

in a spectral representation defined by (6.14, 15). This (deterministic) subject of "generalized harmonic analysis," discussed in summary in [6.7], is sufficient for our purpose, when the transition to the random process framework is made, as *Koopmans* [6.8] discussed in detail. However, we choose to retain the representation of $x(t)$ itself, and write the limit of (6.17) as the Fourier-Stieltjes integral [6.9]

$$x(t)=\int_{-\infty}^{\infty}\exp(\mathrm{j}2\pi ft)X(df), \tag{6.20}$$

where $X(df)$ is an infinitesimal quantity of order $\sqrt{df}$, which is also a function of frequency f. Thus $x(t)$ is decomposed into a linear combination of harmonics $\exp(\mathrm{j}2\pi ft)$, with "phasor" amplitudes $X(df)$. As $T\rightarrow\infty$ the phasors $X(df)$ are heuristically the limit of

$$\frac{1}{\sqrt{T}}X_n=\frac{1}{T}\int_{-T/2}^{T/2}x(t)\exp(-\mathrm{j}2\pi nt/T)dt. \tag{6.21}$$

Then assuming $x(t)$ to be a sample function of a stochastic process, and associating some probability variable ξ to obtain $x(t,\xi)$, the phasor amplitudes $X(df)$ become random variables $X(df,\xi)$ at each fixed frequency.

These remarks are intended only to provide some motivation for acceptance of the fact that any sample function of a weakly stationary stochastic process has a phasor decomposition (6.20), where the phasor amplitudes $X(df)$ are random variables. Here weak stationarity means simply that the mean and autocovariance functions,

$$\mathrm{E}[x(t)]=m_x, \tag{6.22}$$

$$\mathrm{E}[[x(t+\tau)-m_x][x(t)-m_x]^*]=C_x(\tau), \tag{6.23}$$

exist and are independent of t. The spectral decomposition (6.20) of a (weakly) stationary random process is discussed in some depth by *Yaglom* [6.9], and in summary fashion by *Koopmans* [6.8]. The random variables $X(df)$ have the property

$$\mathrm{E}[X(df_1)X^*(df_2)]=0, \qquad df_1\cap df_2=\emptyset, \tag{6.24}$$

where the condition is that the intervals $df_1=(f_1,f_1+df)$ and $df_2=(f_2,f_2+df)$ be disjoint.

Using property (6.24) and the spectral decomposition (6.20), the autocovariance function (6.23) becomes

$$C_x(\tau)=\int_{-\infty}^{\infty}\exp(\mathrm{j}2\pi f\tau)\mathrm{E}[|X(df)|^2]. \tag{6.25}$$

Since $X(df)$ is of order $\sqrt{df}$, $|X(df)|^2$ is of order df, and given the appropriate restrictions, essentially that $x(t)$ be of zero mean and have no discrete harmonic component, we may write

$$\mathrm{E}[|X(df)|^2] = P_x(f)df, \tag{6.26}$$

where $P_x(f)$ is the average power spectral density of the process, i.e., the expected (average) power per Hertz in each frequency component $X(df)\exp(\mathrm{j}2\pi ft)$ of the data $x(t)$. Finally, recognizing the inverse Fourier transform resulting from (6.25, 26),

$$C_x(\tau) = \int_{-\infty}^{\infty} P_x(f)\exp(\mathrm{j}2\pi f\tau)df, \tag{6.27}$$

we may write

$$P_x(f) = \int_{-\infty}^{\infty} C_x(\tau)\exp(-\mathrm{j}2\pi f\tau)d\tau. \tag{6.28}$$

The objective of data processing is to compute an estimate of the spectrum $P_x(f)$ from whatever values of $x(t)$ may be available.

In spatial processing, we have to deal with stochastic time functions $x(t)$ defined over some set of space points with position vectors $\boldsymbol{z}$, i.e., with functions $x(t, \boldsymbol{z})$ of time and three space variables, as well as the implied probability variable defining the stochastic ensemble. If the ensemble $x(t, \boldsymbol{z})$ is (weakly) stationary in both time and space, i.e., if the mean and autocovariance,

$$m_x = \mathrm{E}[x(t, \boldsymbol{z})], \tag{6.29}$$

$$C_x(\tau, \boldsymbol{r}) = \mathrm{E}[[x(t+\tau, \boldsymbol{z}+\boldsymbol{r}) - m_x][x(t, \boldsymbol{z}) - m_x]^*], \tag{6.30}$$

depend neither on t nor $\boldsymbol{z}$, then again there exists a spectral decomposition [6.9]:

$$x(t, \boldsymbol{z}) = \int_{-\infty}^{\infty} \int_{V} \exp[\mathrm{j}2\pi(ft + \boldsymbol{v}\cdot\boldsymbol{z})]X(df, d\boldsymbol{v}), \tag{6.31}$$

where $\boldsymbol{v}$ is the vector wave number, and V indicates all of 3-space. This represents a decomposition of the random "field" $x(t, \boldsymbol{z})$ into plane-traveling-wave components $\exp[\mathrm{j}2\pi(ft + \boldsymbol{v}\cdot\boldsymbol{z})]$ of wavelength

$$\lambda = 1/|\boldsymbol{v}| \tag{6.32}$$

propagating in the direction of $-\boldsymbol{v}$ with speed (phase "velocity")

$$v = f/|\boldsymbol{v}| = f\lambda. \tag{6.33}$$

The phasor amplitudes have the property

$$E[X(df_1, d\boldsymbol{v}_1)X^*(df_2, d\boldsymbol{v}_2)] = 0, \tag{6.34}$$

if either $(f_1, f_1 + df)$, $(f_2, f_2 + df)$ are disjoint, $(\boldsymbol{v}_1, \boldsymbol{v}_1 + d\boldsymbol{v})$, $(\boldsymbol{v}_2, \boldsymbol{v}_2 + d\boldsymbol{v})$ are disjoint, or both. Substituting (6.31) into (6.30) and using (6.34), we obtain

$$C_x(\tau, \boldsymbol{r}) = \int_{-\infty}^{\infty} \int_V \exp[j2\pi(f\tau + \boldsymbol{v}\cdot\boldsymbol{r})] E[|X(df, d\boldsymbol{v})|^2]. \tag{6.35}$$

Defining

$$E[|X(df, d\boldsymbol{v})|^2] = S_x(f, \boldsymbol{v})df\, d\boldsymbol{v}, \tag{6.36}$$

and recognizing the four-dimensional inverse Fourier transform [6.10] in (6.36), we have

$$S_x(f, \boldsymbol{v}) = \int_{-\infty}^{\infty} \int_V C_x(\tau, \boldsymbol{r}) \exp[-j2\pi(f\tau + \boldsymbol{v}\cdot\boldsymbol{r})] d\boldsymbol{r}\, d\tau, \tag{6.37}$$

where V now refers to $\boldsymbol{r}$-space. We shall seek to estimate the frequency–wave-number power spectral density $S_x(f, \boldsymbol{v})$, as it represents the average power density at each frequency of plane waves, propagating through space at each speed and from each direction, into which the data field is decomposable. A frequency–wave-number analysis of the data is needed, to be carried out by some type of time-space filter. In the next section, we will summarize the usual linear filters, and discuss their limitations, before passing on to the nonlinear filters of primary interest.

6.2.2 Conventional Linear Spectrum Analysis

For brevity of notation, let us return to the one-dimensional time-domain problem of estimating the power density spectrum (6.28) of a stationary stochastic process $x(t)$. As discussed by *Yaglom* [6.9], under some weak conditions the stationary process $x(t)$ is ergodic, which is to say that the ensemble average needed in calculating the mean and autocovariance function, namely, (6.22, 23), can be replaced by time averages along a single realization of the process, as for example,

$$m_x = E[x(t)] = \lim_{T\to\infty} \frac{1}{T} \int_{-T/2}^{T/2} x(t)dt. \tag{6.38}$$

Hence in principle the autocovariance (6.23) and its transform, the spectrum (6.28), are exactly computable. However, when we are prevented from passing to the limit in such integrals as (6.38), due to the finite limit of both the available data span and the ability to process it, we must rely on estimates of the power spectrum, computed from a finite data span. The statistical properties of various such estimators then need to be considered.

It has been the universal practice in the past to compute estimates of the power spectrum $P_x(f)$ by linear operations on the available values of $x(t)$. The procedures are fully presented in [6.11, 12]. Two equivalent approaches are in use, the "indirect" method and the "direct" method. In the former, an estimate of the autocovariance is first computed as

$$\hat{m}_x = \frac{1}{T} \int_{-T/2}^{T/2} x(t)dt\,, \tag{6.39}$$

$$\hat{C}_x(\tau) = \frac{1}{T} \int_{-T/2}^{T/2-\tau} [x(t+\tau)-\hat{m}_x][x(t)-\hat{m}_x]^* dt\,, \qquad 0 \leqq \tau \leqq T\,, \tag{6.40}$$

$$\hat{C}_x(\tau) = \hat{C}_x^*(-\tau)\,, \qquad -T \leqq \tau \leqq 0\,, \tag{6.41}$$

$$\hat{C}_x(\tau) = 0\,, \qquad |\tau| > T \tag{6.42}$$

(the alternative divisor $T-|\tau|$ in (6.40) is less desirable [6.12]), and then Fourier-transformed after weighting with an appropriate "lag window":

$$\hat{P}_x(f) = \int_{-T}^{T} \hat{C}_x(\tau) w(\tau) \exp(-j2\pi f\tau) d\tau \tag{6.43}$$

In the "direct" method, the "periodogram" is computed:

$$\mathscr{P}_x(f) = \frac{1}{T} \left| \int_{-T/2}^{T/2} [x(t)-\hat{m}_x] \exp(-j2\pi f t) dt \right|^2\,, \tag{6.44}$$

and then smoothed with an appropriate "frequency window":

$$\hat{P}_x(f) = \int_{-\infty}^{\infty} W(f-\lambda) \mathscr{P}_x(\lambda) d\lambda\,. \tag{6.45}$$

The estimators (6.43, 45) are identical, the choice being largely one of computational convenience in each case, with (6.45) being recently in favor because of the case with which the fast Fourier transform allows the periodogram to be computed. This is easily seen to be the case by carrying out straightforward changes of variable to show that the periodogram (6.44) is just the Fourier transform of the autocovariance estimate defined by (6.40, 41):

$$\mathscr{P}_x(f) = \int_{-\infty}^{\infty} \hat{C}_x(\tau) \exp(-j2\pi f\tau) d\tau\,, \tag{6.46}$$

taking account of the fact that $\hat{C}_x(\tau)=0$, $|\tau|>T$. Then extending the limits of integration in (6.43) to $\pm\infty$, and use of the complex convolution theorem,

$$\mathscr{F}[f(t)g(t)]=F(f)*G(f), \tag{6.47}$$

yields (6.45) where $W(f)$ is the Fourier transform of the lag window $w(\tau)$.

As more and more data are processed ($T\rightarrow\infty$), the estimator $\hat{C}_x(\tau)$ in (6.40, 41) approaches the true autocovariance function $C_x(\tau)$ in (6.23), in the sense that, for every τ, the mean of the random variable $\hat{C}_x(\tau)$ approaches $C_x(\tau)$ while its variance approaches zero. The effect of the window $w(\tau)$ on the spectral estimate $\hat{P}_x(f)$ in (6.43) is then judged in the first instance by its effect on the quantity

$$\tilde{P}_x(f)=\int_{-\infty}^{\infty} C_x(\tau)w(\tau)\exp(-j2\pi f\tau)d\tau, \tag{6.48}$$

where the effect of finite data length is accounted for by taking $w(\tau)=0$, $|\tau|>T$. The corresponding frequency-domain expression is

$$\tilde{P}_x(f)=\int_{-\infty}^{\infty} W(f-\lambda)P_x(\lambda)d\lambda. \tag{6.49}$$

Many possibilities are in use for $w(\tau)$. Provided

$$\begin{aligned} w(0)&=1,\\ w(-\tau)&=w(\tau),\\ w(\tau)&=0, \qquad |\tau|>T, \end{aligned} \tag{6.50}$$

and the window is reasonably smooth, the choices are grossly equivalent. The simplest choice, the rectangular window

$$w(\tau)=1, \qquad |\tau|\leqq T, \tag{6.51}$$

results in a spectrum smoothing function $W(f)$ with the narrowest central lobe [width from peak to first null of $1/2T$)] of all the conventional choices, and is thus capable of resolving more closely spaced peaks in the true spectrum $P_x(f)$. However, the rectangular window has the largest first minor lobe, which is also negative, thus introducing the possibility that a strong peak in the true spectrum may generate a negative power estimate at a nearby frequency. Another possible choice, the Bartlett window,

$$w(\tau)=1-|\tau|/T, \qquad |\tau|\leqq T, \tag{6.52}$$

avoids this latter difficulty, in that its transform $W(f)$ is positive everywhere, and has a considerably smaller minor lobe, but pays the penalty of a main lobe of width from peak to first null of $1/T$. A common general-purpose choice lately has been the Hanning window,

$$w(\tau)=\tfrac{1}{2}[1+\cos(\pi\tau/T)], \qquad |\tau|\leqq T, \tag{6.53}$$

which has a main lobe width (peak to first null) of $1/T$ also. The Kaiser window [6.13] is also widely used, but is not so simply expressed as (6.53).

All the various common window possibilities, in both analog and digital versions, are fully discussed by *Koopmans* [6.8], and *Jenkins* and *Watts* [6.12]. Recently *Harris* [6.13] has given a very extensive comparison of the different possibilities available. Any reasonable window will retain the virtue that, as T becomes large and $E[\hat{C}_x(\tau)]\to C_x(\tau)$, we will have $E[\hat{P}_x(f)]\to P_x(f)$. On the other hand, for finite T, the ability of the estimator $\hat{P}_x(f)$ to resolve peaks in $P_x(f)$ will be limited, in that peaks closer in frequency than $\Delta f=1/T$ are unlikely to be clearly resolved.

A second effect must enter into choice of the window $w(\tau)$. Just as the window used should not disturb the tendency of the mean of $\hat{P}_x$ to approach the true spectrum as more data are processed, the variance of the random variable $\hat{P}_x(f)$ should become small at each f as more data are processed ($T\to\infty$). As is now well appreciated, the fact that $\hat{C}_x(\tau)\to C_x(\tau)$ in an orderly way as $T\to\infty$ (variance $\to 0$) does not imply that

$$\mathscr{P}_x(f)=\mathscr{F}[\hat{C}_x(\tau)]\to\mathscr{F}[C_x(\tau)]=P_x(f)$$

in an equivalent orderly fashion. In fact, the variance of the periodogram $\mathscr{P}_x(f)$ does not generally become small as the data span increases, so that smoothing as in (6.45) is essential if the "direct" method is used. [This is simply to say that (6.44) is not equivalent to (6.43), even if the rectangular window is used.]

Under various simplifying assumptions, including the Gaussian assumption, it turns out [6.12] that the variance of the spectral estimate for fixed f behaves asymptotically for large T as

$$\mathrm{Var}[\hat{P}_x(f)]\sim\left[\frac{1}{T}\int_{-T}^{T} w^2(\tau)d\tau\right]P_x^2(f), \tag{6.54}$$

which does not generally approach zero for large T. The expression (6.54) does argue somewhat against the rectangular window, and in favor of windows which fall off to zero at $\tau=\pm T$, since the variance reduction factor (6.54) is then smaller.

The usual way to improve the variance reduction ratio (6.54) is to narrow the window $w(\tau)$:

$$\begin{aligned} w_1(\tau)&=w(\tau T/T_1), \qquad && |\tau|\leqq T_1,\\ &=0, && \text{elsewhere}. \end{aligned} \tag{6.55}$$

The ratio (6.54) then becomes

$$\mathrm{Var}[\hat{P}_x(f)]_1 \sim \frac{1}{T}\int_{-T_1}^{T_1} w^2(T\tau/T_1)d\tau = (T_1/T)\,\mathrm{Var}[\hat{P}_x(f)]\,, \tag{6.56}$$

a reduction factor of T_1/T. As penalty, the width of the frequency-domain window increases in the same ratio, and resolution deteriorates. In addition, the bias, i.e., the difference between $\mathrm{E}[\hat{P}_x(f)]$ and $P_x(f)$ at fixed f, increases. A compromise must be struck among bias, variance, and resolution. *Jenkins* and *Watts* [6.12] again give a thorough discussion of this issue.

A common procedure used to improve the variance properties of the estimator, as well as to obtain appropriate windowing, is through the use of smoothed periodograms. The data span $(-T/2, T/2)$ is segmented into N blocks of length T_1, and the corresponding periodograms computed, after removing an estimate of the mean, as in (6.44):

$$\mathscr{P}_x^{(i)}(f) = \frac{1}{T_1}\left|\int_{-T/2+(i-1)T_1}^{-T/2+iT_1} x(t)\exp(-\mathrm{j}2\pi f t)dt\right|^2, \qquad i=1,N\,. \tag{6.57}$$

The spectrum estimate is then taken to be

$$\hat{P}_x(f) = \frac{1}{N}\sum_{i=1}^{N}\mathscr{P}_x^{(i)}(f)\,. \tag{6.58}$$

The effect is that of using the Bartlett window with lag width $-T_1 \leqq \tau \leqq T_1$ so that variance reduction occurs in the ratio N, while still obtaining the time-domain averaging properties of the full data span $-T/2 \leqq t \leqq T/2$. *Oppenheim* and *Schafer* [6.14] discuss this procedure for the time-sampled data used in practice.

In dealing with space-time data, we have a whole family $x(t, z)$ of data signals to deal with. In the process of building an estimator of the frequency–wave-number spectrum (6.37), we must deal with the "multivariate" spectral analysis appropriate to multiple data waveforms. Specifically, in considering the random field $x(t, z)$, for each pair $z = z_1$, $z = z_2$, we consider the two time waveforms $x(t, z_1)$, $x(t, z_2)$. Rather than dealing with only one signal, and "univariate" spectral analysis, we now have to consider the cross-covariance function

$$m_x = \mathrm{E}[x(t, z)]\,, \tag{6.59}$$

$$C_x(\tau, z_1, z_2) = \mathrm{E}[[x(t+\tau, z_1) - m_x][x(t, z_2) - m_x]^*]\,, \tag{6.60}$$

and the cross-spectrum

$$P_x(f, z_1, z_2) = \int_{-\infty}^{\infty} C_x(\tau, z_1, z_2)\exp(-\mathrm{j}2\pi f\tau)d\tau\,, \tag{6.61}$$

which may be complex, since the cross-covariance function (6.60), in general, is not an even function of lag τ. All that has been summarized above applies unchanged here [6.12]. We compute

$$\hat{C}_x(\tau, z_1, z_2) = \frac{1}{T} \int_{-T/2}^{T/2-\tau} [x(t+\tau, z_1) - \hat{m}_x][x(t, z_2) - \hat{m}_x]^* dt, \qquad 0 \leqq \tau \leqq T, \tag{6.62}$$

with $\hat{C}_x(\tau, z_1, z_2)$, $-T \leqq \tau \leqq 0$, and $\hat{m}_x$ having the obvious definitions in analogy with (6.39, 41). Then the cross-spectrum is estimated as

$$\hat{P}_x(f, z_1, z_2) = \int_{-T}^{T} \hat{C}_x(\tau, z_1, z_2) w(\tau) \exp(-j2\pi f\tau) d\tau. \tag{6.63}$$

Alternatively, the direct method yields

$$\hat{P}_x(f, z_1, z_2) - \int_{-\infty}^{\infty} W(f-\lambda) \mathscr{P}_{12}(\lambda) d\lambda, \tag{6.64}$$

where

$$\mathscr{P}_{12}(f) = X_1(f) X_2^*(f), \tag{6.65}$$

and

$$X_i(f) = \frac{1}{\sqrt{T}} \int_{-T/2}^{T/2} [x(t, z_i) - \hat{m}_x] \exp(-j2\pi ft) dt. \tag{6.66}$$

Segmenting of the data and averaging of the periodograms calculated on each segment may also be effective in providing variance reduction using the direct method.

6.2.3 Linear Estimation of the Frequency–Wave-Number Spectrum

We seek to estimate the frequency–wave-number spectrum (6.37) of a time–space stationary ("homogeneous") random field $x(t, z)$. We can write down an estimator at once, in analogy with the preceding discussion:

$$\hat{S}_x(f, \boldsymbol{v}) = \int_{-T}^{T} \int_{V} \hat{C}_x(\tau, \boldsymbol{r}) w(\tau, \boldsymbol{r}) \exp[-j2\pi(f\tau + \boldsymbol{v} \cdot \boldsymbol{r})] d\boldsymbol{r} d\tau, \tag{6.67}$$

where V represents the domain in $\boldsymbol{r}$-space over which values $C_x(\tau, \boldsymbol{r})$ are available, and $w(\tau, \boldsymbol{r})$ is a window in four dimensions. Here, we have

$$\hat{C}_x(\tau, \boldsymbol{r}) = \frac{1}{T}\frac{1}{V_L}\int_{-T/2}^{T/2-\tau}\left(\prod_{i=1}^{3}\int_{-L_i/2}^{L_i/2-r_i}\right) \cdot [x(t+\tau, \boldsymbol{z}+\boldsymbol{r})-\hat{m}_x][x(t, \boldsymbol{z})-\hat{m}_x]^* dV_L dt\,, \qquad 0 \leqq \tau \leqq T, \quad 0 \leqq r_i \leqq L_i\,, \tag{6.68}$$

where L_i is the maximum physical dimension of the apparatus in space-coordinate i, and $\hat{m}_x$ and the segments of $\hat{C}_x$ for $-T \leqq \tau \leqq 0$, $-L_i \leqq r_i \leqq 0$, have the obvious definitions in analogy with (6.39, 41). The equivalent direct estimator is

$$\hat{S}_x(f, \boldsymbol{\nu}) = \int_{-\infty}^{\infty}\int_V W(f-\lambda, \boldsymbol{\nu}-\boldsymbol{\xi})\mathscr{P}_x(\lambda, \boldsymbol{\xi})d\boldsymbol{\xi}d\lambda\,, \tag{6.69}$$

with the periodogram being

$$\mathscr{P}_x(f, \boldsymbol{\nu}) = \frac{1}{T}\frac{1}{V_L}\left|\int_{-T/2}^{T/2}\int_{-V_L/2}^{V_L/2}[x(t, \boldsymbol{z})-\hat{m}_x] \cdot \exp[-j2\pi(ft+\boldsymbol{\nu}\cdot\boldsymbol{z})]dVdt\right|^2. \tag{6.70}$$

The smoothing (6.69) may also be done by segmenting the data and smoothing periodograms.

The estimators (6.67, 69) will have variance, bias, and resolution qualities in frequency and wave number which are "good" insofar as the data time span T and apparatus spatial extent L are "large". [L is some measure of the aggregate of L_1, L_2, L_3 in (6.68), the apparatus extent in the three dimensions of physical space.] Our assumption is that T will be adequately large to serve, but L will not. That is, we assume that, for each frequency of interest, many more oscillation periods are contained in the interval T than there are wavelengths across L at the various wave numbers of interest. Thus, linear processing will serve in the time domain, but not necessarily in the spatial domain. Accordingly, in considering nonlinear spectral estimation algorithms, we want to suppress the time domain.

Let us make the reasonable assumption that the time–space window $w(\tau, \boldsymbol{r})$ in (6.67) factors as

$$w(\tau, \boldsymbol{r}) = w_1(\tau)w_2(\boldsymbol{r})\,. \tag{6.71}$$

Then the four-dimensional Fourier transform window in frequency and wave number also factors as

$$W(f, \boldsymbol{\nu}) = W_1(f)W_2(\boldsymbol{\nu})\,. \tag{6.72}$$

Let us write the periodogram (6.70) as

$$\mathscr{P}_x(f,\nu)=\frac{1}{V_L}\left|\int_{-V_L/2}^{V_L/2} X(f,z)\exp(-j2\pi\nu\cdot z)dz\right|^2, \tag{6.73}$$

where

$$X(f,z)=\frac{1}{\sqrt{T}}\int_{-T/2}^{T/2}[x(t,z)-\hat{m}_x]\exp(-j2\pi ft)dt \tag{6.74}$$

is the Fourier coefficient (6.18) of the signal $x(t,z)$, after correcting for the mean. Using (6.72, 73) in (6.69), we obtain

$$\hat{S}_x(f,\nu)=\frac{1}{V_L}\int_V W_2(\nu-\xi)\int_{-V_L/2}^{V_L/2}\int_{-V_L/2}^{V_L/2}\cdot\left[\int_{-\infty}^{\infty} W_1(f-\lambda)X(\lambda,z_1)X^*(\lambda,z_2)d\lambda\right]\cdot\exp[-j2\pi\xi\cdot(z_1-z_2)]dz_1dz_2d\xi. \tag{6.75}$$

We recognize from (6.64–66) that the bracketed quantity in (6.75) is the smoothed-periodogram estimator of the cross-spectrum between $x(t,z_1)$ and $x(t,z_2)$, that is

$$\int_{-\infty}^{\infty} W_1(f-\lambda)X(\lambda,z_1)X^*(\lambda,z_2)d\lambda=\hat{P}_x(f,z_1,z_2)\approx P_x(f,z_1,z_2), \tag{6.76}$$

with $P_x(f,z_1,z_2)$ as in (6.61). Thus we need to compute

$$\hat{S}_x(f,\nu)=\int_V W_2(\nu-\xi)\hat{\mathscr{S}}_x(f,\xi)d\xi, \tag{6.77}$$

where

$$\hat{\mathscr{S}}_x(f,\nu)=\frac{1}{V_L}\int_{-V_L/2}^{V_L/2}\int_{-V_L/2}^{V_L/2}\hat{P}_x(f,z_1,z_2)\exp[-j2\pi\nu\cdot(z_1-z_2)]dz_1dz_2. \tag{6.78}$$

In (6.77), V indicates all of 3-space, while in (6.78), V_L indicates the cube of 3-space which bounds the physical extent of the apparatus.

Henceforth we focus attention on the spatial processing in (6.77, 78). We assume sufficient data span T that, with appropriate windowing and data segmentation, $\hat{P}_x$ in (6.78) may be replaced by the true cross-spectrum P_x

$$\hat{\mathscr{S}}_x(f,\nu)\approx\frac{1}{V_L}\int_{-V_L/2}^{V_L/2}\int_{-V_L/2}^{V_L/2}P_x(f,z_1,z_2)\exp[-j2\pi\nu\cdot(z_1-z_2)]dz_1dz_2. \tag{6.79}$$

In the case of a homogeneous field, we will have

$$P_x(f, z_1, z_2) = P_x(f, z_1 - z_2),$$

in which case (6.79), after the usual change of variable $r = z_1 - z_2$, and interchange of order of integration [6.12], may be rewritten as

$$\hat{\mathscr{P}}_x(f, \nu) \approx \int_{-V_L/2}^{V_L/2} w_B(r) P_x(f, r) \exp(-j2\pi\nu \cdot r) dr, \qquad (6.80)$$

where $w_B(r)$ is the three-dimensional Bartlett window. Thus, if the spatial extent is adequate so that the Bartlett window serves for smoothing, we may bypass the smoothing step (6.77), and take the final spectrum as

$$\hat{S}_x(f, \nu) = \hat{\mathscr{P}}_x(f, \nu) \qquad (6.81)$$

directly, where $\hat{\mathscr{P}}_x(f, \nu)$ is as in (6.78). In case the apparatus spatial extent is inadequate, so that the resolution and bias properties of (6.80) are unsatisfactory, we must seek alternative estimators by moving to nonlinear processing.

6.2.4 A Word on Sampled Data

For convenience of exposition (or perhaps out of personal habit of thought) we have written the above summary in the language of continuous time and space variables. In practice, the sensor signals $x(t, z)$ will almost certainly be time sampled before any computations are performed. Thus, the samples $x(l\Delta t, z)$ will be available for a sensor located at space point z. We assume that suitable prefiltering before sampling is used, and that the sampling rate is adequate to avoid aliasing. These matters are fully discussed in many texts, e.g., [6.15], and we shall not elaborate here. All the time/frequency spectral estimators discussed above will then be computed at discrete frequencies, in the sampled-data versions. The appropriate formulas are written out in [6.12], and discussed with special reference to use of the fast Fourier transform algorithm in [6.14]. The spectral estimators at two different discrete frequencies are uncorrelated, for adequately long data span.

In the space domain, the signal $x(t, z)$ will also be sampled, in that the sensors, be they seismometers, hydrophones, or local antennas in a long-baseline interferometer, are considered to be each located at a discrete space point. Under this assumption, we have available the sampled data $x(l\Delta t, z_i)$. If the sensor array is uniformly spaced along each directional coordinate, the standard formulas of *Jenkins* and *Watts* [6.12], and *Oppenheim* and *Schafer* [6.14] may simply be read off in the spatial domain, and the linear computation of the wave number spectrum proceeds in analogy with the frequency spectrum.

In fact, in some common cases uniformly spaced arrays are used, but by no means always. Hence we will retain as long as possible the assumption of arbitrarily positioned sensors.

One particular formula is worth writing down explicitly in the discrete-space version. Suppose sensor signals $x(t, z_i)$, $i=1, K$, are time analyzed to produce the spectral estimates $\hat{P}_x(f, z_i, z_j)$, $i, j=1, K$, as in (6.78). Typically these are computed using time segmentation to provide (frequency) spectral smoothing:

$$\hat{P}_x(f, z_i, z_j) = \frac{1}{N} \sum_{n=1}^{N} X_i^{(n)}(f) X_j^{(n)*}(f), \qquad i, j = 1, K, \tag{6.82}$$

where f is one of the frequencies of the time-frequency analysis, and

$$X_i^{(n)}(f) = \frac{1}{\sqrt{T_N}} \int_{-T_N/2}^{T_N/2} x^{(n)}(t, z_i) \exp(-\mathrm{j}2\pi f t) dt, \tag{6.83}$$

or the sampled-data version thereof, where $x^{(n)}(t, z_i)$ indicates the segment of $x(t, z_i)$ on the nth time segment, of length $T_N = T/N$, after removal of some estimate $\hat{m}_x$ of the mean. Then the discrete version of the estimator (6.78) is often simply taken as [considering also (6.81)]

$$\hat{S}_x(f, v) = \sum_{i=1}^{K} \sum_{j=1}^{K} \hat{P}_x(f, z_i, z_j) \eta_i^*(v) \eta_j(v), \tag{6.84}$$

where we define

$$\eta_i(v) = w_i \exp(\mathrm{j}2\pi v \cdot z_i), \qquad \mathrm{i} = 1, K, \tag{6.85}$$

with w_i being available for adjustment in some way to improve the estimator properties. Often $w_i = 1$ is used. The estimator (6.84) is the delay-and-sum "beam-former," commonly used in sonar signal processing. *Baggeroer* [6.16] has recently summarized some of its practical implementations, as well as given a brief discussion of the frequency–wave-number spectral analysis procedures considered above. The estimator (6.84) was also discussed in depth by *Lacoss* et al. [6.17].

Let us now turn to the two most popular ways of determining estimates of the wave number spectrum with higher resolution than the linear estimators (6.78) or (6.84).

6.3 The Maximum-Likelihood Processor in the Spatial Domain

The spectral estimator which is now called the maximum-likelihood array processor grew out of data processing work associated with the large aperture seismic array (LASA). In the early 1960s, it became important to the national defense that there exist a capability to discriminate distant underground nuclear blast tests from natural earth tremors. It was necessary to detect the occurrence of an underground pressure wave, to determine the direction from which the wave impinged on the detection site, and to delineate clearly the time waveform of the initial few seconds of the shock wave. Accordingly, the U.S. Defense Advanced Research Projects Agency sponsored the construction, in eastern Montana, of a spatially distributed array of 525 seismometers, spread over an aperture of 200 km. A general description of this Large Aperture Seismic Array (LASA) sensor system, and the problem it was designed to solve, is given by *Green* et al. [6.18] and the general scheme of the associated data processing was reviewed by *Briscoe* and *Fleck* [6.19].

The 525 sensors of LASA were arranged in 21 subarrays of 25 sensors each. Each subarray had an aperture of 7 km. The blast pressure wave, the signal of interest, occurred in a background of reverberation, interfering signals, and more or less spatially organized microseismic noise. The noise background was expected to be essentially incoherent (spatially disorganized) over the full 200 km aperture, but to have significant spatial correlation over the 7 km aperture of a subarray. Hence, simple delay-and-sum (beam-former) processing was proposed for combining the subarray outputs, but more sophisticated processing was proposed for the 25 sensor signals of a given subarray. This latter processing is described by *Capon* et al. [6.20].

The problem treated by *Capon* et al. [6.20] was a standard time-domain problem in communication theory, with the added complexity that multiple data signals were to be processed. However, no spatial-domain processing, other than beam forming, was used at the time. Only later, in the important paper of *Capon* [6.21], did maximum-likelihood processing evolve in the spatial domain, and then not in fact in a maximum-likelihood setting, but rather in the context of the spatial estimation procedure we have been discussing. Here we will summarize the (linear) processing results of [6.20] in order to clarify why the application in [6.21] leads to a nonlinear processor. We will then discuss some important properties of the nonlinear processor. Later on in the section, we will discuss some closely related "adaptive" array processors which evolved independently.

6.3.1 Linear Maximum-Likelihood Processing

The LASA data-processing problem, as initially posed [6.20], was as follows. A deterministic plane traveling pressure wave is assumed to impinge on the sensor subarray from a known direction, described by a unit vector $\boldsymbol{u}$, and to propagate at a known speed v, the direction and speed having been preliminarily determined by linear beam forming over a sufficiently large (up to the full 200 km) aperture. Let the pressure wave signal, at the coordinate origin, be called $s(t)$. Then a subarray sensor at position z relative to the origin receives the signal as $s(t+\tau)$, where $\tau=(\boldsymbol{u}\cdot z)/v$. Let the sensor signals be delayed by this amount in time, so that each sensor with delay then receives just $s(t)$. Assume that this deterministic signal is immersed in a background of additive noise. Then the sensor signals are

$$x(t, z_k)=x_k(t)=s(t)+n_k(t), \qquad k=1, K. \tag{6.86}$$

Assume that these data signals are time sampled at $t_l=l\Delta t$, $l=1$, q, yielding

$$\mathbf{x}=\mathrm{col}\{x_1(t_1), \ldots, x_K(t_1), x_1(t_2), \ldots, x_K(t_q)\}, \tag{6.87}$$

and similarly for the $n_k(t)$. Let the vector $\mathbf{s}$ be the column of the time samples $s(l\Delta t)$, $l=1, q$. We then have the model

$$\mathbf{x}=\mathbf{H}\mathbf{s}+\mathbf{n}, \tag{6.88}$$

where the $Kq\times q$ dimensional "modulator matrix" is

$$\mathbf{H}=\begin{bmatrix}\mathbf{1} & \mathbf{0} \ldots & \mathbf{0}\\ \mathbf{0} & \mathbf{1} & \mathbf{0}\\ & \ldots & \\ \mathbf{0} & \mathbf{0} & \mathbf{1}\end{bmatrix} \tag{6.89}$$

in which $\mathbf{1}$ and $\mathbf{0}$ indicate the K-columns all of whose elements are respectively unity or zero. [The alternative ordering of the $x_k(t_l)$ simply changes the form of the modulator matrix.]

We now want an estimator $\hat{\mathbf{s}}(\mathbf{x})$ of the vector of time samples of the signal $s(t)$. One standard procedure is to use the maximum-likelihood estimator, obtained by maximizing the conditional probability of the data vector:

$$\hat{\mathbf{s}}\Leftarrow \max_{\mathbf{s}} p(\mathbf{x}|\mathbf{s}). \tag{6.90}$$

Deutsch [6.22] has discussed the properties of this maximum-likelihood estimator at this level of generality. The estimator is typically nonlinear, except in the case assumed in [6.20] that the noise vector $\mathbf{n}$ is Gaussian:

$$p_n(\mathbf{n})=(2\pi)^{-Kq/2}|\mathbf{N}|^{-1/2}\exp(-\tfrac{1}{2}\mathbf{n}'\mathbf{N}^{-1}\mathbf{n}). \tag{6.91}$$

Here we assume zero-mean noise, for simplicity. The $\mathbf{N}$ is the covariance matrix of $\mathbf{n}$. The "prime" indicates vector transpose, with conjugation in the case of a complex vector; $|\mathbf{N}|=\det(\mathbf{N})$, which is assumed nonzero. From (6.88, 90) we find that

$$p(\mathbf{x}|\mathbf{s})=p_n(\mathbf{x}-\mathbf{H}\mathbf{s}), \tag{6.92}$$

$$\max_{\mathbf{s}} p(\mathbf{x}|\mathbf{s}) \Rightarrow \min_{\mathbf{s}}(\mathbf{x}-\mathbf{H}\mathbf{s})'\mathbf{N}^{-1}(\mathbf{x}-\mathbf{H}\mathbf{s}), \tag{6.93}$$

$$(\mathbf{x}-\mathbf{H}\hat{\mathbf{s}})'\mathbf{N}^{-1}(-\mathbf{H})=\mathbf{0}, \tag{6.94}$$

$$\hat{\mathbf{s}}=(\mathbf{H}'\mathbf{N}^{-1}\mathbf{H})^{-1}\mathbf{H}'\mathbf{N}^{-1}\mathbf{x}. \tag{6.95}$$

[Since $\mathbf{H}$ is of full rank, and $|\mathbf{N}|\neq 0$, the inverse indicated in (6.95) exists.]

In this latter manipulation, we have used the common formalism in which, in differentiating to determine the singular point of a real scalar function of a real (respectively complex) vector, we assume the vector and its transpose (respectively transpose conjugate) to be independent variables. Thus, in extremizing a real scalar function $J(\mathbf{s},\mathbf{s}')$, we require both $\partial J/\partial\mathbf{s}=0$ and $\partial J/\partial\mathbf{s}'=0$. Since both of these lead to the same necessary conditions on $\mathbf{s}$, we need to consider only $\partial J/\partial\mathbf{s}=0$. This procedure is primarily useful in dealing with a real function of a complex variable and its complex conjugate, and is briefly discussed in that context in the text by *Ahlfors* [6.23]. We shall further use freely the easily verified derivative formulas:

$$\partial(\mathbf{a}'\mathbf{s})/\partial\mathbf{s} \triangleq \underset{i}{\text{row}}\, \partial(\mathbf{a}'\mathbf{s})/\partial s_i=\mathbf{a}', \tag{6.96}$$

$$\partial(\mathbf{a}'\mathbf{S}\mathbf{b})/\partial\mathbf{S} \triangleq \text{matrix}\{\partial(\mathbf{a}'\mathbf{S}\mathbf{b})/\partial s_{ij}\}=\mathbf{b}\mathbf{a}', \tag{6.97}$$

$$\partial(\mathbf{A}\mathbf{s})/\partial\mathbf{s} \triangleq \text{matrix}\,\{\partial(\mathbf{A}\mathbf{s})_i/\partial s_j\}=\mathbf{A}. \tag{6.98}$$

If the assumption of Gaussian noise is not warranted, the maximum-likelihood estimate may be difficult to compute in closed form, and will generally be a nonlinear function of the data. Common practice is then to resort to "minimum-variance unbiased" estimation, which results by changing the criterion of "goodness" of the desired estimator. Rather than requiring that the estimator $\hat{\mathbf{s}}(\mathbf{x})$ maximize $p(\mathbf{x}|\mathbf{s})$, we now require:

1) that $\hat{\mathbf{s}}$ be linear,

$$\hat{\mathbf{s}}=\mathbf{L}\mathbf{x}+\boldsymbol{\alpha}; \tag{6.99}$$

2) that the mean error be zero (the unbiasedness requirement),

$$\mathrm{E}[\hat{\mathbf{s}}]=\mathbf{s}; \quad \text{and} \tag{6.100}$$

3) that, under these constraints, the variance of the error be minimized,

$$\hat{\mathbf{s}} \Leftarrow \min_{\mathbf{s}} \mathrm{E}[(\mathbf{e}-\mathrm{E}[\mathbf{e}])'(\mathbf{e}-\mathrm{E}[\mathbf{e}])], \tag{6.101}$$

where $\mathbf{e}=\mathbf{s}-\hat{\mathbf{s}}(\mathbf{x})$.

The unbiasedness constraint requires

$$\begin{aligned}\mathrm{E}[\hat{\mathbf{s}}] &= \mathbf{L}(\mathrm{E}[\mathbf{x}]) + \boldsymbol{\alpha} = \mathbf{L}\mathrm{E}[\mathbf{Hs} + \mathbf{n}] + \boldsymbol{\alpha} \\ &= \mathbf{LHs} + \mathbf{L}\mathrm{E}[\mathbf{n}] + \boldsymbol{\alpha} = \mathbf{s}. \end{aligned} \tag{6.102}$$

Since (6.102) is to hold for arbitrary **s**, we require (**J** is the unit matrix)

$$\mathbf{LH} = \mathbf{J}, \qquad \boldsymbol{\alpha} = -\mathbf{L}\mathrm{E}[\mathbf{n}] = 0. \tag{6.103}$$

We satisfy the second of these in general by writing

$$\hat{\mathbf{s}} = \mathbf{L}(\mathbf{x} - \mathrm{E}[\mathbf{n}]), \tag{6.104}$$

and we introduce the first via Lagrange multiplier techniques.

At this point, even vector notation becomes somewhat inelegant, in that we must partition the estimator matrix

$$\mathbf{L} = \underset{i}{\mathrm{col}}(\mathbf{l}_i'), \tag{6.105}$$

where $\mathbf{l}_i'$ is row i of the matrix **L**. The unbiasedness constraint then partitions as

$$\mathbf{LH} = \begin{bmatrix} \mathbf{l}_1' \\ \vdots \\ \mathbf{l}_q' \end{bmatrix} \mathbf{H} = \begin{bmatrix} \mathbf{l}_1'\mathbf{H} \\ \vdots \\ \mathbf{l}_q'\mathbf{H} \end{bmatrix} = \mathbf{J} = \begin{bmatrix} \mathbf{e}_1' \\ \vdots \\ \mathbf{e}_q' \end{bmatrix}, \tag{6.106}$$

where $\mathbf{e}_i$ is the column vector with unity in position i and zeros elsewhere. We thus have the constraints

$$\mathbf{l}_i'\mathbf{H} = \mathbf{e}_i', \qquad i = 1, q. \tag{6.107}$$

The error variance can be written

$$\begin{aligned}\sigma_e^2 &= \mathrm{E}[(\mathbf{e} - \mathrm{E}[\mathbf{e}])'(\mathbf{e} - \mathrm{E}[\mathbf{e}])] = \mathrm{E}[\{[\mathbf{s} - \mathbf{L}(\mathbf{x} - \mathrm{E}[\mathbf{n}])] - \mathrm{E}[\text{same}]\}'\{\text{same}\}] \\ &= \mathrm{E}[\{[\mathbf{s} - \mathbf{L}(\mathbf{Hs} + \mathbf{n} - \mathrm{E}[\mathbf{n}])] - [\mathbf{s} - \mathbf{LHs}]\}'\{\text{same}\}] \\ &= \mathrm{E}[\{\mathbf{L}(\mathbf{n} - \mathrm{E}[\mathbf{n}])\}'\{\text{same}\}] = \mathrm{tr}\,\mathbf{L}\{\mathrm{E}[(\mathbf{n} - \mathrm{E}[\mathbf{n}])(\mathbf{n} - \mathrm{E}[\mathbf{n}])']\}\mathbf{L}' \\ &= \mathrm{tr}(\mathbf{LNL}') = \mathrm{tr}\begin{bmatrix} \mathbf{l}_1' \\ \vdots \\ \mathbf{l}_q' \end{bmatrix} \mathbf{N}[\mathbf{l}_1, \ldots, \mathbf{l}_q] \\ &= \sum_{i=1}^{q} \mathbf{l}_i'\mathbf{N}\mathbf{l}_i, \end{aligned} \tag{6.108}$$

where "tr" indicates the matrix trace operation.

Adjoining the constraints (6.106) to the "cost" (6.108), we consider the augmented cost function

$$J=\sum_{i=1}^{q} \mathbf{l}_i'\mathbf{N}\mathbf{l}_i+\sum_{i=1}^{q}(\mathbf{l}_i'\mathbf{H}-\mathbf{e}_i')\boldsymbol{\lambda}_i, \tag{6.109}$$

where $\boldsymbol{\lambda}_i$ is a q-vector of Lagrange multipliers. We then obtain the necessary conditions (also sufficient, since $\mathbf{N}>0$):

$$J_{\mathbf{l}_k}=\mathbf{0}=\mathbf{N}\mathbf{l}_k+\mathbf{H}\boldsymbol{\lambda}_k, \qquad k=1,q. \tag{6.110}$$

Collecting these, we have

$$\begin{aligned}\mathbf{0}&=\mathbf{N}(\mathbf{l}_1,\dots,\mathbf{l}_q)+\mathbf{H}(\boldsymbol{\lambda}_1,\dots,\boldsymbol{\lambda}_q)\\&=\mathbf{N}\mathbf{L}'+\mathbf{H}\boldsymbol{\Lambda}',\end{aligned} \tag{6.111}$$

where we define the matrix

$$\boldsymbol{\Lambda}=\mathrm{col}(\boldsymbol{\lambda}_i'), \qquad i=1,q. \tag{6.112}$$

From (6.111), we have

$$\mathbf{L}=\boldsymbol{\Lambda}\mathbf{H}'\mathbf{N}^{-1}. \tag{6.113}$$

The Lagrange multiplier follows from

$$\begin{aligned}\mathbf{L}\mathbf{H}&=\mathbf{J}=\boldsymbol{\Lambda}\mathbf{H}'\mathbf{N}^{-1}\mathbf{H},\\ \boldsymbol{\Lambda}&=(\mathbf{H}'\mathbf{N}^{-1}\mathbf{H})^{-1},\end{aligned} \tag{6.114}$$

so that finally

$$\mathbf{L}=(\mathbf{H}'\mathbf{N}^{-1}\mathbf{H})^{-1}\mathbf{H}'\mathbf{N}^{-1}, \tag{6.115}$$

$$\hat{\mathbf{s}}=(\mathbf{H}'\mathbf{N}^{-1}\mathbf{H})^{-1}\mathbf{H}'\mathbf{N}^{-1}(\mathbf{x}-\mathrm{E}[\mathbf{n}]), \tag{6.116}$$

which is exactly the maximum-likelihood estimator obtained in the case of Gaussian noise, a well-known result.

The Lagrange multiplier (6.114) is of independent interest, because the attained minimum-error variance is, from (6.108, 115, 114),

$$\sigma_e^2=\mathrm{E}[\mathbf{e}'\mathbf{e}]=\mathrm{tr}(\mathbf{L}\mathbf{N}\mathbf{L}')=\mathrm{tr}(\mathbf{H}'\mathbf{N}^{-1}\mathbf{H})^{-1}=\mathrm{tr}\,\boldsymbol{\Lambda}. \tag{6.117}$$

A computational advantage accompanies the ordering (6.87), as follows. The equations which we need to solve, for the estimator coefficient matrix **L** and the attained error covariance matrix Λ, namely, (6.103, 111), are

$$\begin{aligned} \mathbf{N}\mathbf{L}' + \mathbf{H}\Lambda' &= \mathbf{0}\,, \\ \mathbf{H}'\mathbf{L}' &= \mathbf{J}\,. \end{aligned} \tag{6.118}$$

In the case of the ordering (6.87), with q time samples and K sensors, we write out the rows of the matrix $\mathbf{L}$ as

$$\mathbf{l}_i' = (\mathbf{l}_{i1}', \ldots, \mathbf{l}_{iq}'), \qquad i = 1, q, \tag{6.119}$$

where $\mathbf{l}_{ij}$ is the K-vector of coefficients to be applied to the data $\{x_1(t_j), \ldots, x_K(t_j)\}$ in estimating $s(t_i)$. Then using the partitioning (6.89) of $\mathbf{H}$, the equations (6.118) may be merged as

$$\left[\begin{array}{cc|cc|c|cc} \mathbf{N}_{11} & \mathbf{1} & \mathbf{N}_{12} & \mathbf{0} & \cdots & \mathbf{N}_{1q} & \mathbf{0} \\ \mathbf{1}' & 0 & \mathbf{0}' & 0 & & \mathbf{0}' & 0 \\ \hline \mathbf{N}_{12}' & \mathbf{0} & \mathbf{N}_{22} & \mathbf{1} & \cdots & \mathbf{N}_{2q} & \mathbf{0} \\ \mathbf{0}' & 0 & \mathbf{1}' & 0 & & \mathbf{0}' & 0 \\ \hline & \vdots & & \vdots & & & \vdots \\ \hline \mathbf{N}_{1q}' & \mathbf{0} & \mathbf{N}_{1,q-1}' & \mathbf{0} & \cdots & \mathbf{N}_{qq} & \mathbf{1} \\ \mathbf{0}' & 0 & \mathbf{0}' & 0 & & \mathbf{1}' & 0 \end{array}\right] \left[\begin{array}{cc} \mathbf{l}_{11} & \mathbf{l}_{21} \ldots \mathbf{l}_{q1} \\ \lambda_{11} & \lambda_{21} \ldots \lambda_{q1} \\ \hline \mathbf{l}_{12} & \mathbf{l}_{22} \ldots \mathbf{l}_{q2} \\ \lambda_{12} & \lambda_{22} \ldots \lambda_{q2} \\ \hline & \vdots \\ \hline \mathbf{l}_{1q} & \mathbf{l}_{2q} \ldots \mathbf{l}_{qq} \\ \lambda_{1q} & \lambda_{2q} \ldots \lambda_{qq} \end{array}\right] = \left[\begin{array}{ccc} \mathbf{0} & \mathbf{0} & \mathbf{0} \\ 1 & 0 \ldots & 0 \\ \hline \mathbf{0} & \mathbf{0} \ldots & \mathbf{0} \\ 0 & 1 & 0 \\ \hline & & \\ \hline \mathbf{0} & \mathbf{0} \ldots & \mathbf{0} \\ 0 & 0 \ldots & 1 \end{array}\right] \tag{6.120}$$

Here $\mathbf{N}_{ij}$ is the $K \times K$ matrix with elements

$$\mathbf{N}_{ij}(k, l) = \mathrm{E}[[n_k(t_i) - \mathrm{E}[n_k(t_i)]]\,[n_l(t_j) - \mathrm{E}[n_l(t_j)]]^*] \tag{6.121}$$

$\mathbf{1}$ und $\mathbf{0}$ are the q-columns all of whose elements are, respectively, unity and zero, $\mathbf{l}_{ij}$ are the K-columns in (6.119), and the scalars $\lambda_{ij} = \lambda_{ji}$ are the elements of the $q \times q$ matrix Λ. The point now is that, if the noise field is time stationary, the matrix in the left in (6.120) is block-Toeplitz, and the Levinson algorithm [6.24] may be used to effect a significant computational saving.

The scheme sketched out above proceeds directly in the time domain. For that reason, the matrix $\mathbf{N}$ in (6.116), or its block components (6.121), are not "sparse" in any sense; the element (6.121) is simply the cross-covariance, at lag $t_i - t_j$, of the sensor noise waveforms $x_k(t)$, $x_l(t)$. As *Capon* et al. [6.20] point out, a significant decoupling occurs with passage to the frequency domain, since the power-spectral densities ($k = l$) and cross-spectra ($k \neq l$),

$$P_{kl}(f) = \int_{-\infty}^{\infty} \mathrm{E}[n_k(t+\tau) n_l(t)] \exp(-\mathrm{j}2\pi f \tau) d\tau\,, \tag{6.122}$$

of the noise signals are uncorrelated at different frequencies, as are the corresponding estimators, for adequately large time span of data [6.12].

Suppose then that the sensor signals $x_k(t)$, after time sampling, are Fourier transformed to produce Fourier coefficients $X_k(l\Delta f)$, where $\Delta f = 1/T$, for $l \approx -q/2, q/2$.

Let these be ordered into a Kq column as

$$\mathbf{X} = \mathrm{col}\{X_1(-q\Delta f/2), \ldots, X_K(-q\Delta f/2), \ldots, X_K(q\Delta f/2)\}, \tag{6.123}$$

corresponding to (6.87). Then, we proceed as before, with the model (6.88) being read in terms of Fourier coefficients

$$\mathbf{X} = \mathbf{HS} + \mathbf{N}, \tag{6.124}$$

and the noise covariance matrix $\mathbf{N}$ in (6.121) becoming the noise cross-spectral matrix

$$\mathcal{N} = \mathrm{E}[\mathbf{NN}'], \tag{6.125}$$

where now $\mathbf{N}$ is a Fourier-coefficient vector. Since $\mathcal{N}$ is block diagonal,

$$\mathrm{E}[N_k(m\Delta f)N_l^*(n\Delta f)] = 0, \qquad m \neq n, \tag{6.126}$$

the estimator (6.116) decouples in the frequency domain:

$$\hat{S}_l = (\mathbf{1}'\mathcal{N}_l^{-1}\mathbf{1})^{-1}\mathbf{1}'\mathcal{N}_l^{-1}\mathbf{X}_l, \qquad l \approx -q/2, q/2, \tag{6.127}$$

Here, we have

$$\text{scalar}: \hat{S}_l = \text{estimate of } S(l\Delta f); \tag{6.128}$$

$$K\text{-vector}: \mathbf{X}_l = \underset{k}{\mathrm{col}}\, X_k(l\Delta f),\ k = 1, K; \tag{6.129}$$

$$K\text{-vector}: \mathbf{1}' = \mathrm{col}(1, 1, \ldots, 1); \tag{6.130}$$

$$K \times K\ \text{matrix}: (\mathcal{N}_l)_{ij} = \mathrm{E}[N_i(l\Delta f)N_j^*(l\Delta f)]. \tag{6.131}$$

In scalar form, (6.127) becomes

$$\hat{S}(l\Delta f) = \sum_{i,j=1}^{K} (\mathcal{N}_l^{-1})_{ij} X_j(l\Delta f) \Big/ \sum_{i,j=1}^{K} (\mathcal{N}_l^{-1})_{ij}, \tag{6.132}$$

which is essentially the frequency-domain procedure of [6.20]. The largest inversion involved is now K by K. Levinson's algorithm can be applied to the decoupled problems.

In the LASA application discussed in [6.20] a sufficient time span of signal-free data is available before the arrival of the event of interest, the pressure wave represented above by $s(t)$. During the signal-free data collection interval, the noise matrix $\mathcal{N}$ is estimated, so that the estimator (6.132) can be synthesized. Again, the processing is linear, and it is a solution of the signal-estimation problem of communication theory in a multidimensional data setting. We turn now to the nonlinear estimator which is of primary interest to us.

6.3.2 The (Nonlinear) Maximum-Likelihood Array Processor

In the original discussion of maximum-likelihood processing for the LASA given in [6.20], the signal wave producing $s(t)$ was assumed to be planar, to propagate at a known speed v, and to arrive from direction $\boldsymbol{u}$ with respect to some origin. The data signal at a sensor at position z_k was then delayed by a time $\tau_k = \boldsymbol{u} \cdot z_k / v$, and the result called $x_k(t)$. The modulation matrix $\mathbf{H}$ in (6.89) resulted. Equivalently, we may let the undelayed sensor signals themselves be called $x_k(t)$, and account for the signal propagation delays by redefining the modulator matrix in a simple way, provided we use the frequency domain approach. We simply need to write the components of the Fourier-coefficient vector (6.124) as

$$X_k(l\Delta f) = \eta_k(l\Delta f, \boldsymbol{v}) S(l\Delta f, \boldsymbol{v}), \qquad k = 1, K, \tag{6.133}$$

where

$$\begin{aligned} \eta_k(l\Delta f, \boldsymbol{v}) &= \exp(\mathrm{j}2\pi f_l \boldsymbol{v} \cdot z_k / |\boldsymbol{v}| v) \\ &= \exp(\mathrm{j}2\pi \boldsymbol{v} \cdot z_k), \end{aligned} \tag{6.134}$$

since

$$|\boldsymbol{v}| = 1/\lambda, \qquad f_l \lambda = v, \tag{6.135}$$

with f_l the frequency $l\Delta f$ of interest, and λ_l the associated wavelength. Here $s(t)$ is the desired signal, measured at the coordinate origin. Writing the known direction vector $\boldsymbol{u}$ in terms of wave number simply indicates that, having specified a frequency, direction, and velocity of propagation, we have to deal with a single harmonic traveling wave component.

With the redefinition (6.133), we again have

$$\mathbf{X}_l = \underset{k}{\mathrm{col}}\, X_k(l\Delta f) = \mathbf{H} S_l + \mathbf{N}_l, \tag{6.136}$$

but with

$$\mathbf{H} = \mathbf{H}(l\Delta f, \boldsymbol{v}) = \underset{k}{\mathrm{col}}\, \eta_k(l\Delta f, \boldsymbol{v}) \triangleq \boldsymbol{\eta}(\boldsymbol{v}), \tag{6.137}$$

rather than the earlier

$$\mathbf{H}=(1,1,\ldots,1).\tag{6.138}$$

Nothing changes in any of the developments above, however, and we have at once, from (6.127), that

$$\hat{S}_l=[\boldsymbol{\eta}'(\boldsymbol{v})\mathcal{N}_l^{-1}\boldsymbol{\eta}(\boldsymbol{v})]^{-1}\boldsymbol{\eta}'(\boldsymbol{v})\mathcal{N}_l^{-1}\mathbf{X}_l\mathrel{\hat{=}}\mathcal{L}_l(\boldsymbol{v})\mathbf{X}_l\,.\tag{6.139}$$

The estimator (6.139) realizes the two properties built into it, i.e., it is unbiased:

$$\mathrm{E}[\hat{S}_l]=[\boldsymbol{\eta}'(\boldsymbol{v})\mathcal{N}_l^{-1}\boldsymbol{\eta}(\boldsymbol{v})]^{-1}\boldsymbol{\eta}'(\boldsymbol{v})\mathcal{N}_l^{-1}\mathrm{E}[\mathbf{X}_l]=S_l\,,\tag{6.140}$$

since from (6.136, 137),

$$\mathrm{E}[\mathbf{X}_l]=\mathbf{H}S_l=\boldsymbol{\eta}(\boldsymbol{v})S_l\,,\tag{6.141}$$

and the attained error variance, i.e.,

$$\begin{aligned}\sigma_l^2&=\mathrm{E}[|\hat{S}_l-S_l|^2]=\mathrm{E}[|\mathcal{L}_l(\boldsymbol{v})\mathbf{X}_l-S_l|^2]\\&=\mathrm{E}[|\mathcal{L}_l(\boldsymbol{v})\boldsymbol{\eta}(\boldsymbol{v})S_l-S_l+\mathcal{L}_l(\boldsymbol{v})\mathbf{N}_l|^2]\\&=\mathcal{L}_l(\boldsymbol{v})\mathrm{E}[\mathbf{N}_l\mathbf{N}_l']\mathcal{L}_l'(\boldsymbol{v})=[\boldsymbol{\eta}'(\boldsymbol{v})\mathcal{N}_l^{-1}\boldsymbol{\eta}(\boldsymbol{v})]^{-1}\,,\end{aligned}\tag{6.142}$$

using the unbiasedness property (6.140), is minimized under that constraint. The result (6.142) is of course just (6.117).

Capon [6.21] rephrased the unbiasedness and minimum variance properties of the linear filter (6.139) as follows. The filter (6.139) is "transparent" to the traveling wave

$$S_l\exp[\mathrm{j}2\pi(f_lt+\boldsymbol{v}\cdot\boldsymbol{z})]\,,\tag{6.143}$$

in that the associated Fourier data coefficient is

$$X_l=S_l\exp(\mathrm{j}2\pi\boldsymbol{v}\cdot\boldsymbol{z})\,,\tag{6.144}$$

so that the array Fourier coefficient data vector is

$$\mathbf{X}_l=S_l\boldsymbol{\eta}(\boldsymbol{v})\,.\tag{6.145}$$

Thus

$$\mathcal{L}_l(\boldsymbol{v})\mathbf{X}_l=S_l\mathcal{L}_l(\boldsymbol{v})\boldsymbol{\eta}(\boldsymbol{v})=S_l\,,\tag{6.146}$$

using the definition (6.139). On the other hand, the filter average power response to "everything else," i.e., the noise, is

$$\mathrm{E}[|\mathscr{L}_l(\boldsymbol{v})\mathbf{N}_l|^2] = \sigma_l^2\,, \tag{6.147}$$

as in (6.142), which is just the error variance, which is as small as possible, by design. We need only interpret "everything else" as all traveling waves except those from precisely the specified direction, traveling with precisely the specified phase velocity, i.e., those with the specified $\boldsymbol{v}$, to realize that we have here a filter for wave number spectral analysis of the random field.

In short, we want to return now to the view that the total data field $x(t, \boldsymbol{z})$ is decomposable into plane traveling wave components as in (6.31), i.e.,

$$x(t, \boldsymbol{z}) = \int_{-\infty}^{\infty} \int_{V} \exp[\mathrm{j}2\pi(ft + \boldsymbol{v}\cdot\boldsymbol{z})]X(df, d\boldsymbol{v})\,. \tag{6.148}$$

We seek to estimate the average power $\mathrm{E}[|X(df, d\boldsymbol{v})|^2]$ of each field component to obtain the information about what waves of what frequencies (f) and phase velocities ($v = f/|\boldsymbol{v}|$) are impinging on the array from what directions ($\boldsymbol{v}/|\boldsymbol{v}|$). The answer, as discussed at length above, is to construct an estimator $\hat{S}(f, \boldsymbol{v})$ of the frequency–wave-number spectrum $S(f, \boldsymbol{v})$ in (6.37). There is no signal/noise dichotomy in the data structure. We are not seeking to estimate the structure or power of some special kind of data called "signal" when buried in some other kind of data called "noise"; everything is "noise," and we seek to estimate its structure in time and space. This is the problem dealt with by *Lacoss* et al. [6.17], and differs in concept from the problem treated by *Capon* et al. [6.20], or by *Burg* [6.25], and *Backus* et al. [6.26]. In these latter cases, a deterministic or random signal was assumed present in noise, and the spectrum of the noise was assumed to be available, free of signal, for use in the estimator design.

Capon [6.21] made the step of identifying the totality of data $x(t, \boldsymbol{z})$ with the noise $n(t, \boldsymbol{z})$ in the maximum-likelihood problem. The filter design is then, from (6.139),

$$\mathscr{L}_l(\boldsymbol{v}) = [\boldsymbol{\eta}'(\boldsymbol{v})\boldsymbol{\chi}_l^{-1}\boldsymbol{\eta}(\boldsymbol{v})]^{-1}\boldsymbol{\eta}'(\boldsymbol{v})\boldsymbol{\chi}_l^{-1}\,, \tag{6.149}$$

where $\boldsymbol{\chi}_l$ is the data cross-spectral matrix at frequency $f = l\Delta f$, estimated by some means:

$$(\boldsymbol{\chi}_l)_{ij} \approx P_x(f, \boldsymbol{z}_i, \boldsymbol{z}_j) = \int_{-\infty}^{\infty} C_x(\tau, \boldsymbol{z}_i, \boldsymbol{z}_j)\exp(-\mathrm{j}2\pi f\tau)d\tau\,, \tag{6.150}$$

as in (6.60, 61). The usual estimator used is the average of data-segmented periodograms:

$$(\boldsymbol{\chi}_l)_{ij} = \mathrm{aver}\, X_i(l\Delta f)X_j^*(l\Delta f)\,, \tag{6.151}$$

as in (6.65, 66).

Having designed the filter based on the data, rather than the noise, we then apply the filter to the data, the vector $\mathbf{X}_l$ of data Fourier spectral coefficients (6.123):

$$\mathbf{X}(l\Delta f, n) = \underset{k}{\mathrm{col}} \frac{1}{\sqrt{T_n}} \int_{-T_n/2}^{T_n/2} x^{(n)}(t, z_k) \exp(-\mathrm{j}2\pi f t)\, dt, \tag{6.152}$$

where $x^{(n)}(t, z_k)$ indicates the portion of $x(t, z_k)$ on the nth time segment. Finally, the filter power output on each time segment is averaged:

$$\begin{aligned} \sigma_l^2(v) &= \underset{n}{\mathrm{aver}} |\mathscr{L}_l(v)\mathbf{X}(l\Delta f, n)|^2 \\ &= \mathscr{L}_l(v)\Big[\underset{n}{\mathrm{aver}}\, \mathbf{X}(l\Delta f, n)\mathbf{X}'(l\Delta f, n)\Big]\mathscr{L}_l'(v) \\ &= \mathscr{L}_l(v)\chi_l\mathscr{L}_l'(v) \\ &= [\boldsymbol{\eta}'(v)\chi_l^{-1}\boldsymbol{\eta}(v)]^{-1}. \end{aligned} \tag{6.153}$$

This filter power output is then taken as an estimator of the field frequency–wave-number spectrum

$$\hat{S}_x(f, v) = [\boldsymbol{\eta}'(v)\chi_l^{-1}\boldsymbol{\eta}(v)]^{-1}, \tag{6.154}$$

and this is *Capon*'s high-resolution frequency–wave-number spectral estimator [6.21]. (See also Chap. 5 for another way of deriving this formula, based on a likelihood ratio.)

The estimator (6.154) is nonlinear, because the data itself are used in its design. Thus the filter is data dependent, rather than dependent on some prespecified noise spectrum. Accordingly, "all bets are off" so far as assigning to the estimator (6.154) any of the statistical properties discussed above with regard to linear spectral estimation procedures. Accordingly, *Capon* [6.21], and *Capon* and *Goodman* [6.27], found it necessary to consider the statistical properties of (6.154). The result was that the mean and variance of (6.154) are essentially the same as those of the linear estimator, i.e., $\mathrm{E}[\hat{S}_x]$ is a frequency wave number windowed version of the true spectrum, and the variance of $\hat{S}_x$ decreases as the number of data segments used in the averaging (6.151) increases. The point is that the resolution properties of (6.154) are much superior to those of the usual linear estimator, for an equivalent data span. (A detailed treatment of this issue was presented in Chap. 5.)

The very good success of using the estimator (6.154) in seismic (LASA) data processing was adequately demonstrated by *Capon* [6.27], which is a strong argument for its use. However, the nonlinearity of the estimator gave rise to at least one unexpected disadvantage, which may arise in certain applications. We will return to this point in a later section, before going on to discuss some iterative implementations of estimators in the spirit of (6.154). First, however, let us give a direct derivation of (6.154), which may clarify some of the points made above.

6.3.3 The Maximum-Likelihood Estimator as a Filter in Wave Number Space

The high-resolution estimator (6.154) can be derived directly as a wave number filter, as follows [6.28]. We again regard the field $x(t, z)$ as a superposition of traveling waves (6.148). The sensor data $x(t, z_i)$ are time segmented, into N blocks of length T/N, and on each segment we compute a Fourier coefficient vector (6.152). We seek a wave number filter which will block out all wave number components other than a particular component of interest with wave number $\boldsymbol{\nu}_0$, for instance.

Let the filter input on time segment n be the K-vector $\mathbf{x}_l^{(n)} = \mathbf{x}^{(n)}(l\Delta f)$ of sensor Fourier coefficients at frequency $f_l = l\Delta f$. Let these coefficients be weighted by an arbitrary vector $\mathbf{w}(\boldsymbol{\nu}_0)$, for each analysis wave number $\boldsymbol{\nu}$ of interest, to produce the filter output on segment n:

$$\varrho_n(\boldsymbol{\nu}_0) = \mathbf{w}'(\boldsymbol{\nu}_0)\mathbf{X}_l^{(n)}. \tag{6.155}$$

Let the filter "power" output be the squared magnitude of this quantity, averaged over the N time segments:

$$\varrho(\boldsymbol{\nu}_0) = \underset{n}{\operatorname{aver}} |\mathbf{w}'(\boldsymbol{\nu}_0)\mathbf{X}_l^{(n)}|^2. \tag{6.156}$$

Using the definition (6.151), this becomes

$$\begin{aligned} \varrho(\boldsymbol{\nu}_0) &= \mathbf{w}'(\boldsymbol{\nu}_0)\left(\underset{n}{\operatorname{aver}} \mathbf{X}_l^{(n)}\mathbf{X}_l^{(n)\prime}\right)\mathbf{w}(\boldsymbol{\nu}_0) \\ &= \mathbf{w}'(\boldsymbol{\nu}_0)\chi_l \mathbf{w}(\boldsymbol{\nu}_0) \mathrel{\hat=} \mathbf{w}_0'\chi\mathbf{w}_0. \end{aligned} \tag{6.157}$$

We now make the reasonable requirement that the filter response (6.157) be such that the sensitivity to a plane wave component of the wave number and frequency in question, having Fourier coefficient vector (6.144)

$$\mathbf{X}_l^{(n)} = \underset{k}{\operatorname{col}} \exp(\mathrm{j}2\pi\boldsymbol{\nu}_0 \cdot z_k) \mathrel{\hat=} \boldsymbol{\eta}(\boldsymbol{\nu}_0), \tag{6.158}$$

be fixed at some arbitrary level:

$$\mathbf{w}_0'[\boldsymbol{\eta}(\boldsymbol{\nu}_0)\boldsymbol{\eta}'(\boldsymbol{\nu}_0)]\mathbf{w}_0 = 1. \tag{6.159}$$

Under the sensitivity constraint (6.159), we then require that the filter response to the actual data be minimum:

$$\mathbf{w}_0 \Leftarrow \min_{\mathbf{w}_0} \mathbf{w}_0' \chi \mathbf{w}_0 . \tag{6.160}$$

The constrained minimization problem (6.159, 160) is solved routinely by the introduction of a (scalar) Lagrange multiplier. The augmented "cost" is

$$J = \mathbf{w}_0' \chi \mathbf{w}_0 - \lambda(\mathbf{w}_0' \boldsymbol{\eta}_0 \boldsymbol{\eta}_0' \mathbf{w}_0 - 1), \tag{6.161}$$

where we write

$$\boldsymbol{\eta}_0 = \boldsymbol{\eta}(\boldsymbol{\nu}_0) . \tag{6.162}$$

The necessary condition is

$$J_{\mathbf{w}_0} = 0 = \mathbf{w}_0' \chi - \lambda \mathbf{w}_0' \boldsymbol{\eta}_0 \boldsymbol{\eta}_0' , \tag{6.163}$$

$$\chi \mathbf{w}_0 = \lambda \boldsymbol{\eta}_0 \boldsymbol{\eta}_0' \mathbf{w}_0 . \tag{6.164}$$

This is easily solved, if we assume χ to be nonsingular, which will be the case, barring a degenerate situation. Since then $|\chi| \neq 0$, we must have $\lambda \neq 0$ in (6.164), since by (6.159) we also have $\mathbf{w}_0 \neq 0$. (This follows because χ has no zero eigenvalue.) First, from (6.164), with (6.159), we have the attained processor output as

$$\varrho(\nu_0) = \mathbf{w}_0' \chi \mathbf{w}_0 = \lambda \mathbf{w}_0' \boldsymbol{\eta}_0 \boldsymbol{\eta}_0' \mathbf{w}_0 = \lambda . \tag{6.165}$$

Second, from (6.164) we have

$$\mathbf{w}_0 = \lambda \chi^{-1} \boldsymbol{\eta}_0 \boldsymbol{\eta}_0' \mathbf{w}_0 , \tag{6.166}$$

which, used in (6.165), yields

$$\lambda^2 (\boldsymbol{\eta}' \chi^{-1} \boldsymbol{\eta}) = \lambda , \tag{6.167}$$

or, since $\lambda \neq 0$,

$$\lambda = \varrho(\nu_0) = (\boldsymbol{\eta}_0' \chi^{-1} \boldsymbol{\eta}_0)^{-1} , \tag{6.168}$$

the high-resolution estimator. The filter weight vector, by inspection of (6.164), is

$$\mathbf{w}_0 = \alpha \chi^{-1} \boldsymbol{\eta}_0 , \tag{6.169}$$

where α is a scalar adjusted to satisfy (6.159):

$$\mathbf{w}_0 = (\boldsymbol{\eta}_0' \boldsymbol{\chi}^{-1} \boldsymbol{\eta}_0)^{-1} \boldsymbol{\chi}^{-1} \boldsymbol{\eta}_0 , \tag{6.170}$$

in agreement with (6.149).

We can think of this procedure as one in which the sensitivity of the filter (or data-adaptive beam former, to use the common terminology) on the main lobe center is kept fixed, while the actual response of the filter off the center of the main lobe is minimized. Thus, in directions of low signal power, the spatial processor sensitivity increases (side-lobes grow), while in directions of strong "interference" (signals off the main lobe) the sensitivity is sharply decreased. Strong signals off the "main beam," or off the "look direction," i.e., traveling waves with wave numbers $\boldsymbol{v}_1 \neq \boldsymbol{v}_0$, are sharply discriminated against, weak signals with $\boldsymbol{v}_1 \neq \boldsymbol{v}_0$ are less sharply suppressed, and signals with wave number $\boldsymbol{v}_0$ pass through unaffected. *Cox* [6.29] has given an extensive discussion of the response of the processor (6.168) to signals with wave numbers $\boldsymbol{v}_1 \neq \boldsymbol{v}_0$. *Lacoss* [6.30] has given some computed examples of the ability of the processor (6.168) to resolve spectral peaks.

Before discussing some of the iterative wave number filters which have been developed, following the philosophy of this section, let us set out some results concerning a potential difficulty associated with the maximum-likelihood processor (6.168).

6.3.4 The Effect of Signal Mismatch on the Maximum-Likelihood Processor

The results presented by *Capon* [6.21] clearly indicate the superior resolving power of the maximum-likelihood estimator (6.168), compared with the simple linear delay-and-sum "beam former" (Bartlett window) estimator (6.84). These results refer to the geophysical data received on the large aperture seismic array (LASA), but the relevance to other directional wave analysis problems was clear. Accordingly, the processor (6.168) was investigated for the processing of various other signals received on arrays. It was found that indeed in many cases the "high-resolution" estimator of (6.168) had resolution characteristics much superior to those of the "beam former" of (6.84). However, in some situations the matter was reversed: the beam former had usable resolution, but the high-resolution processor failed to yield useful results.

Seligson [6.31] came to an understanding of the problem, and presented some analytical results. The matter hinges upon the sensitivity of the maximum-likelihood processor (6.168) to deviations of the actual wave fronts propagating over the array from the planar form assumed in the derivation of (6.168) presented in the previous section. This sensitivity may or may not lead to poor behavior of the estimator, depending on the degree of spatial coherence of the true data field. In the presence of a few relatively strong interfering signals, in a background of relatively weak ambient signals, the effects of nonplanar data wave fronts are pronounced. The matter merits some discussion, since the effects in question somewhat limit the applicability of the maximum-likelihood estimator.

In deriving the maximum-likelihood estimator from the point of view of a filter in wave number space, and working the constrained minimization problem (6.159, 160), we assumed the vector of Fourier coefficients of the wave to which the filter was to be "transparent," i.e., $\boldsymbol{\eta}(\boldsymbol{v}_0)$, was as in (6.158), since (6.158) correctly describes plane traveling waves emanating from direction $\boldsymbol{v}_0$. However, it is not plane waves from direction $\boldsymbol{v}_0$ we are interested in, but all waves, planar or not, emanating from direction $\boldsymbol{v}_0$. Now if we know a priori the shape of the wave front of the actual waves from $\boldsymbol{v}_0$ present in the data, we need only use the correct "shape" vector $\boldsymbol{\eta}(\boldsymbol{v}_0)$ in (6.168), and the resulting filter will be transparent to the actual waves, of the specified wave front shape, emanating from direction $\boldsymbol{v}_0$, with all other waves being maximally rejected. However, we may not know the actual data wave front shape, perhaps because of turbulence, multipath propagation, and so on. The practice then is to design the filter for some nominal wave front, e.g., planar, and proceed.

Seligson [6.31] considered a data model leading to

$$\boldsymbol{\chi}=\alpha\mathbf{J}+\boldsymbol{\eta}_1\boldsymbol{\eta}_1', \tag{6.171}$$

which corresponds to a signal wave front described by a Fourier coefficient vector $\boldsymbol{\eta}_1$ across the sensor array, with a background of spatially uncorrelated ("white") noise. The signal wave front vector is arbitrary,

$$(\boldsymbol{\eta}_1)_k=A_k\exp(\mathrm{j}2\pi f\tau_k), \qquad k=1,K, \tag{6.172}$$

where A_k is an amplitude factor, τ_k is the time advance of the wave front at sensor k, relative to the coordinate origin, and f is the frequency of interest. The "data" (6.171) are analyzed by the filter (6.168), designed based on plane wave fronts, with $\boldsymbol{\eta}_0$ as in (6.158):

$$\varrho_{\mathrm{HR}}(\boldsymbol{v}_0)=[\boldsymbol{\eta}_0'(\mathbf{J}+\alpha\boldsymbol{\eta}_1\boldsymbol{\eta}_1')^{-1}\boldsymbol{\eta}_0]^{-1}, \tag{6.173}$$

and also by the simple "beam former" (6.84):

$$\varrho_{\mathrm{BF}}(\boldsymbol{v}_0)=\boldsymbol{\eta}_0'(\mathbf{J}+\alpha\boldsymbol{\eta}_1\boldsymbol{\eta}_1')\boldsymbol{\eta}_0. \tag{6.174}$$

It was found that the inner product

$$B^2\triangleq|\boldsymbol{\eta}_0'\boldsymbol{\eta}_1|^2 \tag{6.175}$$

controlled whether or not a resolution "anomaly" could occur as $\boldsymbol{v}_0$ was varied, and, if it could, the signal-to-noise ratio $1/\alpha$ governed its severity. If the data wave front "matched" the design wave front, i.e., if $\boldsymbol{\eta}_1=\boldsymbol{\eta}_0(\boldsymbol{v})$ occurred for some

$\boldsymbol{\nu}_0$, leading to $B^2 = K^2$, then no anomaly could occur. However, if the data wave front was so distorted that B^2 never rose above a certain threshold as $\boldsymbol{\eta}_0(\boldsymbol{\nu}_0)$ varied through the family of planar waves, then an anomaly occurred, of severity governed by $1/\alpha$, being more severe the higher the signal-to-noise ratio.

This question was further pursued by *McDonough* [6.28], where much the same conclusions resulted from a different analysis. It was assumed that data were received, and a spatial correlation matrix $\boldsymbol{\chi}$ computed. An estimator of the general form

$$\varrho(\boldsymbol{\nu}_0) = \mathbf{w}'(\boldsymbol{\nu}_0, \boldsymbol{\chi})\boldsymbol{\chi}\mathbf{w}(\boldsymbol{\nu}_0, \boldsymbol{\chi}) \tag{6.176}$$

is to be constructed, with $\mathbf{w}$ being a "weight" vector, whose functional dependence on $\boldsymbol{\chi}$ is at our pleasure. The vector $\boldsymbol{\nu}_0$ indicates the true geometric direction from which the waves of interest are assumed to emanate, i.e., we want the filter to be "transparent" to waves from direction $\boldsymbol{\nu}_0$. The key question to be answered is, how do we describe these waves from direction $\boldsymbol{\nu}_0$, as they arrive at the array?

We may be willing to assume waves from sources in direction $\boldsymbol{\nu}_0$ propagate undisturbed through the medium, and arrive as plane waves, producing a Fourier-coefficient vector as in (6.158), i.e.,

$$\boldsymbol{\eta}_0 = \boldsymbol{\eta}(\boldsymbol{\nu}_0) = \operatorname*{col}_k \exp(\mathrm{j}2\pi\boldsymbol{\nu}_0 \cdot \boldsymbol{z}_k), \qquad k = 1, K. \tag{6.177}$$

The "transparency" constraint is then (6.159), i.e.,

$$\mathbf{w}'(\boldsymbol{\nu}_0, \boldsymbol{\chi})\boldsymbol{\eta}_0\boldsymbol{\eta}_0'\mathbf{w}(\boldsymbol{\nu}_0, \boldsymbol{\chi}) = 1. \tag{6.178}$$

If we minimize (6.176), holding (6.178), the functional form (6.170) results, as before. On the other hand, we may assume (6.177), whereas the actual data field contains waves from $\boldsymbol{\nu}_0$ with wave fronts which differ from (6.177). The filter response to such waves is not constrained, and these waves will be discriminated against by the filter. We emphasize that here we are not speaking only of wave fronts which differ from (6.177) by virtue of emanating from a geometric direction other than $\boldsymbol{\nu}_0$, but rather we want to consider the effects of waves from $\boldsymbol{\nu}_0$ with perturbed wave fronts.

If we want to constrain the filter to pass some variety of perturbed wave fronts from $\boldsymbol{\nu}_0$, we should require (6.178) for a family of vectors $\boldsymbol{\eta}_0$ "near" the nominal (6.177). *McDonough* [6.28] did not do this, but rather investigated the extent to which the constraint equation (6.178) varies as $\boldsymbol{\eta}_0$ varies around the nominal plane wave (6.177). Specifically, a family

$$\boldsymbol{\eta} = \operatorname*{col}_k A_k(1 + \alpha_k)\exp[\mathrm{j}(\phi_k + \xi_k)], \qquad k = 1, K \tag{6.179}$$

was considered, where the nominal constraint vector is

$$\boldsymbol{\eta}_0 = \operatorname*{col}_k A_k \exp(\mathrm{j}\phi_k). \tag{6.180}$$

We let the amplitude perturbations α_k and phase perturbations ξ_k be uncorrelated zero-mean Gaussian random variables, with variances σ_α^2 and σ_ξ^2 which were the same at each sensor. Then the expectation

$$\begin{aligned} E[\mathbf{w}_0'\boldsymbol{\eta}\boldsymbol{\eta}'\mathbf{w}_0] &= \mathbf{w}_0'(E[\boldsymbol{\eta}\boldsymbol{\eta}'])\mathbf{w}_0 \\ &= \exp(-\sigma_\xi^2)\mathbf{w}_0'\boldsymbol{\eta}_0\boldsymbol{\eta}_0'\mathbf{w}_0 \\ &\quad + [1-\exp(-\sigma_\xi^2)+\sigma_\alpha^2]\mathbf{w}_0'\mathbf{w}_0 \end{aligned} \tag{6.181}$$

was calculated. Thus, if we design the filter using the constraint (6.178), thereby obtaining the maximum-likelihood estimator, the constraint is violated for perturbed wave fronts insofar as the right side of (6.181) differs from unity.

First, note that if we have no perturbations in the design nominal vector $\boldsymbol{\eta}_0$, i.e., $\sigma_\xi^2 = \sigma_\alpha^2 = 0$, the right side of (6.181) is just the constrained quantity (6.178). However, if either σ_ξ^2 or σ_α^2 is nonzero, or both, then the right side of (6.181) will differ from the design constraint (6.178). The amount of difference is largely governed by the sensitivity factor, or "perturbation gain,"

$$g = \mathbf{w}'(\nu_0, \boldsymbol{\chi})\mathbf{w}(\nu_0, \boldsymbol{\chi}). \tag{6.182}$$

In the case of the maximum-likelihood processor weights (6.170), this sensitivity measure is

$$g = (\boldsymbol{\eta}_0'\boldsymbol{\chi}^{-2}\boldsymbol{\eta}_0)/(\boldsymbol{\eta}_0'\boldsymbol{\chi}^{-1}\boldsymbol{\eta}_0)^2. \tag{6.183}$$

It turns out that this factor can become arbitrarily large, causing the maximum-likelihood processor design to be extremely sensitive to the choice of nominal "steering" vector $\boldsymbol{\eta}_0$. If we choose this vector inappropriately, anomalous results can be expected. Unfortunately, an appropriate choice of the vector may not be known.

In general, the sensitivity measure (6.183) is found to increase as the condition of the data spatial correlation matrix $\boldsymbol{\chi}$ deteriorates [6.28]. Specifically, the effects in question here worsen as the ratio of the largest eigenvalue of $\boldsymbol{\chi}$ to its smallest eigenvalue increases. This may explain why perturbation effects, if they occur, tend to be worse in the presence of strong signals in the data field, since then $\boldsymbol{\chi}$ will tend to have a relatively few strongly dominant eigenvectors.

In an important paper, *Cox* [6.29] made all this quite precise. He considered three processors of the form (6.176), and their responses to data corresponding to

$$\boldsymbol{\chi} = \sigma_0^2\mathbf{Q} + \sigma_1^2\boldsymbol{\eta}_1\boldsymbol{\eta}_1', \tag{6.184}$$

i.e., a "signal" $\boldsymbol{\eta}_1$ in a background of "noise" with spatial covariance matrix $\mathbf{Q}$. The three processors considered are: the simple beam former, with weights

$$\mathbf{w}_1 = \boldsymbol{\eta}_0, \tag{6.185}$$

the "maximum-likelihood" processor (6.170), with weights

$$\mathbf{w}_3 = (\boldsymbol{\eta}_0' \boldsymbol{\chi}^{-1} \boldsymbol{\eta}_0)^{-1} \boldsymbol{\chi}^{-1} \boldsymbol{\eta}_0, \tag{6.186}$$

and the maximum-likelihood processor designed on the basis of the true noise background, with weights

$$\mathbf{w}_2 = (\boldsymbol{\eta}_0' \mathbf{Q}^{-1} \boldsymbol{\eta}_0)^{-1} \mathbf{Q}^{-1} \boldsymbol{\eta}_0. \tag{6.187}$$

Cox computed the signal-to-noise ratio improvement due to use of these three filters, as it depends on the deviation of the assumed steering vector $\boldsymbol{\eta}_0$ from the actual signal vector $\boldsymbol{\eta}_1$. He found that the processors (6.186, 187) have equivalent behavior when perfectly matched ($\boldsymbol{\eta}_0 = \boldsymbol{\eta}_1$), but that the processor (6.186) is subject to the difficulties we have been discussing if $\boldsymbol{\eta}_0 \neq \boldsymbol{\eta}_1$.

Cox [6.29] calculated the resolving power of the processor (6.186), the "maximum-likelihood" processor, when the data consists of uncorrelated background noise and two signals:

$$\boldsymbol{\chi} = \sigma_0^2 \mathbf{J} + \sigma_1^2 \boldsymbol{\eta}_1 \boldsymbol{\eta}_1' + \sigma_2^2 \boldsymbol{\eta}_2 \boldsymbol{\eta}_2'. \tag{6.188}$$

Various results were obtained, both in the case of "on-target" steering in (6.186) ($\boldsymbol{\eta}_0 = \boldsymbol{\eta}_1$ or $\boldsymbol{\eta}_0 = \boldsymbol{\eta}_2$), or "off-target" steering, in which case $\boldsymbol{\eta}_0$ was chosen such that

$$|\boldsymbol{\eta}_0' \boldsymbol{\eta}_1| / (\boldsymbol{\eta}_1' \boldsymbol{\eta}_1)^{1/2} = |\boldsymbol{\eta}_0' \boldsymbol{\eta}_2| / (\boldsymbol{\eta}_2' \boldsymbol{\eta}_2)^{1/2}. \tag{6.189}$$

If the processor responses for on-target and off-target steering differ sufficiently, the sources $\boldsymbol{\eta}_1, \boldsymbol{\eta}_2$ are resolved. Conditions are presented under which the maximum-likelihood processor (6.186) resolves the two signals $\boldsymbol{\eta}_1$, $\boldsymbol{\eta}_2$ in (6.188).

We may conjecture that an environment with a relatively stable propagation medium will be suited to use of the "maximum-likelihood" processor, but that some modifications may be needed in a more random environment. We will now discuss some processors which, being adaptive, may be more successful in the latter cases.

6.3.5 Adaptive Wave Number Filters

We have discussed above the fact that a wave number spectral estimator may be regarded as a filter in wave number space, that has fixed sensitivity to wave numbers of some specified value $\boldsymbol{\nu}_0$, and which maximally rejects traveling waves with wave numbers other than $\boldsymbol{\nu}_0$. The analysis of the ambient traveling wave field then amounts to measuring the actual output of the filter when "tuned" to the various wave numbers of interest. This led to the constrained minimization problem in (6.159, 160), viz.,

$$\mathbf{w}_0'\boldsymbol{\eta}(\nu_0)\boldsymbol{\eta}'(\nu_0)\mathbf{w}_0=1\,, \tag{6.190}$$

$$\mathbf{w}_0 \Leftarrow \min_{\mathbf{w}_0} \mathbf{w}_0'\boldsymbol{\chi}\mathbf{w}_0\,, \tag{6.191}$$

where $\boldsymbol{\chi}$ is the spatial cross-covariance matrix (6.151) at some frequency of interest, the latter selected by conventional frequency filtering of the sensor signals. The solution to this problem. with the "hard" constraint (6.190), is the "maximum-likelihood" processor, as we discussed above.

The same philosophy of filter design has been independently developed as the theory and practice of "adaptive array processing". The relationship to the method we have been discussing is so close, seen in hindsight, that a discussion of adaptive processing will not be out of context.

In an important paper, *Widrow* et al. [6.32] worked out an iterative solution to a problem closely related to (6.190, 191). Let the sensor data as before be Fourier analyzed at some frequency $l\Delta f$ of interest to produce a K-vector of Fourier coefficients $\mathbf{X}(l\Delta f)$. Suppose this vector of Fourier coefficients to have been calculated on some data subinterval $(n-1)T \leqq t \leqq nT$, where T is the Fourier analysis interval. Accordingly, let the vector be called $\mathbf{X}_n(l\Delta f)$, and finally drop the frequency dependence. In the problem defined by (6.190, 191), we sought a wavenumber filter weight vector $\mathbf{w}$ such that

$$\mathbf{w}'\boldsymbol{\eta}_0\boldsymbol{\eta}_0'\mathbf{w}=|\mathbf{w}'\boldsymbol{\eta}_0|^2=1\,, \tag{6.192}$$

$$\begin{aligned}\mathbf{w}'\boldsymbol{\chi}\mathbf{w}&=\mathbf{w}'(\mathrm{E}[\mathbf{X}_n\mathbf{X}_n'])\mathbf{w}\\&=\mathrm{E}[|\mathbf{w}'\mathbf{X}_n|^2]\Rightarrow\min,\end{aligned} \tag{6.193}$$

where we drop the subscript zero on $\mathbf{w}$, and let $\boldsymbol{\eta}_0$ represent $\boldsymbol{\eta}(\nu_0)$. *Widrow* et al. [6.32] replaced this problem with the related problem

$$\mathbf{w}'\boldsymbol{\eta}_0 \sim 1\,, \tag{6.194}$$

$$\mathbf{w}'\mathbf{X}_n \sim 0\,, \tag{6.195}$$

where the approximations are in the statistical least-squares sense. For convenience, let $\mathbf{y}$ be either of $\boldsymbol{\eta}_0$ or $\mathbf{X}_n$, and let d be correspondingly 1 or 0. The problem defined by (6.194, 195) then becomes

$$\mathbf{w}'\mathbf{y} \sim d\,. \tag{6.196}$$

The ensemble least-squares problem (6.196) can be solved by iterative gradient search techniques. The ensemble mean-squared error is

$$\varepsilon=\mathrm{E}[|\mathbf{w}'\mathbf{y}-d|^2]\,. \tag{6.197}$$

Descent to the minimum of ε by gradient search leads to the iteration

$$\mathbf{w}'_j = \mathbf{w}'_{j-1} - \mu_j(\partial\varepsilon/\partial\mathbf{w})_{j-1}, \tag{6.198}$$

where the partial derivative is evaluated at the current weight vector $\mathbf{w}_{j-1}$; μ_j is a scalar convergence factor. Using (6.197) in (6.198) yields

$$\mathbf{w}'_j = \mathbf{w}'_{j-1} - \mu_j \mathrm{E}[(\mathbf{w}'_{j-1}\mathbf{y} - d)\mathbf{y}']. \tag{6.199}$$

The difficulty, of course, is that we cannot compute the expected value in (6.199), since we presumably don't know the probability density of the data vector $\mathbf{X}_n$. However, following the procedures of the theory of stochastic approximation [6.33], we simply drop the expectation sign in (6.199), and replace $\mathbf{y}$ by its time-sequential realizations. We also synchronize the steps of the gradient search with the data flow, to obtain the iteration

$$\mathbf{w}_j = \mathbf{w}_{j-1} - \mu_j \mathbf{y}_j(\mathbf{y}'_j \mathbf{w}_{j-1} - d_j), \qquad j = 0, 1, \dots . \tag{6.200}$$

In operation, the algorithm (6.200) is cycled with $\mathbf{y}_j = \boldsymbol{\eta}_0$, $d_j = 1$, or alternatively $\mathbf{y}_j = \mathbf{X}_n$, $d_j = 0$, with some scheme for determining what proportion of "pilot signal" cycles to use compared to data cycles.

Roughly speaking, the iterations (6.199, 200) converge to the same limit, under suitable restrictions on the sequence μ_j and the data sequence $\mathbf{X}_n$ (the latter should be independent). Even lacking independence of the sequence elements $\mathbf{X}_n$, which is in fact the case in our application, the algorithm (6.200) generally converges, and represents a useful data processing scheme. *Griffiths* [6.34] has given a variant of (6.200), and discussed some convergence questions. *Widrow* et al. [6.35] have recently given an extensive discussion of convergence of the adaptive process under various conditions of use.

The weights determined using the iteration (6.200) cannot be expected to be those computed using the "maximum-likelihood" procedure, since the least-squares problem (6.194, 195) does not impose the hard constraint (6.192). The two methods yield different solutions to the same problem, following the same philosophy of an optimal filter in wave number space. It may be conjectured that the filter designed using (6.200) would be less sensitive to the problem of data wave front distortion which we discussed above, but this appears not to have been investigated.

A second approach to iterative solution of the constrained minimization problem (6.190, 191) has also been investigated by a number of writers. This amounts to carrying out a gradient search to work the minimization problem (6.191), while adjusting the filter weight vector at each iteration to maintain (6.190). *Lacoss* [6.36] applied this technique to the LASA data. In brief, we seek to minimize $\mathbf{w}'\chi\mathbf{w}$ by an iterative adjustment of the filter weight vector:

$$\mathbf{w}_{n+1} = \mathbf{w}_n - (\Delta\mathbf{w})_n, \qquad \mathrm{n} = 1, 2, \dots . \tag{6.201}$$

The change $\Delta\mathbf{w}$ is to be perpendicular to the "steering" vector $\boldsymbol{\eta}_0$:

$$\boldsymbol{\eta}_0'(\Delta\mathbf{w})_n = 0. \tag{6.202}$$

This is brought about by taking $\Delta\mathbf{w}$ to be proportional to the portion $\mathbf{g}^\perp$ of the gradient $\nabla_{\mathbf{w}'}(\mathbf{w}'\chi\mathbf{w})$ which is orthogonal to $\boldsymbol{\eta}_0$,

$$\mathbf{g}^\perp \triangleq \chi\mathbf{w} - (\boldsymbol{\eta}_0'\chi\mathbf{w})\boldsymbol{\eta}_0/K^2, \tag{6.203}$$

$$\Delta\mathbf{w} = \mu\mathbf{g}^\perp. \tag{6.204}$$

Applying the ideas of stochastic approximation, we obtain the "projected gradient" search algorithm ($\mathbf{J}$ is the unit matrix)

$$\mathbf{w}_{n+1} = \mathbf{w}_n - \mu_n(\mathbf{J} - \boldsymbol{\eta}_0\boldsymbol{\eta}_0'/K^2)\mathbf{X}_n\mathbf{X}_n'\mathbf{w}_n, \tag{6.205}$$

where μ_n is a variable step-size control. The initial weight vector is taken such as to satisfy the constraint (6.192), e.g., $\mathbf{w}_0 = \boldsymbol{\eta}_0/K$. *Lacoss* [6.36] gave examples, and discussed some convergence questions for the algorithm (6.205). He particularly considered versions which are computationally efficient, and which are robust to a certain extent against "wild" (invalid) data points. In particular, he recommended choosing

$$\mu_n = \mu/\|\mathbf{X}_n\mathbf{X}_n'\mathbf{w}_n\|, \tag{6.206}$$

so that an inappropriately large "burst" in the magnitude of the vector $\mathbf{X}_n$ will not cause the weights to change too much in one step of the algorithm. [In (6.206), we use the notation $\|\mathbf{v}\| = (\mathbf{v}'\mathbf{v})^{1/2}$.]

Kobayashi [6.37] also used a projected gradient method in processing some LASA data. He arranged that the proportion of the projected estimated negative gradient taken to form $\Delta\mathbf{w}_n$ was such that, were the covariance χ_n to remain unchanged, the minimum of $\mathbf{w}'\chi_n\mathbf{w}$ would be reached at the next step. Thus, we again choose $\Delta\mathbf{w}$ as in (6.203, 204):

$$\mathbf{w}_{n+1} = \mathbf{w}_n - \mu_n\mathbf{g}_n^\perp, \tag{6.207}$$

where $\mathbf{g}_n^\perp$ indicates that (6.203) is evaluated for $\mathbf{w} = \mathbf{w}_n$, $\chi = \chi_n$. Choosing μ_n in (6.207) to attain

$$\mu_n \Leftarrow \min_{\mu_n} \mathbf{w}_{n+1}'\chi_n\mathbf{w}_{n+1} \tag{6.208}$$

results in

$$\begin{aligned}\mu_n &= \mathbf{g}_n^{\perp\prime}\chi_n\mathbf{w}_n/(\mathbf{g}_n^{\perp\prime}\chi_n\mathbf{g}_n^\perp)\\ &= \mathbf{g}_n^{\perp\prime}\mathbf{g}_n^\perp/(\mathbf{g}_n^{\perp\prime}\chi_n\mathbf{g}_n^\perp),\end{aligned} \tag{6.209}$$

where the last step follows because $\mathbf{g}_n^{\perp}$ is orthogonal to $\boldsymbol{\eta}_0$, hence whether we write $\boldsymbol{\chi}_n \mathbf{w}_n$ or $\mathbf{g}_n^{\perp}$ in the numerator is irrelevant, the two differing only by a term proportional to $\boldsymbol{\eta}_0$.

Using (6.209) in (6.207) yields Kobayashi's algorithm. In fact, *Kobayashi* was concerned with a single fixed value $\boldsymbol{\chi}_n$, call it $\hat{\boldsymbol{\chi}}$, and used his algorithm simply as an efficient way of computing the maximum-likelihood filter weights, avoiding the inversion of $\hat{\boldsymbol{\chi}}$ in the usual expression (6.154). In that case, the vectors $\mathbf{g}^{\perp}$ obey the recursion

$$\begin{aligned}\mathbf{g}_{n+1}^{\perp} &= (\mathbf{J} - \boldsymbol{\eta}_0 \boldsymbol{\eta}_0'/K^2)\hat{\boldsymbol{\chi}} \mathbf{w}_{n+1} \\ &= \mathbf{g}_n^{\perp} - \mu_n (\mathbf{J} - \boldsymbol{\eta}_0 \boldsymbol{\eta}_0'/K^2)\hat{\boldsymbol{\chi}} \mathbf{g}_n^{\perp} .\end{aligned} \tag{6.210}$$

In this application, with fixed spatial covariance matrix estimate, *Kobayashi* gave some examples of applying his method to LASA data. He also developed a similar algorithm using minimization by the conjugate-gradient method.

Frost [6.38] worked out another approach to the "hard constraint" problem (6.192, 193). A constraint equivalent to (6.192) is adjoined to the cost (6.193) with a Lagrange multiplier to obtain the augmented cost

$$J = \mathbf{w}' \boldsymbol{\chi} \mathbf{w} + \lambda(\boldsymbol{\eta}_0' \mathbf{w} - 1) . \tag{6.211}$$

An iterative gradient descent algorithm is used to minimize this cost, yielding

$$\mathbf{w}_{n+1} = \mathbf{w}_n - \mu_n (\boldsymbol{\chi}_n \mathbf{w}_n + \lambda \boldsymbol{\eta}_0) . \tag{6.212}$$

At each step, the Lagrange multiplier is adjusted to satisfy the constraint

$$\boldsymbol{\eta}_0' \mathbf{w}_{n+1} = \boldsymbol{\eta}_0' \mathbf{w}_n - \mu_n (\boldsymbol{\eta}_0' \boldsymbol{\chi}_n \mathbf{w}_n + \lambda_n K) = 1 , \tag{6.213}$$

$$\lambda_n = -\boldsymbol{\eta}_0' \boldsymbol{\chi}_n \mathbf{w}_n / K , \tag{6.214}$$

where we assume $\boldsymbol{\eta}_0' \mathbf{w}_n = 1$. Using (6.214) in (6.212) yields Frost's algorithm, with $\boldsymbol{\chi}_n = \mathbf{X}_n \mathbf{X}_n'$, or some average of such terms:

$$\mathbf{w}_{n+1} = \mathbf{w}_n - \mu(\mathbf{J} - \boldsymbol{\eta}_0 \boldsymbol{\eta}_0'/K) \mathbf{X}_n \mathbf{X}_n' \mathbf{w}_n . \tag{6.215}$$

This is seen to be essentially the gradient projection algorithm (6.205) of *Lacoss* [6.36]. However, *Frost* actually used an alternate algorithm, which has some error-correcting features, designed to provide robustness against computational errors:

$$\begin{aligned}\mathbf{w}_{n+1} &= \mathbf{w}_n - \mu(\mathbf{J} - \boldsymbol{\eta}_0 \boldsymbol{\eta}_0'/K) \mathbf{X}_n \mathbf{X}_n' \mathbf{w}_n \\ &\quad - (\boldsymbol{\eta}_0/K)(\boldsymbol{\eta}_0' \mathbf{w}_n - 1) \\ &= (\mathbf{J} - \boldsymbol{\eta}_0 \boldsymbol{\eta}_0'/K)[\mathbf{w}_n - \mu(\mathbf{X}_n' \mathbf{w}_n) \mathbf{X}_n] + \boldsymbol{\eta}_0/K .\end{aligned} \tag{6.216}$$

If indeed $\boldsymbol{\eta}_0'\mathbf{w}_n = 1$, (6.215, 216) are identical. Otherwise, (6.216) tends to correct the error.

We have written (6.216) for the case of one directional sensitivity constraint $\boldsymbol{\eta}'(\boldsymbol{v}_0)\mathbf{w}_n = 1$. In fact, *Frost* [6.38] presented a more general form, in which a number of constraints $\boldsymbol{\eta}'(\boldsymbol{v}_l)\mathbf{w}_n = 1$, $l = 1, L$ may be held simultaneously. This can be useful in widening the main beam of the processor, or in simultaneously steering multiple beams.

Recently *Gangi* and *Byun* [6.39] have discussed Frost's algorithm, in reference to processing of LASA data.

Finally, mention should be made of the survey paper by *Widrow* et al. [6.40] on adaptive noise cancellation, and that of *Gabriel* [6.41] on adaptive array processors for radio-frequency use.

6.4 The Maximum-Entropy Method in Array Processing

In a presentation not widely noticed at the time, *Burg* [6.42] discussed a nonlinear spectral estimation technique which he called the maximum-entropy spectral analysis (MESA). *Burg* observed that the true spectrum (6.28) of a random process is in principle computable from the infinite span $-\infty \leqq \tau \leqq \infty$ of values $C(\tau)$ of the process autocovariance function. With a finite data span $x(t)$, $-T/2 \leqq t \leqq T/2$, only values of $C(\tau)$ for $-T \leqq \tau \leqq T$ are available. The conventional spectral estimator (6.43) uses estimates of $C(\tau)$ only over that span, and implicitly estimates $C(\tau) = 0$, $|\tau| > T$, thereby incurring the decreased resolution capabilities which were discussed above. *Burg* proposed to use the available nonzero estimates $\hat{C}(\tau)$ of $C(\tau)$ for $|\tau| \leqq T$ to construct a better, nonzero estimate of $C(\tau)$ for $|\tau| > T$. These "artificial" covariance values, which in reality are no more artificial than the values $\hat{C}(\tau) = 0$, $|\tau| > T$, could then be used to extend the correlation span used in the spectrum estimator, thereby improving its resolution properties. *Burg* chose to compute these artificial covariance values by a technique based on concepts of information theory, involving entropy considerations; hence the name, maximum-entropy spectral analysis.

In subsequent presentations, *Burg* [6.43, 44] related his method more strongly to the use of prediction-error filters, which was familiar ground for geophysical workers. The very good results, presented in [6.44], in application of the method to wave number analysis of seismic data, stimulated other investigators to consider the method. In particular, *Lacoss* [6.30] compared the maximum-likelihood and maximum-entropy methods as to resolution, both analytically and by computation, for some idealized cases, and found the maximum-entropy method to be superior. In an elegant note, *Burg* [6.45] supported this result analytically, in the general case. A comprehensive survey paper [6.46] subsequently appeared, summarizing various relevant developments in closely related fields of study, and presenting essentially the complete picture of the method in application to time-domain processing.

The maximum-entropy spectral analysis method has strong ties to many areas of very active work. *Van den Bos* [6.47] pointed out the link to autoregressive modeling in statistics, shortly after Burg's original presentations. *Burg* [6.44] himself worked directly from the ideas of predictive deconvolution used in geophysical processing [6.48]. In particular, the link to linear predictive coding in speech research [6.49] is very strong, and the intense recent activity in that area has led to much related literature. With the appearance of survey papers such as [6.46, 50, 51] there is little need to devote space here to detailing these links. Further, with the appearance of Burg's thesis [6.52], the original method has been fully detailed.

Accordingly, we will deal here only with the spatial domain version of maximum-entropy processing, a topic to which some preliminary attention has been given by *McDonough* [6.53]. We will first discuss the information-theoretic aspects of the problem, in dealing with an arbitrarily spaced sensor array. We will then specify the results to the case of a uniformly spaced line array, to which the standard results of the theory apply. Finally we will make some suggestions as to how to deal with the case of a nonuniformly spaced array.

6.4.1 The Information-Theoretic Basis of Maximum-Entropy Analysis

Shannon in 1948 published an important paper, reprinted as [6.55], which became the foundation of the discipline of information theory. The portion of that theory of interest here deals with the "information" conveyed by specifying one of several alternatives. In brief, if some number n of alternatives are possible in some situation, and if the alternatives have probabilities of occurrence p_i, known a priori, then the information conveyed by specifying alternative i is $-\log p_i$. Thus the average information conveyed, $\mathrm{E}[-\log p_i]$, is

$$H = -\sum_{i=1}^{n} p_i \log p_i , \tag{6.217}$$

where the logarithm can be taken as the natural logarithm (to the base e) by suitable definition of the units of H. This definition is essentially the only one which satisfies a resonable intuitive feeling as to how "information" should behave. Specifically, we are led uniquely to (6.217) by requiring only that H be a continuous function of the p_i; that, in the case of equally likely alternatives, the information carried by specification of any one of them be a monotonically increasing function of the number of possible alternatives n; and that the average information (6.217) be the numerical sum of the average information in the two component choices, if the specification of one alternative is done in two stages, weighted by the probability of each component choice being made. *Shannon* called the quantity in (6.217) the "entropy" of the collection p_i, $i=1, n$. In the case that the p_i are the probabilities of values of a random variable x,

$p_i = \text{Prob}(x = x_i)$, H in (6.217) is written $H(x)$, the "entropy of the random variable x."

In applying entropy ideas to spectral estimation, we need to deal with random variables which can take on a continuum of values. If x is a continuous-valued random variable, with probability density $p(x)$, we can write

$$p_i = \text{Prob}(x_i \leqq x \leqq x_i + \Delta x) = p(x_i)\Delta x\,, \tag{6.218}$$

so that (6.217) becomes

$$\begin{aligned} H(x) &= -\lim_{\Delta x \to 0} \sum_i p(x_i)\Delta x \log[p(x_i)\Delta x] \\ &= -\lim_{\Delta x \to 0} \sum_i p(x_i)[\log p(x_i)]\Delta x \\ &\quad -\lim_{\Delta x \to 0} \sum_i p(x_i)[\log(\Delta x)]\Delta x\,. \end{aligned} \tag{6.219}$$

The first term on the right in (6.219) has a limit, but the second term diverges. However, since the second term behaves for small Δx as

$$-\log(\Delta x) \int_{-\infty}^{\infty} p(x)dx = -\log(\Delta x)\,, \tag{6.220}$$

which is independent of $p(x)$, it will cancel in any calculation involving the difference in the entropy of two random variables. Since such calculations are the ones of interest, as it turns out, for a continuous-valued random variable the entropy is defined as

$$H(x) = -\int_{-\infty}^{\infty} p(x)[\log p(x)]dx\,, \tag{6.221}$$

corresponding to the limit of the first term on the right in (6.219).

In dealing with a zero-mean stationary stochastic process $x(t)$, $-T/2 \leqq t \leqq T/2$, we can initially consider the process to be band limited to B Hz and time-sampled with $\Delta t = 1/(2B)$, yielding a collection of random variables $x(i\Delta t)$, $i = -BT, BT$, which may be arranged into a vector $\mathbf{x}$. The entropy of the vector $\mathbf{x}$, which we may call the entropy of the stochastic process $x(t)$, is then just (6.221), where now $p(\mathbf{x})$ is the joint probability density of the samples $x(i\Delta t)$ and the integral is multidimensional. In our discussion, however, we have to deal with the spectrum of the process $x(t)$, rather than with the time samples themselves. Therefore, we introduce the discrete spectrum through the discrete form of (6.18):

$$X(n\Delta f) = (\Delta t/\sqrt{T}) \sum_{i=-BT}^{BT} x(i\Delta t)\exp(-jn2\pi i\Delta t/T)$$

$$= 1/(2B\sqrt{T}) \sum_{i=-BT}^{BT} x(i\Delta t)\exp(-jn2\pi i/2BT), \quad n = -BT, BT. \quad (6.222)$$

We are then interested in the entropy of the vector random variable **X** of the Fourier coefficients $X(n\Delta f)$. This is given by [6.55]:

$$H(\mathbf{X}) = H(\mathbf{x}) - \int_{-\infty}^{\infty} p(\mathbf{x}) \log|\det(\partial\mathbf{x}/\partial\mathbf{X})| d\mathbf{x}, \quad (6.223)$$

where we indicate the absolute value of the determinant of the Jacobian matrix $\partial\mathbf{x}/\partial\mathbf{X}$ of the transformation (6.222). With the particular choice of constant in (6.222), we have $|\partial\mathbf{x}/\partial\mathbf{X}| = \sqrt{T}$, so that [6.56]

$$H(\mathbf{X}) = H(\mathbf{x}) - \log\sqrt{T}. \quad (6.224)$$

By a different choice of constants in (6.17, 18), we may arrange $H(\mathbf{X}) = H(\mathbf{x})$, but since we will be dealing only with maximization problems, the additive constant in (6.222) is of no concern.

In (6.224), the entropy of the sampled stochastic process $x(i\Delta t)$ is expressed in terms of the Fourier coefficients $X(n\Delta f)$ of the process at discrete frequencies:

$$H(\mathbf{x}) = H(\mathbf{X}) + \log\sqrt{T}. \quad (6.225)$$

Here

$$H(\mathbf{X}) = -\int_{-\infty}^{\infty} p(\mathbf{X})[\log p(\mathbf{X})] d\mathbf{X} \quad (6.226)$$

is a $2BT$-fold multiple integral, corresponding to the $2BT$ discrete Fourier coefficients. What we would like to have is an expression for $H(\mathbf{x})$ in terms of a one-dimensional integral along the power density spectrum $P_x(f)$ of the stochastic process $x(t)$, with $x(t)$ not necessarily band limited. Such an expression can be found provided we assume $x(t)$ to be a Gaussian stochastic process, which suffices for our purposes.

As a preliminary, consider a linear time-invariant filter with transfer function $G(f)$. Let $n(t)$ be a band-limited process at the input, with Fourier coefficients $N(n\Delta f)$, and let $x(t)$ be the corresponding band-limited output, with Fourier coefficients $X(n\Delta f)$. Then

$$X(n\Delta f)=G(n\Delta f)N(n\Delta f), \tag{6.227}$$

which is a linear transformation relating the random variables $X(n\Delta f)$ and $N(n\Delta f)$. The Jacobian matrix $\partial\mathbf{X}/\partial\mathbf{N}$ is diagonal with determinant having the absolute value

$$|\det(\partial\mathbf{X}/\partial\mathbf{N})|=\prod_{n=-BT}^{BT}|G(n\Delta f)|. \tag{6.228}$$

Then, from (6.223)

$$\begin{aligned} H(\mathbf{X})&=H(\mathbf{N})+\int_{-\infty}^{\infty}p(\mathbf{N})\sum_{n=-BT}^{BT}\log|G(n\Delta f)|d\mathbf{N} \\ &=H(\mathbf{N})+\sum_{n=-BT}^{BT}\log|G(n\Delta f)|. \end{aligned} \tag{6.229}$$

In this form, we can heuristically pass to the limit $T\to\infty$, taking account of the fact that $\Delta f=1/T$, provided we work with the entropy "per degree of freedom" of the band-limited signals:

$$H'(\mathbf{x})\triangleq H(\mathbf{x})/2BT. \tag{6.230}$$

Then we obtain from (6.225, 229) that

$$H'(\mathbf{x})=H'(\mathbf{n})+\frac{1}{2B}\int_{-B}^{B}\log|G(f)|df. \tag{6.231}$$

In (6.231), we may take the filter input $n(t)$ and output $x(t)$ to be band-limited Gaussian stochastic processes, with power spectral densities say $P_n(f)$, $P_x(f)$. Then we have [6.7]

$$P_x(f)=|G(f)|^2P_n(f). \tag{6.232}$$

Finally, let us assume that $n(t)$ is band-limited Gaussian white noise, with power spectral density N_0, $|f|\leqq B$. Then, we find that [6.55]

$$H'(n)=\log(4\pi eBN_0)^{1/2}/\log 2. \tag{6.233}$$

Finally choosing $N_0=1$, we have from (6.232) that

$$|G(f)|^2=P_x(f), \tag{6.234}$$

and in turn from (6.231) that

$$H'(x) = \log(4\pi eB)^{1/2}/\log 2 + \frac{1}{2B}\int_0^B \log P_x(f)df. \tag{6.235}$$

This is the formula we need, expressing the entropy per degree of freedom of a band-limited random process $x(t)$ in terms of its power spectral density $P_x(f)$.

Let us now proceed to indicate the use of these formulas in the estimation of power spectral densities.

6.4.2 The Principle of Maximum Entropy

The first use of entropy ideas for calculating probability density functions appears to occur in a paper by *Jaynes* [6.4], who was concerned with providing a firm basis for the procedures of statistical mechanics. In statistical mechanics, we hypothesize a certain probability structure for the states of some physical system, and then calculate therefrom the average values of the various thermodynamic quantities of interest. *Jaynes* proposed to choose as the underlying probability distribution of the microscopic states of the system, that distribution which corresponded to minimum assumptions about the system, beyond what is surely known by measurement (the values of certain macroscopic variables). Accordingly, he chose as underlying distribution that function which had maximum Shannon entropy, among all those distributions which were in accord with the observed macroscopic values, these latter being the values of various ensemble average quantities. Specifically, *Jaynes* maximized the entropy of the discrete probabilities p_i of the microscopic states x_i, under the constraint that certain averages $E[f_l(x)]$, $l=1, L$, be contrained to have fixed values, say β_l. Adjoining constraints to the "cost" (6.217) leads to the augmented cost

$$J = -\sum_i p_i \log p_i - \sum_{l=1}^{L} \lambda_l \left(\sum_i p_i f_l(x_i) - \beta_l\right)$$

$$-\lambda_0\left(\sum_i p_i - 1\right), \tag{6.236}$$

where the final constraint is included to insure that we obtain a proper set of probabilities. The necessary conditions for a maximum are

$$\partial J/\partial p_i = 0 = -1 - \log p_i - \sum_{l=1}^{L} \lambda_l f_l(x_i) - \lambda_0, \tag{6.237}$$

from which

$$p_i = \exp\left[-1 - \lambda_0 - \sum_{l=1}^{L} \lambda_l f_l(x_i)\right] \tag{6.238}$$

are the desired probabilities. The $f_l(x_i)$ represent the values of certain quantities f_l if the system is in state x_i. The Lagrange multipliers are found by substituting (6.238) into the constraints indicated in (6.236)

$$1=\sum_i \exp\left[-1-\lambda_0-\sum_l \lambda_l f_l(x_i)\right], \tag{6.239}$$

$$\lambda_0=-1+\log Z(\lambda_1,\ldots,\lambda_L), \tag{6.240}$$

where

$$Z(\lambda_1,\ldots,\lambda_L)\triangleq\sum_i \exp\left[-\sum_l \lambda_l f_l(x_i)\right]. \tag{6.241}$$

Further, the $\lambda_1,\ldots,\lambda_L$ are determined by

$$\sum_i f_l(x_i)\exp\left[-1-\lambda_0-\sum_l \lambda_l f_l(x_i)\right]=\beta_l, \qquad l=1,L, \tag{6.242}$$

which, with (6.240, 241), is equivalent to

$$-\partial(\log Z)/\partial\lambda_l=\beta_l, \qquad l=1,L, \tag{6.243}$$

which is, in general, a nonlinear system of equations.

These developments of *Jaynes* in regard to thermodynamics and statistical mechanics were followed up by *Tribus* [6.57] in a thermodynamics text based on this new approach, and in a text [6.58] on design and decision making in the face of uncertainty. However, the application to spectral estimation seems to have first been made by *Burg* [6.42]. The precise nature of the link was pointed out by *Edward* and *Fitelson* [6.59], and at about the same time by *Ables* [6.60]. We suppose there to be available time samples $x(i\Delta t)$, $i=-BT, BT$, of a zero-mean (for convenience) stationary stochastic process $x(t)$, band limited to $|f|\leqq B$ Hz, and extending over a time span $-T/2\leqq t\leqq T/2$. Accordingly, estimates $\hat{C}_x(l\Delta\tau)$, $l=-L, L$, of the autocovariance function of $x(t)$ are computed over the span $-T\leqq\tau\leqq T$, with $\Delta\tau=\Delta t=1/2B$. We seek an estimator $\hat{P}_x(f)$ of the power density spectrum $P_x(f)$. Following the maximum-entropy philosophy, we take $\hat{P}_x(f)$ to be the spectrum corresponding to the joint-probability density of the $x(i\Delta t)$ which has maximum entropy, under the constraint that the observed covariance values $\hat{C}_x(l\Delta\tau)$ be attained.

First, the desired probability density is Gaussian, since among all probability densities having the same second-order moments, that with maximum entropy is Gaussian [6.55]. Thus we may use (6.235) for the entropy expression, and since we are concerned only with a maximization problem, we may drop the constant first term on the right. Thus we seek

$$\hat{P}_x(f)\Leftarrow\max_{P_x(f)}\int_{-B}^{B}\log P_x(f)df. \tag{6.244}$$

The constraints are, from (6.27),

$$\hat{C}_x(l\Delta\tau) = \int_{-B}^{B} P_x(f)\exp(\mathrm{j}2\pi f l\Delta\tau)df\,, \qquad l = -L, L\,, \tag{6.245}$$

where we assume that the covariance estimates are in fact exact, as a first-order approximation. Adjoining the constraints (6.245) to the cost (6.244) with Lagrange multipliers, we obtain the augmented cost

$$J = \int_{-B}^{B} \log P_x(f)df - \sum_{l=-L}^{L} \lambda_l \left[\int_{-B}^{B} P_x(f)\exp(\mathrm{j}2\pi f l\Delta\tau)df - \hat{C}_x(l\Delta\tau)\right]. \tag{6.246}$$

This is a routine variational problem, with necessary condition for an extremum:

$$\hat{P}_x(f) = 1\Big/\sum_{l=-L}^{L} \lambda_l \exp(\mathrm{j}2\pi f l\Delta\tau)\,. \tag{6.247}$$

The Lagrange multipliers are found as the solution to the nonlinear system resulting by using (6.247) in (6.245)

$$\int_{-B}^{B}\left[\sum_{m=-L}^{L} \lambda_m \exp(\mathrm{j}2\pi f m\Delta\tau)\right]^{-1} \exp(\mathrm{j}2\pi f l\Delta\tau)df = \hat{C}_x(l\Delta\tau), \qquad l = -L, L\,. \tag{6.248}$$

This procedure has been used by *Frieden* [6.61] in a picture-restoration application.

It is interesting to find that $\hat{P}_x(f)$, with the λ_l being the solution of (6.248), is exactly [6.59]

$$\hat{P}_x(f) = 1\Big/\left|\sum_{l=0}^{L} \alpha_l \exp(-\mathrm{j}2\pi f l\Delta\tau)\right|^2, \tag{6.249}$$

where the numbers α_l are related to the Lagrange multipliers $\lambda_l = \lambda^*_{-l}$ by (here λ_l is in fact real)

$$\sum_{m=0}^{L-l} \alpha_m \alpha^*_{l+m} = \lambda_l\,, \qquad l = 0, 1, \ldots, L\,. \tag{6.250}$$

[This is seen by expanding the denominator of (6.249) as a double sum, making standard changes of variable, and recognizing that (6.250) is necessary to obtain

(6.247).] Further, taking account of the fact that sampling is at the Nyquist rate, so that $\Delta\tau = \Delta t = 1/2B$, following [6.59] it can be shown that the constraints in (6.246) are satisfied provided the α_m satisfy

$$\sum_{m=0}^{L} \hat{C}_x[(l-m)\Delta\tau]\alpha_m = \delta_{l,0}/\alpha_0, \qquad l=0,L. \tag{6.251}$$

Defining $\hat{\mathbf{C}}_x$ as the matrix with elements

$$(\hat{\mathbf{C}}_x)_{lm} = \hat{C}_x[(l-m)\Delta\tau], \qquad l,m=0,L, \tag{6.252}$$

and defining the vectors

$$\boldsymbol{\alpha} = \operatorname*{col}_m(\alpha_m), \qquad m=0,L, \tag{6.253}$$

$$\boldsymbol{\delta} = \operatorname{col}(1,0,\ldots,0), \tag{6.254}$$

$$\boldsymbol{\varepsilon} = \operatorname*{col}_m[\exp(\mathrm{j}2\pi f m\Delta\tau)], \qquad m=0,L, \tag{6.255}$$

the maximum-entropy spectrum (6.249) becomes

$$\hat{P}_x(f) = 1/|\boldsymbol{\varepsilon}'\boldsymbol{\alpha}|^2, \tag{6.256}$$

where the prime indicates the conjugate transpose. Further, from (6.251).

$$\boldsymbol{\alpha} = \hat{\mathbf{C}}_x^{-1}\boldsymbol{\delta}/\alpha_0. \tag{6.257}$$

Using this result in (6.256), we get

$$\hat{P}_x(f) = \alpha_0^2/|\boldsymbol{\varepsilon}'\hat{\mathbf{C}}_x^{-1}\boldsymbol{\delta}|^2. \tag{6.258}$$

Finally, (6.257) yields

$$\alpha_0 = (\hat{\mathbf{C}}_x^{-1})_{00}/\alpha_0, \tag{6.259}$$

so that (6.258) becomes

$$\hat{P}_x(f) = (\hat{\mathbf{C}}_x^{-1})_{00}/|\boldsymbol{\varepsilon}'\hat{\mathbf{C}}_x^{-1}\boldsymbol{\delta}|^2. \tag{6.260}$$

The nonlinearity of (6.248) has thus been "concentrated" into (6.259), by the use of the factorization of the denominator of (6.247) as indicated in (6.249).

These developments lead at once to all the interpretations of maximum-entropy spectra in terms of autoregressive modeling and linear prediction. Writing out (6.256), for example, after dividing the numerator and denominator by α_0^2, we have

$$\hat{P}_x(f)=(1/\alpha_0^2)\left|1+\sum_{l=1}^{L}(\alpha_l/\alpha_0)\exp(-j2\pi fl\Delta\tau)\right|^{-2}. \tag{6.261}$$

Defining

$$\sigma_n^2=1/\alpha_0^2, \tag{6.262}$$

$$a_l=\alpha_l/\alpha_0, \qquad l=1,L, \tag{6.263}$$

(6.261) becomes

$$\sigma_n^2=\left|1+\sum_{l=1}^{L}a_l\exp(-j2\pi fl\Delta\tau)\right|^2\hat{P}_x(f), \tag{6.264}$$

which is the expression for the output power spectrum of a digital filter which whitens the sampled input signal $x(i\Delta t)$. Thus, if we can determine such a filter, its coefficients and its output variance lead at once to the maximum-entropy spectrum.

Such a whitening filter is described in the frequency domain by [6.15]

$$\left[1+\sum_{l=1}^{L}a_l\exp(-j2\pi fl\Delta\tau)\right]X(z)=N(z), \tag{6.265}$$

where $X(z)$ and $N(z)$ are the z transforms of the sampled input and output signals. In the time domain, this becomes

$$x(i\Delta t)+\sum_{l=1}^{L}a_lx[(i-l)\Delta t]=n(i\Delta t), \tag{6.266}$$

where $n(i\Delta t)$ is to be discrete white noise. We thus seek to model the data as an autoregressive sequence. It turns out that exactly the correct filter coefficients a_l result, if we interpret (6.266) as a one-step prediction-error filter along the data, and seek a_l as the coefficients in the least-squares fit

$$x(i\Delta t)\sim-\sum_{l=1}^{L}a_lx[(i-l)\Delta t]. \tag{6.267}$$

The well-known Levinson algorithm [6.24] serves well in all these problems, since the matrix to be inverted is (6.252), which is Toeplitz.

Two procedures are common in calculating the spectrum. The covariances $\hat{C}_x(l\Delta\tau)$, $l=0, L$, can be estimated in some way, perhaps using the usual linear estimator (6.40). These can then be used in (6.251), or some equivalent expression, and the resulting equations solved sequentially for increasing L, using the Levinson recursion. Alternatively, a filter such as (6.266) can be fitted directly to the time-domain data, and the order of the filter L increased sequentially, again using a Levinson algorithm. The former procedure implicitly assumes data off the end of the observation interval vanish, or repeat periodically in the digital version. The latter procedure, which is the one developed by *Burg* originally, and described by *Andersen* [6.62] makes no such implicit assumption. For short data spans, Burg's choice would seem more in the spirit of entropy processing. The choice may involve a bias/variance trade-off, with Burg's procedure having better resolution properties, but larger variance.

All these matters related to (6.264), including the important question of how large to take the maximum model order, have been fully treated in [6.46, 51–53]. (See also Chaps. 2, 3 for a detailed treatment of maximum-entropy spectral estimation and autoregressive modeling.)

We now turn to a consideration of the use of these techniques in array processing.

6.4.3 Maximum-Entropy Spatial Processing

Again we assume a signal field $x(t, z)$, with sensors at some positions z_i, $i=1, K$. Assuming space–time stationarity, the desired frequency–wave-number spectrum is as in (6.37), which we rewrite as

$$S_x(f, \boldsymbol{v}) = \int_V P_x(f, \boldsymbol{r}) \exp(-\mathrm{j}2\pi \boldsymbol{v}\cdot\boldsymbol{r}) d\boldsymbol{r}, \tag{6.268}$$

where

$$P_x(f, \boldsymbol{r}) = \int_{-\infty}^{\infty} C_x(\tau, \boldsymbol{r}) \exp(-\mathrm{j}2\pi f\tau) d\tau, \tag{6.269}$$

with

$$C_x(\tau, \boldsymbol{r}) = \mathrm{E}[[x(t+\tau, z+\boldsymbol{r}) - m_x][x(t, z) - m_x]^*] \tag{6.270}$$

being the space-time autocovariance function. As before, we assume adequate time span of data that a good estimate of the cross-spectrum between any two sensors is available:

$$\hat{P}_x(f, z_i, z_j) \approx \int_{-\infty}^{\infty} C_x(\tau, z_i, z_j) \exp(-\mathrm{j}2\pi f\tau) d\tau, \tag{6.271}$$

where the cross-covariance is

$$C_x(\tau, z_i, z_j) = \mathrm{E}[[x(t+\tau, z_i) - m_{x_i}][x(t, z_j) - m_{x_j}]^*]. \tag{6.272}$$

The estimate (6.271) may be obtained by conventional linear processing, by smoothing periodograms over data segments, as discussed earlier. Given the cross-spectra $\hat{P}_x(f, z_i, z_j)$, $i, j=1, K$, we then seek an estimator $\hat{S}_x(f, \boldsymbol{\nu})$ of the spectrum (6.268), using entropy ideas.

It may be pointed out explicitly here that we will not be concerned with the "multichannel" or multivariate maximum-entropy spectral analysis problem. In that case, we have available the K time series with values $x(i\Delta t, z_k)=x_k(i\Delta t)$, $k=1, K$, and seek an estimator of the K by K dimensional cross-spectral matrix, with elements

$$P_{kl}(f)=P_x(f, z_k, z_l). \tag{6.273}$$

The multivariate version of time-domain maximum-entropy spectral analysis has been worked out for this case by *Burg* [6.52], *Nuttall* [6.63], and *Strand* [6.64], and can be used to determine the estimates (6.271) which we assume here to be available; *Baggeroer* [6.65] has considered such a multivariate case also (see also Sect. 2.12). On the other hand, we are concerned with spectral analysis over multiple independent variables (space and time).

The expression for the entropy per degree of freedom of the sensor signals can easily be written down, following the same procedures used above in the time domain. Let us assume the space–time field $x(t, z)$ to be band limited in frequency to $|f|\leqq B$, as above, and also in wave number. Suppose each component ν_i of the wave number vector $\boldsymbol{\nu}$ to be limited as $|\nu_i|\leqq W$, where we can take W to be the largest of the three limits of ν_i, $i=1, 2, 3$. Then just as we considered the signals $x(t)$ above to be time sampled, to yield $x(i\Delta t)$, $\Delta t=1/2B$, we can consider $x(t, z)$ to be space sampled as well, yielding $x(i\Delta t, l_1\Delta z, l_2\Delta z, l_3\Delta z)$, where $\Delta z=1/2W$. Initially we consider a time extent $-T/2\leqq t\leqq T/2$, and signals in the cube $-L/2\leqq z_l\leqq L/2$. We thus have $2BT$ time samples at each of $(2WL)^3$ space points (at most of which there will not, in fact, be a sensor). We then consider $\mathbf{x}$ to be the vector of the totality of all these samples, of dimension $16BT(WL)^3$, each sample being a random variable. Some joint probability function $p(\mathbf{x})$ can be considered. The developments presented earlier then apply, and we can write the entropy of the field as in (6.221).

The shift to spectral representation is made through the four-dimensional Fourier transformation, yielding

$$\begin{aligned}
&X(n\Delta f, m_1\Delta\nu, m_2\Delta\nu, m_3\Delta\nu)\\
&\quad=(1/16BW^3T^{1/2}L^{3/2})\sum_{i=-BT}^{BT}\sum_{l_1=-WL}^{WL}\cdots\sum_{l_3=-WL}^{WL}x(i\Delta t, l_1\Delta z, \ldots, l_3\Delta z)\\
&\qquad\cdot\exp(-\mathrm{j}2\pi ni/2BT)\\
&\qquad\cdot\exp[-\mathrm{j}2\pi(m_1l_1+\ldots+m_3l_3)/2WL],\\
&\qquad\qquad n=-BT, BT;\ m_1, m_2, m_3=-WL, WL.
\end{aligned} \tag{6.274}$$

Each component $X(n, \boldsymbol{m})$ can be written, as in (6.227), as

$$X(n, \boldsymbol{m}) = G(n\Delta f, \boldsymbol{m} \cdot \Delta \boldsymbol{v}) N(n, \boldsymbol{m}), \tag{6.275}$$

a further change of variable. Finally, assuming $x(t, z)$ to be Gaussian, and letting $T, L \to \infty$, maintaining the band-limited assumptions, we obtain from (6.235) the result

$$H'(x) = \log(32\pi e B W^3)^{1/2}/\log 2 + \frac{1}{16BW^3} \int_0^B \int_0^W \dots \int_0^W \log S_x(f, \boldsymbol{v}) d\boldsymbol{v} df, \tag{6.276}$$

which, except for constants, has been written down also for one spatial dimension by *Ables* [6.60].

We thus wish to maximize the integral term on the right in (6.276). We assume the time-frequency processing to have been done elsewhere, and accordingly treat frequency as a fixed parameter. We thus seek

$$\hat{S}_x(f, \boldsymbol{v}) \Leftarrow \max_{S_x(f, \boldsymbol{v})} \int_0^W \dots \int_0^W \log S_x(f, \boldsymbol{v}) d\boldsymbol{v}. \tag{6.277}$$

The constraints are that the observed estimated cross-power spectra be compatible with the estimated wave number spectrum:

$$\hat{P}_x(f, \boldsymbol{z}_i, \boldsymbol{z}_j) = \int_V S_x(f, \boldsymbol{v}) \exp[\mathrm{j} 2\pi \boldsymbol{v} \cdot (\boldsymbol{z}_i - \boldsymbol{z}_j)] d\boldsymbol{v}, \qquad i, j = 1, K, \tag{6.278}$$

where we have used the inverse relationship to (6.268):

$$P_x(f, \boldsymbol{r}) = \int_V S_x(f, \boldsymbol{v}) \exp(\mathrm{j} 2\pi \boldsymbol{v} \cdot \boldsymbol{r}) d\boldsymbol{v}, \tag{6.279}$$

evaluated for lag $\boldsymbol{r} = \boldsymbol{z}_1 - \boldsymbol{z}_2$.

The solution to the variational problem (6.277, 278) is

$$\hat{S}_x(f, \boldsymbol{v}) = 1 \Big/ \sum_{i,j=1}^{K} \lambda_{ij} \exp[\mathrm{j} 2\pi \boldsymbol{v} \cdot (\boldsymbol{z}_i - \boldsymbol{z}_j)], \tag{6.280}$$

where

$$\int_{V_L} \left\{ \sum_{i,j=1}^{K} \lambda_{ij} \exp[\mathrm{j} 2\pi \boldsymbol{v} \cdot (\boldsymbol{z}_i - \boldsymbol{z}_j)] \right\}^{-1} \cdot \exp[\mathrm{j} 2\pi \boldsymbol{v} \cdot (\boldsymbol{z}_k - \boldsymbol{z}_l)] d\boldsymbol{v} = \hat{P}_x(f, \boldsymbol{z}_k, \boldsymbol{z}_l), \qquad k, l = 1, K. \tag{6.281}$$

More compactly, these can be written

$$\hat{S}_x(f, \boldsymbol{v}) = [\boldsymbol{\eta}'(\boldsymbol{v})\boldsymbol{\Lambda}(f)\boldsymbol{\eta}(\boldsymbol{v})]^{-1}, \tag{6.282}$$

$$\int_{V_L} [\boldsymbol{\eta}'(\boldsymbol{v})\boldsymbol{\Lambda}(f)\boldsymbol{\eta}(\boldsymbol{v})]^{-1}\boldsymbol{\eta}(\boldsymbol{v})\boldsymbol{\eta}'(\boldsymbol{v})d\boldsymbol{v} = \hat{\mathbf{P}}_x(f), \tag{6.283}$$

where, as in (6.134), we have

$$\boldsymbol{\eta}(\boldsymbol{v}) = \underset{k}{\mathrm{col}} \exp(\mathrm{j}2\pi\boldsymbol{v}\cdot\boldsymbol{z}_k), \qquad k = 1, K, \tag{6.284}$$

$\boldsymbol{\Lambda}$ is the matrix of the λ_{ij}, and $\hat{\mathbf{P}}_x(f)$ is the estimated frequency cross-spectral matrix of the sensor signals, with elements

$$(\hat{\mathbf{P}}_x)_{kl} = \hat{P}_x(f, \boldsymbol{z}_k, \boldsymbol{z}_l), \tag{6.285}$$

and V_L indicates the finite region of $\boldsymbol{v}$-space which is the signal "band". As always, the "prime" indicates conjugate transpose.

The form (6.282) for the maximum-entropy estimator should be compared to that of the maximum-likelihood estimator (6.154). Were we to have $\boldsymbol{\Lambda} = \hat{\mathbf{P}}_x^{-1}$, then (6.282) would be the maximum-likelihood estimator. However, this is not the case. *Burg* [6.45] has written down the formula (6.280) for the case of a uniformly spaced line array of sensors, and proved that its resolution properties are superior to those of the maximum-likelihood processor for the same case.

In the general case of the problem defined by (6.282, 283), no simplification equivalent to that used by *Edward* and *Fitelson* [6.59] in treating the one-dimensional case defined by (6.247, 248) seems to have been worked out. We shall not pursue the point, but rather turn now to the case of the line array.

6.4.4 Maximum-Entropy Processing for a Uniformly Spaced Line Array

Let us assume the sensors of the array to be located along a straight line in space, which we can take to be the coordinate axis z, the third component of the general position vector $\boldsymbol{z}$. Then the basic expressions (6.280, 281) become

$$\hat{S}_x(f, \boldsymbol{v}) = 1 \bigg/ \sum_{k,l=1}^{K} \lambda_{kl} \exp[\mathrm{j}2\pi v_z(z_k - z_l)], \tag{6.286}$$

where v_z is the z component of $\boldsymbol{v}$, and

$$\int_{V_L} \left\{ \sum_{k,l=1}^{K} \lambda_{kl} \exp[\mathrm{j}2\pi v_z(z_k - z_l)] \right\}^{-1} \cdot \exp[\mathrm{j}2\pi v_z(z_m - z_n)] d\boldsymbol{v} = \hat{P}_x(f, z_m, z_n), \qquad m, n = 1, K. \tag{6.287}$$

No particular simplification is workable here, unless we further assume the sensors to be uniformly spaced, with $z_k = k\Delta z$. Then the matrix $\hat{\mathbf{P}}_x$ is in principle Toeplitz,

$$\hat{P}_x(f, z_m, z_n) = \hat{P}_x[f, (m-n)\Delta z]. \tag{6.288}$$

Since then all quantities in (6.287) depend only on index differences, we can seek a solution $\lambda_{kl} = \lambda_{k-l}$. With this assumption, and the usual changes of variable, (6.286, 287) become

$$\hat{S}_x(f, v) = 1 \Big/ \sum_{m=-(K-1)}^{K-1} \mu_m \exp(\mathrm{j}\, 2\pi v_z m \Delta z), \tag{6.289}$$

where we define

$$\mu_m = (\mathrm{K} - |\mathrm{m}|)\lambda_m, \tag{6.290}$$

and

$$4W^2 \int_{-W}^{W} \left[\sum_{m=-(K-1)}^{K-1} \mu_m \exp(\mathrm{j}\, 2\pi v_z m \Delta z) \right]^{-1} \cdot \exp[\mathrm{j}\, 2\pi v_z (k-l)\Delta z] dv_z = \hat{P}_x[f, (k-l)\Delta z], \qquad k, l = 1, K, \tag{6.291}$$

where $|v_x| \leqq W$, and similarly for v_y and v_z, is the band limit in wave number. Finally, we can collapse the K^2 equations of (6.291) into $2K-1$ equations, by eliminating duplicates, and obtain

$$4W^2 \int_{-W}^{W} \left[\sum_{m=-(K-1)}^{K-1} \mu_m \exp(\mathrm{j}\, 2\pi v_z m \Delta z) \right]^{-1} \cdot \exp(\mathrm{j}\, 2\pi v_z n \Delta z) dv_z = \hat{P}_x(f, m\Delta z), \qquad m = -(K-1), K-1. \tag{6.292}$$

Equations (6.289, 292) are identical to (6.247, 248), except for constants, and we at once conclude that (6.260) holds:

$$4W^2 \hat{S}_x(f, v_z) = (\hat{\mathbf{P}}_x^{-1})_{00} / |\boldsymbol{\varepsilon}' \hat{\mathbf{P}}_x^{-1} \boldsymbol{\delta}|^2, \tag{6.293}$$

where $\hat{\mathbf{P}}_x$ is the Toeplitz matrix formed from the numbers $\hat{P}_x(f, m\Delta z)$, i.e.,

$$(\hat{\mathbf{P}}_x)_{kl} = \hat{P}_x[f, (k-l)\Delta z], \qquad k, l = 1, K, \tag{6.294}$$

$$\boldsymbol{\varepsilon} = \underset{k}{\mathrm{col}} \exp(\mathrm{j}\, 2\pi v_z k \Delta z), \qquad k = 1, K, \tag{6.295}$$

$$\boldsymbol{\delta} = \mathrm{col}(1, 0, \ldots, 0). \tag{6.296}$$

All the implications about spatial autoregressive models, prediction-error filters in the spatial domain, and so forth, follow at once, as *Burg* [6.45] points out. In particular, the spectrum (6.293) may be computed by sliding prediction-error filters forward and backward over the data $\hat{P}_x(f, m\Delta z)$, sequentially increasing the model order using a Levinson algorithm until a stopping criterion, such as that of *Akaike* [6.66] is satisfied.

In all the manipulations along this line, the complex conjugate operations implied by the "prime" symbol are important, since the matrix $\hat{\mathbf{P}}_x$ is in general complex. This follows because the sensor cross-covariance functions are not generally even functions of the time-lag variable, leading to complex-valued sensor cross-power spectra.

Because of the importance of the special case of a uniformly spaced line array, we shall write out the corresponding algorithm in some detail.

Let us consider first the so-called Yule–Walker procedure [6.46]. Let the sensor cross-power spectral density estimates $\hat{P}_x(f, z_i, z_j)$, $i,j=1, K$, be available, where $z_i = i\Delta z$.

Define

$$\hat{P}_x(f, m) = \text{aver}\, \hat{P}_x(f, z_i, z_j), \qquad m=0, K-1, \tag{6.297}$$

where the averaging is over all $i.j \leqq i$ such that $i-j=m$. Define the matrix $\hat{\mathbf{P}}_x(f)$ by

$$\begin{aligned}
&[\hat{\mathbf{P}}_x(f)]_{ij} = \hat{\mathbf{P}}_x(f, m), \qquad i,j=1, K;\ j \leqq i;\ i-j=m;\ m=0, K-1, \\
&[\hat{\mathbf{P}}_x(f)]_{ji} = [\hat{\mathbf{P}}_x(f)]^*_{ij}, \qquad i,j=1, K.
\end{aligned} \tag{6.298}$$

Thus $\hat{\mathbf{P}}_x(f)$ is Toeplitz.

We now need to solve the equations corresponding to (6.251), which it is convenient to write in the form involving the coefficients a_l of the autoregressive model (6.266). Dividing (6.251) through by α_0, and using (6.262, 263), we obtain

$$\sum_{m=0}^{K-1} \hat{P}_x(f, l-m) a_m = \delta_{l,0} \sigma_n^2, \qquad a_0 = 1, \qquad l=0, K-1. \tag{6.299}$$

Writing out the first equation of this set separately, we obtain

$$\sigma_n^2 = \hat{P}_x(f, 0) + \sum_{m=1}^{K-1} a_m \hat{\mathbf{P}}^*_x(f, m), \tag{6.300}$$

$$\sum_{m=1}^{K-1} a_m \hat{P}_x(f, l-m) = -\hat{P}_x(f, l), \qquad l=1, K-1. \tag{6.301}$$

This can be solved sequentially by considering a general value of $K-1$, call it M, with $M=1, K-1$. Then we have the family

$$\sigma_M^2 = \hat{P}_x(0) + \sum_{m=1}^{M} a_m^{(M)} \hat{P}^*_x(m), \tag{6.302}$$

$$\sum_{m=1}^{M} a_m^{(M)} \hat{P}_x(l-m) = -\hat{P}_x(l), \qquad l=1, M. \tag{6.303}$$

For $M=1$, we have at once

$$a_1^{(1)} = -\hat{P}_x(1)/\hat{P}_x(0),$$
$$\sigma_1^2 = \hat{P}_x(0) + a_1^{(1)}\hat{P}_x^*(1) = \hat{P}_x(0)(1-|a_1^{(1)}|^2). \tag{6.304}$$

For $M=2,3\ldots$ we have [6.53, 67]

$$a_M^{(M)} = \frac{-1}{\sigma_{M-1}^2}\left[\sum_{m=1}^{M-1} a_m^{(M-1)}\hat{P}_x(M-m) + \hat{P}_x(M)\right], \tag{6.305}$$

$$a_m^{(M)} = a_m^{(M-1)} + a_M^{(M)} a_{M-m}^{(M-1)*}, \qquad m=1, M-1, \tag{6.306}$$

$$\sigma_M^2 = \sigma_{M-1}^2(1-|a_M^{(M)}|^2). \tag{6.307}$$

This is essentially the Levinson algorithm, and allows sequential refinement of the spectral estimate as M increases.

The maximum possible value of M in (6.302, 303) is $K-1$, else we exhaust the available covariance estimates, due to "running off the end of the array". However, good stability of the spectral estimates $\hat{P}_x(f, m)$ dictates that the maximum lag $m\Delta z$ used be considerably less than the array length. The Akaike final prediction-error criterion [6.66] balances resolution with variance, and prescribes that the largest M be taken such that the criterion

$$(\mathrm{FPE})_M = \frac{K+(M+1)}{K-(M+1)}\sigma_M^2 \tag{6.308}$$

is minimum. (See Sects. 2.11, 3.6 for a detailed assessment of this criterion.)

Having selected M in accord with (6.308), the final spectrum estimate is, from (6.264)

$$\hat{S}_x(f, v_z) = (\sigma_M^2/4W^2)\left|1 + \sum_{m=1}^{M} a_m^{(M)} \exp(-\mathrm{j}2\pi v_z m\Delta z)\right|^{-2}. \tag{6.309}$$

In the Burg procedure [6.52, 62] based on prediction-error filters, the spectral coefficients $a_m^{(M)}$ in (6.309) are calculated directly from the data, without the intermediate step of finding the cross-spectral matrix $\hat{\mathbf{P}}_x(f)$. Here the data are the sensor Fourier coefficients (6.66), $X_i(f)$, $i=1, K$, at the frequency of interest. We set up an M-step least-squares predictor along the X_i sequence

$$X_i \sim \hat{X}_i = -\sum_{m=1}^{M} a_m^{(M)} X_{i-m}, \qquad i=M+1, \ldots, K, \tag{6.310}$$

and an M-step "backwards predictor":

$$X_i \sim \hat{X}_i = -\sum_{m=1}^{M} b_m^{(M)} X_{i+m}, \qquad i=1, \ldots, K-m. \tag{6.311}$$

The normal equations for (6.310) can be found using the orthogonality principle [6.68]

$$\mathrm{E}[(X_i-\hat{X}_i)X_{i-l}^*]=0, \qquad l=1,m, \tag{6.312}$$

which leads to

$$P_x(l)+\sum_{m=1}^{M} a_m^{(M)}P_x(l-m)=0, \qquad l=1,M, \tag{6.313}$$

where

$$P_x(l-m)=\mathrm{E}[X_l(f)X_m^*(f)] \tag{6.314}$$

is the cross-spectral power density between $x(t, z_l)$ and $x(t, z_m)$. Similarly, (6.311) requires

$$P_x(-l)+\sum_{m=1}^{M} b_m^{(M)}P_x(m-l)=0, \qquad l=1,M, \tag{6.315}$$

or, conjugating,

$$P_x(l)+\sum_{m=1}^{M} b_m^{(M)*}P_x(l-m)=0, \qquad l=1,M. \tag{6.316}$$

From (6.313, 316) it is clear that, ideally,

$$b_m^{(M)}=a_m^{(M)*}, \qquad m=1,M, \tag{6.317}$$

and we use that relation in (6.311). The attained minimum mean-squared errors for (6.310, 311) are then the same:

$$\begin{aligned}\sigma_M^2&=\mathrm{E}[(X_i-\hat{X}_i)^*(X_i-\hat{X}_i)]\\&=\mathrm{E}[X_i(X_i-\hat{X}_i)^*]=P_x(0)+\sum_{m=1}^{M} a_m^{(M)}P_x^*(m),\end{aligned} \tag{6.318}$$

which agrees with (6.302) in form.

Since the errors in (6.310, 311) are ideally the same, we may as well consider both simultaneously, and seek to minimize the average summed squared error

$$\varepsilon=\frac{1}{2}\frac{1}{K-M}\sum_{i=1}^{K-M}\left\{\left|X_{i+M}+\sum_{m=1}^{M} a_m^{(M)}X_{i+M-m}\right|^2+\left|X_i+\sum_{m=1}^{M} a_m^{(M)*}X_{i+m}\right|^2\right\}. \tag{6.319}$$

The normal equations are

$$\sum_{m=1}^{M} P_x(l-m)a_m^{(M)} = -P_x(l,0), \qquad l=1,m, \tag{6.320}$$

where

$$P_x(l,m) = \frac{1}{2}\frac{1}{K-M}\sum_{i=1}^{K-M}(X_{i+M-l}^{*}X_{i+M-m}+X_{i+l}X_{i+m}^{*}), \quad l,m=0,M. \tag{6.321}$$

The attained minimum error is

$$\begin{aligned}\varepsilon=\sigma_M^2 &= P_x(0,0)+\sum_{m=1}^{M}a_m^{(M)}P_x(0,m)\\ &= P_x(0,0)+\sum_{m=1}^{M}a_m^{(M)}P_x^{*}(m,0).\end{aligned} \tag{6.322}$$

Equations (6.320, 322) are of the form of (6.302, 303).

The matrix with elements (6.321) is not Toeplitz, however. Nonetheless, the recursion (6.306, 307) and the solution (6.305) are still correct, because of the special form (6.321) of the matrix elements $P_x(l,m)$. This allows the Eqs. (6.320) for the prediction-error coefficients to be solved easily as a sequence of one-dimensional problems, as *Burg* [6.43] long ago pointed out. This can be seen easily by noting from (6.321) that

$$P_x^{*}(M-l,M-m) = P_x(l,m), \qquad l,m=0,M. \tag{6.323}$$

Then we assume inductively that (6.320) holds with M replaced by $M-1$. Thus, writing (6.320) with use of (6.306), we obtain

$$\sum_{m=1}^{M-1} P_x(l,m)(a_m^{(M-1)}+a_m^{(M)}a_{M-m}^{(M-1)*})+P_x(l,M)a_M^{(M)} = -P_x(l,0), \qquad l=1,M \tag{6.324}$$

which, because of the inductive hypothesis, becomes

$$a_M^{(M)}\left[\sum_{m=1}^{M-1} P_x(l,m)a_{M-m}^{(M-1)*}+P_x(l,M)\right]=0, \qquad l=1,M-1. \tag{6.325}$$

This is true for $l=1, M-1$ because the term in parentheses vanishes, due to the inductive hypothesis

$$\begin{aligned}&\sum_{m=1}^{M-1} P_x^{*}(l,m)a_{M-m}^{(M-1)}+P_l^{*}(l,M)\\ &=\sum_{m=1}^{M-1} P_x^{*}(l,M-m)a_m^{(M-1)}+P_x^{*}(l,M), \qquad l=1,M-1\end{aligned} \tag{6.326}$$

or

$$\sum_{m=1}^{M-1} P_x^*(M-l, M-m) a_m^{(M-1)} = -P_x^*(M-l, M), \qquad l=1, M-1, \tag{6.327}$$

which, because of (6.323), is the inductive hypothesis. Continuing, we satisfy (6.324) for $l=M$ by simply requiring

$$\begin{aligned} a_M^{(M)} &= -\left[\sum_{m=1}^{M-1} P_x(M,m) a_m^{(M-1)} + P_x(M,0)\right] \Big/ \left[\sum_{m=1}^{M-1} a_{M-m}^{(M-1)*} P_x(M,m) + P_x(M,M)\right] \\ &= -\left[\sum_{m=1}^{M-1} P_x(M,m) a_m^{(M-1)} + P_x(M,0)\right] \Big/ \sigma_{M-1}^2, \end{aligned} \tag{6.328}$$

the last step following from using (6.322) with M replaced by $M-1$, conjugating, and using (6.323) after a change of variable. Finally, assuming inductively that (6.322) holds with M replaced by $M-1$, and using (6.306), we have

$$\begin{aligned} \sigma_M^2 &= P_x(0,0) + \sum_{m=1}^{M-1} a_m^{(M)} P_x^*(m,0) + a_M^{(M)} P_x^*(M,0) \\ &= P_x(0,0) + \sum_{m=1}^{M-1} a_m^{(M-1)} P_x^*(m,0) \\ &\quad + a_M^{(M)} \left[\sum_{m=1}^{M-1} a_{M-m}^{(M-1)*} P^*(m,0) + P^*(M,0)\right] \\ &= \sigma_{M-1}^2 + a_M^{(M)}(-a_M^{(M)*} \sigma_{M-1}^2) \\ &= \sigma_{M-1}^2 (1 - |a_M^{(M)}|^2), \end{aligned} \tag{6.329}$$

again making changes of variables and using (6.323, 328). Thus the recursion is fully verified in this case also, allowing the Yule–Walker and Burg procedures to be used with equal convenience.

6.4.5 Remarks on Nonuniform Sample Spacing

In the case of an array with nonuniformly spaced sensors, the general expression (6.287) may be used to determine the wave number spectrum. However, no recursion such as discussed above has been worked out for the nonuniformly spaced analog of the Yule–Walker approach (6.299), and if Burg's procedure is used, there is no apparent way to pass from the model (6.310) to a spectrum, although the basic relation (6.323) still holds. Accordingly, we may wish to interpolate the actual sensor data, the Fourier coefficients $X_i(f)$, $i=1, K$, in the spatial domain in some appropriate way, and then "read off" uniformly spaced "data" for use in the formulas appropriate to the uniformly spaced case.

A large literature exists in interpolation of band-limited signals. Since we have assumed both frequency ($|f| \leqq B$) and wave number ($|v_{x,y,z}| \leqq W$) band limiting, we may as well speak of time-domain signals. Accordingly, let $x(t)$ be a band-limited stochastic process

$$P_x(f)=0, \qquad |f|>B, \tag{6.330}$$

where $P_x(f)$ is the power spectral density function. The basic result is the stochastic process form of Shannon's sampling theorem [6.54, 69]

$$x(t)=\sum_{n=-\infty}^{\infty} x(t_n)\phi(t,t_n), \qquad \text{all } t, \tag{6.331}$$

where $t_n = n\Delta t$, $\Delta t = 1/2B$,

$$\phi(t,t_n)=[\sin 2\pi B(t-t_n)]/2\pi B(t-t_n), \tag{6.332}$$

and the equality is with probability one.

The objective now is to obtain (6.331) also in the case that the t_n are not uniformly spaced in time. Some interesting results exist, which may be used for this purpose. For example, *Beutler* [6.69] uses some results of *Levinson* [6.70] to show that, provided the t_n do not differ too much from $n\Delta t$, exact reconstruction is still possible. Specifically, if

$$|t_n - n\Delta t| < (\log 2)/(2\pi B), \qquad \text{all } n, \tag{6.333}$$

then, for appropriate $\phi(t,t_n)$, (6.331) still holds. This is the kind of result we want, except that the infinite sum in (6.331) can only be approximated, and the bound in (6.333) is generally too restrictive for use with typical spatial arrays. *Higgins* [6.71] has recently presented a similar result.

Thus we must have recourse to some weaker results than (6.331), typically involving least-squares approximation of the true $x(t)$ by some approximation $\hat{x}(t)$ based on the available $x(t_n)$, $n=1, N$:

$$x(t) \sim \hat{x}(t) = \sum_{n=1}^{N} x(t_n)\phi(t,t_n), \tag{6.334}$$

for some appropriate function set $\phi_n(t)=\phi(t,t_n)$. *Petersen* and *Middleton* [6.72] considered this class of problems, and pointed out among other things that the approximation (6.334) passes exactly through the given data points:

$$x(t_n)=\hat{x}(t_n), \qquad n=1,N, \tag{6.335}$$

when the expansion coefficients $\phi_n(t)$ satisfy the corresponding normal equations

$$\mathrm{E}\left[\left[x(t)-\sum_{m=1}^{N} x(t_m)\phi(t,t_m)\right]x^*(t_n)\right]=0, \qquad n=1,N, \tag{6.336}$$

so that we do have a least-squares approximation.

In a closely related situation, *Levi* [6.73] showed that the band-limited signal of least energy passing exactly through given sample points $x(t_n)$, $n=1$, N, is

$$\hat{x}(t)=\sum_{n=1}^{N} \lambda_n\phi(t,t_n), \tag{6.337}$$

with $\phi(t,t_n)$ as in (6.332), and where the λ_n satisfy

$$\sum_{m=1}^{N} \lambda_m\phi(t_n,t_m)=x(t_n), \qquad n=1,N. \tag{6.338}$$

Further, *Brown* [6.74] showed that, if one seeks a least-squares approximation (6.334), with the error integrated over all t, not simply summed at the given sample points, and if one uses the "basis" functions (6.332), then the result is just (6.337, 338), which is thus, in addition, the function of least energy which passes exactly through all the given sample points.

The procedure of *Brown* [6.74] can be used to interpolate the Fourier coefficients $X_i(f)$, $i=1$, K, of (6.166), if Burg's approach (6.319) through the prediction-error filter is used, or, if the Yule–Walker approach (6.301) is used, the estimated cross-power spectra $\hat{P}_x(f,z_i,z_j)$ can be interpolated to find uniformly spaced values $\hat{P}_x(f,m)$, $m=1$, K.

6.4.6 Direction Finding with a Line Array

As is apparent from (6.286), only the z component ν_z of the vector wave number $\boldsymbol{\nu}$ can be sensed with a line array along the z axis of coordinate space. This is physically clear, since we can then sense waves of various wavelengths propagating through space only in terms of their passage down the z axis. In general, there then exists a fundamental ambiguity. For direction finding, we need to determine the vector $\boldsymbol{\nu}$ itself, since this points in the direction we associate with wave components $\exp[\mathrm{j}2\pi(ft+\boldsymbol{\nu}\cdot\boldsymbol{z})]$ impinging on the array. However, a particular ν_z value can arise from waves with many different $\boldsymbol{\nu}$ values. If we consider the usual spherical coordinate system (r,θ,ϕ), with polar angle θ measured from the positive z axis (the array line), we are insensitive to differing values of ϕ, and further we confuse waves having the same value of $\nu_z=|\boldsymbol{\nu}|\cos\theta=(f/v)\cos\theta$. This latter ambiguity is removed provided we know (or specify) the velocity of propagation v in the medium [v is conveniently a constant in some applications, such as sonar (approximate) and radio astronomy]. This is because the frequency f is fixed by the frequency filters applied to the individual sensors. Thus f/v is known, and v_z is "locked" to the bearing angle $\cos\theta$.

In brief, the estimators discussed above will produce the power density associated with waves of each value of v_z. The manner in which the experimenter makes use of this result is another matter. We are simply pointing out that the observed power results from the superposition of all waves from azimuthal angles $0 \leqq \phi \leqq 2\pi$, and for all waves with velocity and bearing such that $(f/v)\cos\theta = \text{const}$. At best, we may have physical reasons to assume a number for v, in which case we obtain a bearing analysis over $\cos\theta$ for the azimuthal superposition of actual waves. Only a two-dimensional array can remove the azimuthal ambiguity, and a full three-dimensional array is needed for complete analysis. This would return us to the general case (6.282, 283) of maximum-entropy analysis.

6.5 Conclusions

We have described in considerable detail the extension of the conventional linear algorithm for time-domain power spectral density analysis to the case of multiple independent variables (time and space). This makes evident the relevance of time series analysis to the problem of spatial analysis of a traveling wave field.

The maximum-likelihood processor was derived using a reasonably compact vector – matrix notation. This allowed us to summarize easily some of the adaptive algorithms which also realize nonlinear filtering in wave number space.

The maximum-entropy algorithm was derived from first principles, and its general form for spatial processing was written down. The special case of a line array of uniformly spaced sensors was considered in detail. The appropriate recursions on the filter order were presented for the "covariance" and "prediction filter" versions of the method. Some suggestions were made as to the possibility of dealing with nonuniformly spaced sensors, but more analysis and experimentation need to be done in that line.

It is hoped that this summary of work in nonlinear spatial processing will lead to further applications of the methods to diverse fields of study, and to broader evaluation of their potential.

Acknowledgments. The author's awareness of the matters discussed herein developed through association with students at the University of Delaware, particularly Ms. Rosemary Lafrance, and through various consulting and other affiliations, particularly with Bell Laboratories, Systems Control, Inc., and the Scripps Institution of Oceanography.

It is a pleasure to make special mention of the contributions of Peter Hirsch, of Bell Laboratories, to the author's technical growth through a very pleasant collaboration extending over the years.

References

6.1 K. Tomiyasu: Proc. IEEE **66** (5), 563–583 (1978)
6.2 L.J. Cutrona: J. Acoust. Soc. Am. **58** (2), 336–348 (1975)
6.3 M.H. Cohen: Proc. IEEE **61** (9), 1192–1197 (1973)
6.4 E.T. Jaynes: Phys. Rev. **106** (4), 620–630 (1957)
6.5 W.H. Hayt, Jr., J.E. Kemmerly: *Engineering Circuit Analysis* (McGraw-Hill, New York 1971)
6.6 N. Wiener: *The Fourier Integral and Certain of Its Applications* (Cambridge Univ. Press, Cambridge 1933); (Reprint, Dover, New York)
6.7 W.B. Davenport, Jr., W.L. Root: *An Introduction to the Theory of Random Signals and Noise* (McGraw-Hill, New York 1958)
6.8 L.H. Koopmans: *The Spectral Analysis of Time Series* (Academic Press, New York 1974)
6.9 A.M. Yaglom: *An Introduction to the Theory of Stationary Random Functions* (Prentice-Hall, Englewood Cliffs, NJ 1962)
6.10 P.M. Morse, H. Feshbach: *Methods of Theoretical Physics* (McGraw-Hill, New York 1953)
6.11 R.B. Blackman, J.W. Tukey: *The Measurement of Power Spectra* (Dover, New York 1959)
6.12 G.M. Jenkins, D.G. Watts: *Spectral Analysis and Its Applications* (Holden-Day, San Francisco 1968)
6.13 F.J. Harris: Proc. IEEE **66** (1), 51–83 (1978)
6.14 A.V. Oppenheim, R.W. Schafer: *Digital Signal Processing* (Prentice-Hall, Englewood Cliffs, NJ 1975)
6.15 W.D. Stanley: *Digital Signal Processing* (Reston Publ., Reston, VA 1975)
6.16 A.B. Baggeroer: "Sonar Signal Processing," in *Applications of Digital Signal Processing*, ed. by A.V. Oppenheim (Prentice-Hall, Englewood Cliffs, NJ 1978)
6.17 R.T. Lacoss, E.J. Kelley, M.N. Toksöz: Geophys. **34** (1), 21–38 (1969)
6.18 P.E. Green, Jr., R.A. Frosch, C.F. Romney: Proc. IEEE **53** (12), 1821–1833 (1965)
6.19 H.W. Briscoe, P.L. Fleck: Proc. IEEE **53** (12), 1852–1859 (1965)
6.20 J. Capon, R.J. Greenfield, R.J. Kolker: Proc. IEEE **55** (2), 192–211 (1967)
6.21 J. Capon: Proc. IEEE **57** (8), 1408–1418 (1969)
6.22 R. Deutsch: *Estimation Theory* (Prentice-Hall, Englewood Cliffs, NJ 1965)
6.23 L.V. Ahlfors: *Complex Analysis* (McGraw-Hill, New York 1953)
6.24 N. Levinson: "The Wiener RMS (root mean square) Error Criterion in Filter Design and Prediction," in *Extrapolation, Interpolation, and Smoothing of Stationary Time Series*, (Wiley and Sons, New York 1960)
6.25 J.P. Burg: Geophys. **29**, 693–713 (1964)
6.26 M. Backus, J. Burg, D. Baldwin, E. Bryan: Geophys. **29**, 672–692 (1964)
6.27 J, Capon, N.R. Goodman: Proc. IEEE **59** (1), 112 (1971)
6.28 R.N. McDonough: J. Acoust. Soc. Am. **51** (4), 1186–1193 (1972)
6.29 H. Cox: J. Acoust. Soc. Am. **54** (3), 771–785 (1973)
6.30 R.T. Lacoss: Geophys. **36** (4), 661–675 (1971)
6.31 C.D. Seligson: Proc. IEEE **58** (6), 947–949 (1970)
6.32 B. Widrow, P.E. Mantey, L.J. Griffiths, B.B. Goode: Proc. IEEE **55** (12), 2143–2159 (1967)
6.33 D.J. Sakrison: "Stochastic Approximation: A Recursive Method for Solving Regression Problems," in *Advances in Communication Systems*, Vol. 2, ed. by A.V. Balakrishnan (Academic Press, New York 1966)
6.34 L.J. Griffiths: Proc. IEEE **57** (10), 1696–1704 (1969)
6.35 B. Widrow, J.M. McCool, M.G. Larimore, C.R. Johnson, Jr.: Proc. IEEE **64** (8), 1151–1162 (1976)
6.36 R.T. Lacoss: IEEE Trans. GE-**6** (2), 78–86 (1968)
6.37 H. Kobayashi: IEEE Trans. GE-**8** (3), 169–178 (1970)
6.38 O.L. Frost, III: Proc. IEEE **60** (8), 926–935 (1972)
6.39 A.F. Gangi, B.S. Byun: Geophys. **41** (5), 970–984 (1976)
6.40 B. Widrow, J.R. Glover, Jr., J.M. McCool, J. Kaunitz, C.S. Williams, N.H. Hearn, J.R. Zeidler, E. Dong, Jr., R.C. Goodlin: Proc. IEEE **63** (12), 1692–1716 (1975)

6.41 W.F.Gabriel: Proc. IEEE **64** (2), 239–272 (1976)
6.42 J.P.Burg: "Maximum Entropy Spectral Analysis," 37th Ann. Intern. Meeting, Soc. Exploration Geophysicists, Oklahoma City (1967)
6.43 J.P.Burg: "A New Analysis Technique for Time Series Data," NATO Advanced Study Inst. in Signal Processing with Emphasis on Underwater Acoustics, Enschede, Netherlands (1968)
6.44 J.P.Burg: "New Concepts in Power Spectra Estimation," 40th Ann. Intern. Meeting, Soc. Exploration Geophysicists, New Orleans (1970)
6.45 J.P.Burg: Geophys. **37** (2), 375–376 (1972)
6.46 T.J.Ulrych, T.N.Bishop: Rev. Geophys. Space Phys. **13** (1), 183–200 (1975)
6.47 A. van den Bos: IEEE Trans. IT-**17** (4), 493–494 (1971)
6.48 K.L.Peacock, S.Treitel: Geophys. **34** (2), 155–169 (1969)
6.49 B.S.Atal, S.L.Hanauer: J. Acoust. Soc. Am. **50** (2), 637–655 (1971)
6.50 G.J.Fryer, M.E.Odegard, G.H.Sutton: Geophys. **40** (3), 411–425 (1975)
6.51 J.Makhoul: Proc. IEEE **63** (4), 561–580 (1975)
6.52 J.P.Burg: "Maximum Entropy Spectral Analysis"; Ph. D. Dissertation, Dept. of Geophys., Stanford University (1975)
6.53 R.N.McDonough: Geophys. **39** (6), 843–851 (1974)
6.54 C.E.Shannon: Proc. IEEE **37** (1), 20–21 (1949)
6.55 C.E.Shannon, W.Weaver: *The Mathematical Theory of Communication* (University of Illinois Press, Urbana 1962)
6.56 S.Goldman: *Information Theory* (Prentice-Hall, Englewood Cliffs, NJ 1953)
6.57 M.Tribus: *Thermostatics and Thermodynamics* (Van Nostrand, Princeton 1961)
6.58 M.Tribus: *Rational Descriptions, Decisions, and Designs* (Pergamon, New York 1969)
6.59 J.A.Edward, M.M.Fitelson: IEEE Trans. IT-**19** (2), 232–234 (1973)
6.60 J.G.Ables: Astron. Astrophys. Suppl. Ser. **15**, 383–393 (1974)
6.61 B.R.Frieden: J. Opt. Soc. Am. **62** (4), 511–518 (1972)
6.62 N.Andersen: Geophys. **39** (1), 69–72 (1974)
6.63 A.H.Nuttall: "Multivariate Linear Predictive Spectral Analysis Employing Weighted Forward and Backward Averaging: A Generalization of Burg's Algorithm," Naval Underwater Systems Center Tech. Rpt. 5501, New London, Conn. (Oct. 1976)
6.64 O.N.Strand: IEEE Trans. AC-**22** (4), 634–640 (1977)
6.65 A.B.Baggeroer: IEEE Trans. IT-**22** (5), 534–545 (1976)
6.66 H.Akaike: Ann. Inst. Stat. Math. (Tokyo) **22**, 203–217 (1970)
6.67 S.Haykin, S.Kesler: Proc. IEEE **64** (5), 822–823 (1976)
6.68 A.Papoulis: *Probability, Random Variables, and Stochastic Processes* (McGraw-Hill, New York 1965)
6.69 F.J.Beutler: Inf. Control **4** ,97–117 (1961)
6.70 N.Levinson: "Gap and Density Theorems," Coll. Publ. No. 26, Am. Math. Soc., New York (1940)
6.71 J.R.Higgins: IEEE Trans. IT-**22** (5), 621–622 (1976)
6.72 D.P.Petersen, D.Middleton: IEEE Trans. IT-**11** (1), 18–30 (1965)
6.73 L.Levi: IEEE Trans. IT-**11**, 372–376 (1965)
6.74 J.L.Brown, Jr.: Inf. Control **10**, 409–418 (1967)

7. Recent Advances in Spectral Estimation

S. Haykin, S. Kesler, and E. A. Robinson

In this chapter, we describe three algorithms intended for different applications, which have been developed recently. The first one represents a generalization of Burg's algorithm (developed by Kesler and Haykin) for computing the maximum entropy wavenumber spectrum of a spatial-time series. This algorithm is useful for beam-forming applications. The second algorithm, developed by Tufts and Kumareson, offers an improvement in forward-backward linear prediction type of computation. The algorithm uses a singular value decomposition of the covariance matrix of the given time series. The third algorithm is known as the CDE ARMA algorithm, developed by Robinson. It is useful for estimating the parameters of an ARMA process from a given time series.

Finally, in Sect. 7.4, brief mention is made of two other new procedures that are useful in spectral estimation applications, of which the reader should be aware. Note the notation used in any section of this chapter is not to be confused with that in another.

7.1 Generalized Burg Algorithm

The algorithm presented in this section is a generalization of the Burg algorithm (BA) for maximum entropy spectral estimation. The algorithm is suitable for spatial processing of nonstationary (i.e., nonhomogeneous) array data. The cross-spectral (CS) matrix of such a process is non-Toeplitz and the BA is not directly applicable.

Unlike the least-squares (LS) methods [7.1–3], the generalized Burg algorithm (GBA) presented here, retains the feature that the magnitude of the reflection coefficient is smaller than unity at every stage of the computation.

Consider a linear array of N uniformly spaced omnidirectional sensors. At each sensor, signals are received, sampled and Fourier analyzed via the discrete Fourier transform (DFT) [7.4]. Let $X(n,f,i)$ denote the complex Fourier coefficient for the nth sensor, at frequency f and for the ith DFT. The cross-spectrum (CS) between sensors n and m is given by

$$\begin{aligned} R(n,m,f,i,j) &= \mathrm{E}[X(n,f,i)X^*(m,f,j)] \\ &= R(n,m,f)\delta(i-j)\,, \end{aligned} \tag{7.1}$$

where $\delta(i-j)$ is the Kronecker delta function and $\mathrm{E}[\cdot]$ is the expectation in the temporal sense. It is assumed in (7.1) that the sensor signals are temporally stationary. On the other hand, we allow for spatial nonstationarity, i.e., inhomogeneity, as indicated by the explicit dependence of CS on both sensor indices n and m, and not only on their difference $n-m$.

We now derive the GBA for the array beam pattern estimate when the N by N CS matrix $\boldsymbol{R}(f)$ has the (n,m)-th element as given by (7.1). The CS matrix is computed for each frequency of interest, and the array beam pattern is estimated for that frequency. In most applications, the sensor signals are narrow-band, so that the beam-pattern estimate at only one frequency is needed. For convenience, we shall drop the dependence on the operating frequency f and on the DFT numbers i and j from here on.

The maximum-entropy (ME) beam pattern as a function of the elevation angle θ, measured with respect to the normal of the array, is given by [7.5, 6]

$$S(\theta)=\frac{P_M d/\lambda}{\left|\sum_{m=0}^{M} a_m^{(M)}\exp[-jm\psi(\theta)]\right|^2}, \tag{7.2}$$

where λ is the wavelength, d is the intersensor spacing, and $\psi(\theta)=2\pi(d/\lambda)\sin\theta$. Parameters $a_m^{(M)}$, $m=0,1,\ldots,M$. $(a_0^{(M)}=1)$ and P_M are the Mth order prediction error filter (PEF) coefficients and output power, respectively. The choice of the order M depends on the application of interest. If we are interested only in field mapping, in which we assume that the CS properties of the field are known over the array aperture, we ultimately choose the maximum PEF order, $M=N-1$ [7, 5]. In the detection problem, where CS must be estimated from a finite data collection, a lower-order filter may be necessary to produce reasonable stability in the ME estimate.

The PEF coefficients and power are determined in the following recursive manner. For PEF of order $M-1$, the forward prediction error at the sensor n and the backward prediction error at the sensor $n-1$, are defined by the following convolutional sums, respectively,

$$e_{f,n}^{(M-1)}=\sum_{m=0}^{M-1} a_m^{(M-1)}X_{n-m}=\boldsymbol{a}^{\mathrm{T}}\boldsymbol{X}_n=\boldsymbol{X}_n^{\mathrm{T}}\boldsymbol{a}, \tag{7.3}$$

$$e_{b,n-1}^{(M-1)}=\sum_{m=0}^{M-1} a_m^{(M-1)*}X_{n-M+k}=\boldsymbol{a}^{\mathrm{H}}\tilde{\boldsymbol{X}}_{n-1}=\tilde{\boldsymbol{X}}_{n-1}^{\mathrm{T}}\boldsymbol{a}^*=\boldsymbol{X}_{n-1}^{\mathrm{T}}\tilde{\boldsymbol{a}}^*. \tag{7.4}$$

The summations in (7.3, 4) are those shown in (2.72, 74) and used in the derivation of the original Burg algorithm. The dot products in (7.3, 4) are between the vectors of length M defined as follows:

$$\boldsymbol{a}=[1, a_1^{(M-1)},\ldots,a_{M-1}^{(M-1)}]^{\mathrm{T}}, \tag{7.5}$$

$$\boldsymbol{X}_n=[X_n, X_{n-1},\ldots,X_{n-M+1}]^{\mathrm{T}},\quad n\geqq M+1 \tag{7.6}$$

$$\boldsymbol{X}_{n-1}=[X_{n-1}, X_{n-2},\ldots,X_{n-M}]^{\mathrm{T}},\quad n\geqq M+1 \tag{7.7}$$

In (7.3, 4), the asterisk signifies complex conjugation; the superscripts T and H signify transposition and Hermitian transposition, respectively; and the tilde over a vector signifies the fact that the components of that vector appear in reversed order. The requirement is that, given the coefficients of PEF of order $M-1$, to determine the corresponding coefficients of PEF of order M. Substituting the dot products in (7.3, 4) into the Burg formula for the Mth-order filter reflection coefficient $a_M^{(M)}$ in (7.4, 5), we get

$$a_M^{(M)} = \frac{-2 \sum_{n=M+1}^{N} E[e_{f,n}^{(M-1)} e_{b,n-1}^{(M-1)*}]}{\sum_{n=M+1}^{N} E[e_{f,n}^{(M-1)} e_{f,n}^{(M-1)*} + e_{b,n-1}^{(M-1)} e_{b,n-1}^{(M-1)*}]} . \tag{7.8}$$

Next, using (7.1), we obtain the generalized Burg formula [7.5, 6]

$$\begin{aligned} a_M^{(M)} &= \frac{-2 \sum_{n=M+1}^{N} E[\boldsymbol{a}^{\mathrm{T}} \boldsymbol{X}_n \boldsymbol{X}_{n-1}^{\mathrm{H}} \tilde{\boldsymbol{a}}]}{\sum_{m=M+1}^{N} E[\boldsymbol{a}^{\mathrm{H}} \boldsymbol{X}_n^* \boldsymbol{X}_n^{\mathrm{T}} \boldsymbol{a} + \boldsymbol{a}^{\mathrm{H}} \tilde{\boldsymbol{X}}_{n-1} \tilde{\boldsymbol{X}}_{n-1}^{\mathrm{H}} \boldsymbol{a}]} \\ &= \frac{-2\boldsymbol{a}^{\mathrm{T}} \left\{ \sum_{n=M+1}^{N} \hat{\boldsymbol{R}}(n, n-1) \right\} \tilde{\boldsymbol{a}}}{\boldsymbol{a}^{\mathrm{H}} \left\{ \sum_{n=M+1}^{N} [\hat{\boldsymbol{R}}(n,n) + \hat{\boldsymbol{R}}(n-M, n-M)] \right\} \boldsymbol{a}} . \end{aligned} \tag{7.9}$$

The expectation operators in (7.8, 9) signify temporal expectations. The matrices in the numerator and the denominator of (7.9) are all of dimensions M by M, and are given by

$$\hat{\boldsymbol{R}}(n, n-1) = \begin{bmatrix} R(n, n-1) & \dots & R(n, n-M) \\ \vdots & \ddots & \vdots \\ R(n-M+1, n-1) & \dots & R(n-M+1, n-M) \end{bmatrix} , \tag{7.10}$$

$$\hat{\boldsymbol{R}}(n, n) = \begin{bmatrix} R(n, n) & \dots & R(n-M+1, n) \\ \vdots & \ddots & \vdots \\ R(n, n-M+1) & \dots & R(n-M+1, n-M+1) \end{bmatrix} , \tag{7.11}$$

$$\hat{\boldsymbol{R}}(n-M, n-M) = \begin{bmatrix} R(n-M, n-M) & \dots & R(n-M, n-1) \\ \vdots & \ddots & \vdots \\ R(n-1, n-M) & \dots & R(n-1, n-1) \end{bmatrix} . \tag{7.12}$$

These matrices are non-Toeplitz; moreover, $\boldsymbol{R}(n, n-1)$ is not even Hermitian.

Having computed the reflection coefficient $a_M^{(M)}$, we may next compute the remaining PEF coefficients and the prediction power by using the Levinson recursion in the usual way, as given by (2.66, 70). The recursion starts with the trace of the CS matrix, namely,

$$P_0 = \frac{1}{N} \operatorname{tr}[\hat{\boldsymbol{R}}(n,n)] = \frac{1}{N} \sum_{n=1}^{N} R(n,n). \tag{7.13}$$

Equation (7.9) represents a generalization of the original Burg algorithm for the case when the CS matrix is available. The algorithm developed here is the complex space-time or wavenumber-frequency extension of the so-called covariance lattice method derived by *Makhoul* [7.7] for the case of real-valued time series. Recently, two related algorithms have been published [7.8], one using the CS matrix, and the other one computing the reflection coefficient directly from sensor Fourier coefficients.

For the case when the CS matrix is Toeplitz, we find that the generalized Burg algorithm of (7.10) yields an identical result to the original Burg algorithm.

The definition given in (7.1) for the (n,m)-th element $R(n,m)$ of the CS matrix is basic to the formulation of matrices $\hat{\boldsymbol{R}}(n,n-1)$, $\hat{\boldsymbol{R}}(n,n)$, and $\hat{\boldsymbol{R}}(n-M,n-M)$, which are needed to compute the reflection coefficient $a_M^{(M)}$ according to (7.9). Using the exact analytical expressions for the CS matrix of a particular spatial field, it has been demonstrated that the resolution capabilities of the GBA in the presence of two spatially correlated plane-wave signals, are comparable to those of the original Burg algorithm for two uncorrelated plane waves [7.5, 6]. It has also been demonstrated that the statistical properties of the GBA [7.9] are similar to those of the original Burg algorithm [7.10].

Comparing (7.8 and 9), we see that in GBA, instead of updating the forward and backward prediction errors at each recursion, we evaluate two quadratic forms, one in the numerator and the other in the denominator of the formula for the reflection coefficient. Such a procedure preserves the absolute stability of PEF at each recursion, since $|a_M^{(M)}| \leqq 1$ for all M.

7.2 Singular Valve Decomposition in Spectral Analysis

Another major development in the last few years is the improvement in forward-backward linear prediction (FBLP) type of algorithms, brought about by the use of singular value decomposition (SVD) of the covariance matrix. This improvement was first reported by *Kumaresan* and *Tufts* [7.11]. By modifying the FBLP procedure with SVD of the covariance matrix, they achieved much better performance of FBLP at moderate and low signal-to-noise ratios (SNR) than conventional methods. The threshold effect in the FBLP procedure is improved, and the frequency estimates of complex sinusoids in additive noise approach the Cramer-Rao bound. Since the procedure uses prediction error filters (PEF) of large orders (beyond the traditional limits of $N/3$ to $N/2$, where N is the number of data samples used conventionally), the resolution capability of the new procedure is also improved with respect to the FBLP procedure.

The singular value decomposition algorithm for linear prediction works as follows [7.11a]. Suppose we are given a sequence of complex valued N data points, x_n, $n=1,2,\ldots,N$, known to be composed of narrow-band signals in noise. The forward-backward linear prediction equations can be written down by using PEF of order M with impulse response given by the column vector $\boldsymbol{a}'=[1,a_1^{(M)},\ldots,a_M^{(M)}]^{\mathrm{T}}=[1,\boldsymbol{a}]^{\mathrm{T}}$. The vector $-\boldsymbol{a}$ consists of the prediction filter coefficients. Based on an input sequence of N data samples, we may write

$$\begin{bmatrix} X_M & X_{M-1} & \cdots & X_1 \\ X_{M+1} & X_M & \cdots & X_2 \\ \vdots & \vdots & & \vdots \\ X_{N-1} & X_{N-2} & \cdots & X_{N-M} \\ \hline X_2^* & X_3^* & \cdots & X_{M+1}^* \\ X_3^* & X_4^* & \cdots & X_{M+2}^* \\ \vdots & \vdots & & \vdots \\ X_{N-M+1}^* & X_{N-M+2}^* & \cdots & X_N^* \end{bmatrix} \begin{bmatrix} a_1^{(M)} \\ a_2^{(M)} \\ \vdots \\ a_M^{(M)} \end{bmatrix} = - \begin{bmatrix} X_{M+1} \\ X_{M+2} \\ \vdots \\ X_N \\ \hline X_1^* \\ X_2^* \\ \vdots \\ X_{N-M}^* \end{bmatrix} \tag{7.14}$$

or

$$\boldsymbol{X}\boldsymbol{a}=-\boldsymbol{h} \tag{7.15}$$

where $\boldsymbol{X}$ is the $2(N-M)$ by M data matrix on the left of (7.14), and $\boldsymbol{h}$ is the $2(N-M)$ by 1 data vector on the right. The least-squares solution $\boldsymbol{a}$ is found by minimizing the Euclidean norm $\|\boldsymbol{X}\boldsymbol{a}+\boldsymbol{h}\|$, since the linear system of equations is usually overdetermined. Then $\boldsymbol{a}$ is given by

$$\boldsymbol{a}=-[\boldsymbol{X}^{\mathrm{H}}\boldsymbol{X}]^{-1}\boldsymbol{X}^{\mathrm{H}}\boldsymbol{h}, \tag{7.16}$$

where the superscript H denotes Hermitian transposition. The matrix $\boldsymbol{X}^{\mathrm{H}}\boldsymbol{X}$ represents an M by M estimated correlation matrix $\boldsymbol{R}$ and $-\boldsymbol{X}^{\mathrm{H}}\boldsymbol{h}$ represents a cross-correlation vector $\boldsymbol{r}$. This method of setting up the linear equations and finding the vector $\boldsymbol{a}$, which is called the FBLP method, is a straightforward extension of the standard covariance method of linear prediction [7.7, 12], which itself is a variation of the Prony method [7.13].

The above least-squares solution is now related to the singular value decomposition (SVD) of the data matrix $\boldsymbol{X}$. In terms of its SVD, the data matrix $\boldsymbol{X}$ is written as [7.14, 15]

$$\boldsymbol{X}=\boldsymbol{U}\boldsymbol{S}\boldsymbol{V}^{\mathrm{H}}, \tag{7.17}$$

where U is a $2(N-M)$ by $2(N-M)$ unitary matrix, V is an M by M unitary matrix, and S is a real $2(N-M)$ by M diagonal matrix with non-negative numbers $s_1, s_2, \ldots, s_M$ (or $s_{2(N-M)}$ if $M \geqq 3N/2$), arranged in nonincreasing order. The columns of the matrices U and V are orthonormal eigenvectors of XX^H and $X^H X$, respectively. The diagonal elements of S are called the singular values and are the square roots of the eigenvalues of XX^H (or $X^H X$). Note that $X^H X$ and XX^H have common non-zero eigenvalues. Now, in terms of SVD of X, the least-squares solution (or the minimum norm solution depending on whether the system is overdetermined or underdetermined) can be written as [7.14, 15]

$$a = -VS^{(-1)}U^H h, \tag{7.18}$$

where $S^{(-1)}$ is the pseudoinverse of S. The matrix $S^{(-1)}$ has along its diagonal the reciprocal of the nonzero singular values of X, and the rest of its elements are zeros. Then a can be explicitly written as a linear combination of the eigenvectors of $X^H X$, namely, v_i, $i = 1, 2, \ldots, M$ (columns of V) as

$$a = c_1 v_1 + c_2 v_2 + \ldots + c_M v_M \tag{7.19}$$

(or $c_{2(N-M)} v_{2(N-M)}$ for the last element if $M < 3N/2$), where $c_i = -u_i^* h / s_i$, where u_i, $i = 1, 2, \ldots, 2(N-M)$, (columns of U) are the eigenvectors of XX^H. if the data consists of L complex sinusoidal signals only and no noise, then the X matrix will have only L non-zero singular values, and (7.18) will give the minimum norm solution [7.11b]. However, in the presence of noise, X is generally of full rank and all the singular values will be non-zero. Although X is of full rank, there is an underlying effective rank Q for the matrix X, depending on the number of narrow-band or sinusoidal signals in the data. We decide on the value of Q by looking at the size of the singular values. Singular values are known to be very useful in determining the numerical rank of a matrix [7.15]. Then, we set the singular values smaller than s_Q to zero, effectively removing large perturbations introduced into the a vector by the eigenvectors with subscripts $Q+1$ to $2(N-M)$. Thus, our new a vector, now called a_Q, is

$$a_Q = -VS_Q^{(-1)}U^H h, \tag{7.20}$$

where $S_Q^{(-1)}$ has along its diagonals $1/s_1, 1/s_2, \ldots, 1/s_Q, 0, \ldots, 0$. Therefore, a_Q can be written explicitly as

$$a_Q = c_1 v_1 + c_2 v_2 + \ldots + c_Q v_Q . \tag{7.21}$$

This method of computing prediction error filter coefficients is called the principal eigenvectors method, since the coefficients are formed as linear combinations of the Q principal eigenvectors of $X^H X$. The above step is equivalent to approximating the matrix X by a matrix X_Q of rank Q in the least-squares sense [7.15] and using its SVD in (7.20, 21). It should be noted that in the noiseless data case, with L complex sinusoids, Q will be equal to L, since only L singular values will be non-zero, and in this case, the frequencies of the sinusoids can be determined exactly from the PEF coefficients.

The step of replacing $\boldsymbol{X}$ by $\boldsymbol{X}_Q$ can be viewed as improving SNR in the data by using prior information regarding the nature of the signal in the data. This is so, because the Q principal eigenvectors are considerably more robust to noise perturbations than the rest. When the signal is a sum of sinusoids, the eigenvectors corresponding to the noiseless data of $\boldsymbol{X}^H\boldsymbol{X}$ are very susceptible to even small noise perturbations. This is due to the corresponding eigenvalues of $\boldsymbol{X}^H\boldsymbol{X}$ being equal to zero in the noiseless data case. From a numerical analysis point of view [7.14, 15], the truncation of the singular values to s_Q is intended to alleviate extreme ill-conditioning caused by the close dependency of the columns of the $\boldsymbol{X}$ in (7.14).

The best choice of Q is the number L of complex sinusoids present in the data. Of course, L is not known a priori and has to be estimated. The magnitudes of the singular values give an indication of the effective rank of $\boldsymbol{X}$ or the rank of the underlying noiseless data matrix. Thus, the estimate of the number of signals in the data is a by-product in the principal-eigenvectors method. It has been demonstrated by computer simulation that the method is relatively insensitive to the choice of Q which is equal to or somewhat larger than the number of complex sinusoids L [7.11].

The choice of the PEF order M is governed, as in the FBLP method, by a trade-off between the resolution and stability of the estimate, i.e., the number and magnitudes of the spurious peaks. Unlike the FBLP method, where the recommended range of M is between $N/3$ and $N/2$, the principal-eigenvectors method improves steadily with M increasing up to the value of $3N/4$ [7.11b] since statistical fluctuations of the eigenvectors with indexes greater than Q are eliminated. As M is increased to a maximum value of $M=N-L/2$ (beyond which frequencies of even noiseless sinusoids cannot be found using the FBLP method), the number of equations in (7.14) is equal to the number of complex sinusoids in the data. The rank of $\boldsymbol{X}$ is also L (for $L \leqq N/2$) and $\boldsymbol{X}$ has only L non-zero singular values. Therefore, $\boldsymbol{a}_Q$ would be the same as $\boldsymbol{a}$ (minimum norm solution) since $c_{L+1}, c_{L+2}, \ldots$ would be all zero. Thus, the additional eigenvectors are all automatically eliminated from $\boldsymbol{a}$. Note that $(\boldsymbol{a}=\boldsymbol{a}_Q)$ can be calculated without SVD as

$$\boldsymbol{a}_Q = -\boldsymbol{X}^H(\boldsymbol{X}\boldsymbol{X}^H)^{-1}\boldsymbol{h}. \tag{7.22}$$

This case is called the Kumaresan-Prony case [7.11]. It is computationally simple and its performance is essentially the same as the FBLP method at $M=N/3$.

The significance of the method described above is in improving the FBLP method which in itself has a higher resolution capability than the procedure based on the Levinson recursion. The improvement consists of increasing the stability of the FBLP frequency estimates with high-order PEF's and effectively improving SNR by an order of 10 dB.

7.3 CDE ARMA Algorithm

The CDE ARMA algorithm is a new algorithm developed by one of the authors (E.A.R.). It works as follows.

Given a guess of the ARMA parameters, the CDE method generates the prediction error series, and the derivatives of the prediction error series with respect to each parameter. These derivative series form a multichannel actual input for a multichannel shaping operation, and the prediction error series forms the desired output. The resulting least-squares shaping filter represents the corrections to be made to the guess of the ARMA parameters. With the corrected guess, this process is then repeated. The iterations are stopped when the value of the ARMA parameters stabilize.

The value of this algorithm rests on the fact that it uses the Levinson-Wiggins-Robinson (LWR) algorithm and thus is computationally more efficient than existing nonlinear regression methods for ARMA parameter estimation. Also, it makes use of the engineering input-output approach which makes modification and extension of the algorithm easier than the conventional theoretical approach.

The engineering approach recognizes that there are certain fundamental input-output operations that keep recurring in time-series theory. One such fundamental operation is the shaping filter. The shaping filter is embodied in the form of subroutine SHAPE for the single-channel case and subroutine WIENER for the multi-channel case. These subroutines and their descriptions and examples of their use are documented [7.16], so we will not repeat them here. The essential points to remember about the shaping filter are these. We are given an *actual input signal* and a *desired output signal*. These two signals go into subroutine SHAPE (or subroutine WIENER). What comes out is the *shaping filter coefficients*. These shaping filter coefficients are determined by requiring that the approximation

$$(\text{actual input signal}) * (\text{shaping filter}) \approx (\text{desired output signal})$$

holds in the least-squares sense. We call the shaping filter so computed the least-squares shaping filter. Note that the asterisk used above indicates convolution. In the computation of the least-squares shaping filter, both SHAPE and WIENER use the LWR algorithm.

The letters CDE stand for *corrigitur-dum-erigitur*, which means "corrected while erected". The reason for this name will be given later. In this section, we want to treat the *corrigitur-dum-erigitur* method (CDE method) to estimate the parameters of a MA time series. As an introduction to the CDE method, let us suppose that the time series $x(k)$ is generated by the invertible MA(1) model

$$x(k) = \varepsilon(k) + \beta\varepsilon(k-1), \tag{7.23}$$

where the parameter β is less than one in magnitude and $\varepsilon(k)$ is a white noise series. If we write this equation in the form

$$\varepsilon(k) = -\beta\varepsilon(k-1) + x(k), \tag{7.24}$$

we find by recursive deductions that

$$\varepsilon(k) = \sum_{j=0}^{\infty} (-\beta)^j x(k-j). \tag{7.25}$$

This equation represents the AR(∞) form of the MA(1) model. It shows that $\varepsilon(k)$ is a nonlinear function of β. Let us expand the function $\varepsilon(k)$ in a Taylor series around the point b. We obtain

$$\varepsilon(k) = \varepsilon(k)|_{\beta=b} + \left.\frac{\partial\varepsilon(k)}{\partial\beta}\right|_{\beta=b} (\beta - b) + \text{remainder}. \tag{7.26}$$

Let us define the time series $v(k)$ and $n(k)$ as

$$v(k) = -\frac{\partial\varepsilon(k)}{\partial\beta}, \quad n(k) = v(k)|_{\beta=b}. \tag{7.27}$$

That is, $n(k)$ is the time series obtained from $v(k)$ by replacing the parameter β by the value b. If we differentiate (7.23), we obtain

$$\frac{\partial\varepsilon(k)}{\partial\beta} + \beta\frac{\partial\varepsilon(k-1)}{\partial\beta} + \varepsilon(k-1) = 0$$

which is

$$v(k) + \beta v(k-1) = \varepsilon(k-1). \tag{7.28}$$

Because $\varepsilon(k-1)$ is white, it follows that $v(k)$ is an AR(1) time series. Let us define $e(k)$ as the time series obtained by replacing β with b in $\varepsilon(k)$; that is,

$$e(k) = \varepsilon(k)|_{\beta=b} = \sum_{j=0}^{\infty} (-b)^j x(k-j). \tag{7.29}$$

Thus, (7.23 and 28) become

$$\begin{aligned} x(k) &= e(k) + be(k-1) \\ n(k) + bn(k-1) &= e(k-1). \end{aligned} \tag{7.30}$$

The Taylor series (7.26) is seen to be

$$e(k)=(\beta-b)n(k)+\text{error} \tag{7.31}$$

where the error is $\varepsilon(k)$ minus the remainder. Equation (7.31) is in the form of the shaping filter equation, where $n(k)$ is the actual input signal and $e(k)$ is the desired output signal. We can thus use the subroutine SHAPE to find the filter coefficient g_0. By the construction of the Taylor series, we know that

$$g_0=\beta-b.$$

Thus, the estimate of the parameter β is equal to the value b plus the filter coefficient g_0; that is,

$$\beta=b+g_0. \tag{7.32}$$

Let us now give the *corrigitur-dum-erigitur* algorithm for the MA(1) model. Let the numerical values of the given time series be $x(0), x(1), \ldots, x(N)$. Let b be the guess of parameter β and let $e(-1)$ and $n(-1)$ be known initial values. (These initial values can be taken to be zero, or some extrapolation procedure can be used to estimate them. However, we will not dwell on how initial values are obtained, as this is the type of information which is so highly valued and closely guarded by practitioners of this art.) By use of the equations

$$\begin{aligned} e(k)&=x(k)-be(k-1)\\ n(k)&=e(k-1)-bn(k-1) \end{aligned} \tag{7.33}$$

we generate the values of the time series

$$\begin{aligned} &e(0), e(1), \ldots, e(N) \quad \text{(the desired output)}\\ &n(0), n(1), \ldots, n(N) \quad \text{(the actual output)}. \end{aligned} \tag{7.34}$$

We call subroutine SHAPE, and thus obtain the filter coefficient g_0. The corrected guess for the parameter β is thus $b+g_0$. We now repeat the algorithm with this corrected guess. Thus, in an iterative fashion, we correct the model as we erect it, which explains the name *corrigitur-dum-erigitur*.

We shall now state the *corigitur-dum-erigitur* algorithm for the MA(q) model

$$x(k)=\varepsilon(k)+\beta_1\varepsilon(k-1)+\ \ldots\ +\beta_q\varepsilon(k-q). \tag{7.35}$$

Let the given numerical time series be $x(0), x(1), \ldots, x(N)$. Let $b_1, b_2, \ldots, b_q$ be the respective guesses of the parameters $\beta_1, \beta_2, \ldots, \beta_q$.

Let $e(-q)$, $e(-q+1), \ldots, e(-1)$ and $n(-q)$, $n(-q+1), \ldots, n(-1)$ be known initial values. By use of the equations,

$$\begin{aligned} e(k) &= x(k) - b_1 e(k-1) - \ldots - b_q e(k-q) \\ n(k) &= e(k-1) - b_1 n(k-1) - \ldots - b_q n(k-q) \end{aligned} \tag{7.36}$$

generate the values of the time series

$$\begin{aligned} &e(0), e(1), \ldots, e(N) \quad \text{(the desired output)} \\ &n(0), n(1), \ldots, n(N) \quad \text{(the actual input)}. \end{aligned} \tag{7.37}$$

Call subroutine SHAPE, and thus obtain the numerical values of the filter coefficients $g_0, g_1, \ldots, g_{q-1}$. We now repeat the algorithm with the corrected guesses

$$b_1 + g_0, b_2 + g_1, \ldots, b_q + g_{q-1}. \tag{7.38}$$

We continue repeating the algorithm until the guesses converge to the moving average parameters $\beta_1, \beta_2, \ldots, \beta_q$.

We now want to give the *corrigitur-dum-erigitur* (i.e., CDE) ARMA algorithm. As an introduction, let us suppose that the time series $x(k)$ is generated by the invertible ARMA(2, 2) model

$$x(k) + \alpha_1 x(k-1) + \alpha_2 x(k-2) = \varepsilon(k) + \beta_1 \varepsilon(k-1) + \beta_2 \varepsilon(k-2), \tag{7.39}$$

where the operators $(1, \alpha_1, \alpha_2)$ and $(1, \beta_1, \beta_2)$ are each minimum-delay and $\varepsilon(k)$ is white noise. If we replace k by $k+1$, this equation becomes

$$x(k+1) + \alpha_1 x(k) + \alpha_2 x(k-1) = \varepsilon(k+1) + \beta_1 \varepsilon(k) + \beta_2 \varepsilon(k-1). \tag{7.40}$$

The time series $\varepsilon(k)$ is a function of the AR parameters α_1, α_2 and the MA parameters β_1, β_2. If we differentiate (7.39 and 40) with respect to α_1 and α_2, respectively, we obtain

$$\begin{aligned} x(k-1) &= \frac{\partial \varepsilon(k)}{\partial \alpha_1} + \beta_1 \frac{\partial \varepsilon(k)}{\partial \alpha_1} + \beta_2 \frac{\partial \varepsilon(k-2)}{\partial \alpha_1} \\ x(k-1) &= \frac{\partial \varepsilon(k+1)}{\partial \alpha_2} + \beta_1 \frac{\partial \varepsilon(k)}{\partial \alpha_2} + \beta_2 \frac{\partial \varepsilon(k-1)}{\partial \alpha_2}. \end{aligned} \tag{7.41}$$

We see that both of these difference equations have the same left-hand side and the same coefficients $1, \beta_1, \beta_2$ on the right. Suppose they both have the same initial conditions in the remote past. We then see that they are equivalent if we define $\mu(k)$ as

$$\mu(k)=-\frac{\partial\varepsilon(k)}{\partial\alpha_1}=-\frac{\partial\varepsilon(k+1)}{\partial\alpha_2}. \qquad (7.42)$$

The equivalent difference equation is thus

$$-x(k-1)=\mu(k)+\beta_1\mu(k-1)+\beta_2\mu(k-2). \qquad (7.43)$$

Likewise, we may write

$$\varepsilon(k-1)+\frac{\partial\varepsilon(k)}{\partial\beta_1}+\beta_1\frac{\partial\varepsilon(k-1)}{\partial\beta_1}+\beta_2\frac{\partial\varepsilon(k-2)}{\partial\beta_1}=0, \qquad (7.44)$$

$$\varepsilon(k-1)+\frac{\partial\varepsilon(k+1)}{\partial\beta_2}+\beta_1\frac{\partial\varepsilon(k)}{\partial\beta_2}+\beta_2\frac{\partial\varepsilon(k-1)}{\partial\beta_2}=0. \qquad (7.45)$$

If we define

$$v(k)=-\frac{\partial\varepsilon(k)}{\partial\beta_1}=-\frac{\partial\varepsilon(k+1)}{\partial\beta_2}, \qquad (7.46)$$

then each of (7.44 and 45) is equivalent to

$$\varepsilon(k-1)-v(k)-\beta_1 v(k-1)-\beta_2 v(k-2)=0. \qquad (7.47)$$

When the ARMA parameters $\alpha_1, \alpha_2, \beta_1, \beta_2$ are replaced by the values a_1, a_2, b_1, b_2, then the time series $\varepsilon(k), \mu(k), v(k)$ are denoted by $e(k), m(k), n(k)$, respectively.

Let us now expand $\varepsilon(k)$ in a Taylor series around the point a_1, a_2, b_1, b_2. We have

$$\begin{aligned}\varepsilon(k)=e(k)&-m(k)(\alpha_1-a_1)-m(k-1)(\alpha_2-a_2)\\&-n(k)(\beta_1-b_1)-n(k-1)(\beta_2-b_2)+R\end{aligned} \qquad (7.48)$$

where R is the remainder. If we let the error be $\varepsilon(k)-R$, and if we define the filter coefficients as

$$\begin{aligned}f_{10}&=\alpha_1-a_1, \quad f_{11}=\alpha_2-a_2\\ f_{20}&=\beta_1-b_1, \quad f_{21}=\beta_2-b_2,\end{aligned} \qquad (7.49)$$

then the Taylor series becomes

$$e(k)=f_{10}m(k)+f_{11}m(k-1)+f_{20}n(k)+f_{21}n(k-1)+\text{error}. \qquad (7.50)$$

This equation is in the form of a multichannel shaping filter (two input channels and one output channel). The two input signals are $m(k)$ and $n(k)$, and the desired output signal is $e(k)$. The filter coefficients are f_{10}, f_{11} on the first input channel, and f_{20}, f_{21} on the second input channel. The multichannel shaping filter WIENER is used to compute the filter coefficients. Then the estimates of the ARMA parameters are given by

$$\begin{aligned} \alpha_1 &= a_1 + f_{10}, \quad \alpha_2 = a_2 + f_{11} \\ \beta_1 &= b_1 + f_{20}, \quad \beta_2 = b_2 + f_{21}. \end{aligned} \tag{7.51}$$

With this background, we can now describe the *corrigitur-dum-erigitur* algorithm for finding the parameters of the ARMA(p, p) model

$$x(k) + \alpha_1 x(k-1) + \ldots + \alpha_p x(k-p) = \varepsilon(k) + \beta_1 \varepsilon(k-1) + \ldots + \beta_p \varepsilon(k-p). \tag{7.52}$$

Let the given numerical time series be $x(0), x(1), \ldots, x(N)$. Let $a_1, \ldots, a_p, b_1, \ldots, b_p$ be the respective guesses of the parameters $\alpha_1, \ldots, \alpha_p, \beta_1, \ldots, \beta_p$. Let sufficient initial values of the various time series be given, so that we can use the equations

$$\begin{aligned} e(k) &= x(k) + a_1 x(k-1) + \ldots + a_p x(k-p) - b_1 \varepsilon(k-1) - \ldots - b_p \varepsilon(k-p) \\ m(k) &= -x(k-1) - b_1 m(k-1) - \ldots - b_p m(k-p) \\ n(k) &= e(k-1) - b_1 n(k-1) - \ldots - b_p n(k-p) \end{aligned} \tag{7.53}$$

to generate the values

$$\begin{aligned} &e(0),\ e(1), \ldots,\ e(N) && \text{(desired output)} \\ &m(0), m(1), \ldots, m(N) && \text{(first channel actual input)} \\ &n(0),\ n(1), \ldots,\ n(N) && \text{(second channel actual input)}. \end{aligned} \tag{7.54}$$

We call subroutine WIENER to obtain the filter coefficients $f_{10}, f_{11}, \ldots, f_{1,p-1}$, $f_{20}, f_{21}, \ldots, f_{2,p-1}$. Subroutine WIENER uses the LWR algorithm for solving the block Toeplitz (Yule-Walker) multichannel normal equations, and so is computationally efficient. We now report the CDE algorithm with the corrected guesses

$$\begin{aligned} &a_1 + f_{10}, a_2 + f_{11}, \ldots, a_p + f_{1,p-1} \\ &b_1 + f_{20}, b_2 + f_{21}, \ldots, b_p + f_{2,p-1} \end{aligned} \tag{7.55}$$

for the ARMA parameters. We iterate until the process converges. We thus correct the ARMA model as we erect it. This procedure makes up the CDE method.

Both the iterative algorithm and the CDE algorithm use the LWR recursion, and so they are computationally more efficient than existing nonlinear regression methods used to estimate ARMA parameters. The CDE ARMA algorithm, in fact, makes use of the multichannel LWR recursion. Although we have treated only single channel time series, the iterative and CDE algorithm can be extended to the case of multichannel time series.

Once the values of the ARMA(p,q) coefficients have been determined, the estimate of the spectral density is given by

$$\Phi(\omega)=\left|\frac{1+b_1\,\mathrm{e}^{-\mathrm{j}\omega}+\ldots+b_q\,\mathrm{e}^{-\mathrm{j}\omega q}}{1+a_1\,\mathrm{e}^{-\mathrm{j}\omega}+\ldots+a_p\,\mathrm{e}^{-\mathrm{j}\omega p}}\right|^2 v\,,$$

where v is the variance of the white noise process.

In the use of the CDE algorithm care must be taken to remove the mean values of all the time series. Also one must start with a good initial guess. The critical step in the CDE algorithm is the truncation of the Taylor series to a first-order approximation. A major problem could occur when this approximation is no longer valid. The classical problem of nonlinear regression, which is the method used in the CDE algorithm, is that, if the update terms yield estimates which are not near the initial values, then the neglecting of second and higher order terms in the Taylor series may invalidate the process. One solution of this problem is called damped least squares. The idea is to avoid very large update terms by providing a constant derived from the algorithm parameters to damp the updates. The damped least squares idea has been applied to the CDE algorithm by *Cupo* [7.17]. In addition to demonstrating the excellent empirical results of the CDE algorithm on time series data, he developed other important theoretical and empirical methods, which include some new algorithms for spectral analysis. Computer programs are available with the thesis.

7.4 Other Algorithms

Finally, special mention should be made of two other new procedures designed to improve the performance of spectral estimates, as described below:

1) *Burg* [7.18] described a new iterative algorithm for estimating the covariance matrix of specified structure, using a given time series. Basically, the method assumes that the time series is derived from a zero-mean multivariate Gaussian process, and a maximum likelihood procedure is used to find the covariance matrix with the specified structure. *Burg's* new algorithm appears to be interrelated with the concept of minimum cross-entropy [7.19].

It is shown in [7.18] that the use of this new algorithm overcomes the problem of line splitting. This phenomena arises in the maximum-entropy spectral estimate of a time series consisting of a sine wave and additive white noise, particularly when the initial phase of the sine wave is unfavourable and the conventional Burg technique (based on the multistage lattice filter) is used to estimate the prediction-error filter coefficients. Another possible application of the algorithm is in the estimation of multichannel and multidimensional covariance matrices, perhaps followed by maximum-entropy spectral analysis. Nevertheless, much research needs to be done on this new algorithm to fully establish its capabilities and limitations.

2) *Thomson* [7.20] described a new non-parametric spectrum estimation procedure that appears to overcome many of the shortcomings of the conventional spectrum estimation techniques based on the use of the periodogram.

Thomson's approach is designed to provide a solution to the basic equation of spectrum estimation, namely

$$X(f)=\int_{-1/2}^{1/2}\frac{\sin[N\pi(f-v)]}{\sin[\pi(f-v)]}dZ(v)\,, \tag{7.56}$$

where $X(f)$ is the finite discrete Fourier transform of the time series $x_0, x_1, \ldots, x_{N-1}$, which (for notational convenience), is defined by

$$X(f)=\sum_{n=0}^{N-1}\exp\left[-\mathrm{j}2\pi f\left(n-\frac{N-1}{2}\right)\right]x_n\,. \tag{7.57}$$

For zero-mean processes, the random orthogonal-increments measure $dZ(v)$ has, likewise, zero mean. Its second moment, the spectral density of the process, is defined by

$$S_x(f)df=E[|dZ(f)|^2]\,. \tag{7.58}$$

Thomson regarded (7.56) as a linear Fredholm integral equation of the first kind for $dZ(v)$, with the aim of obtaining approximate solutions whose statistical properties are, in some sense, close to those of $dZ(f)$. He presented a detailed solution to this basic problem by using prolate spheroidal wave functions that are fundamental to the study of time and frequency limited systems. Computationally, *Thomson*'s method is equivalent to using a weighted average of a series of direct spectrum estimates based on orthogonal data windows in the form of discrete prolate spheroidal sequences, thereby treating the problems of bias smoothing.

Finally, the reader's attention is directed to an excellent survey paper on spectrum estimation techniques [7.21] and two special publications on spectral estimation [7.22, 7.23].

Mention should also be made of a two-volume book by Priestley on spectral analysis [7.24]. The first volume of this study is devoted to univariate series, while the second volume is devoted to multivariate series, prediction and control.

References

7.1 T.J. Ulrych, R.W. Clayton: Phys. Earth Planet. Inter. **12**, 188–200 (1976)
7.2 M. Morf, B. Dickinson, T. Kailath, A. Viveira: IEEE Trans. ASSP-**25**, 429–433 (1977)
7.3 L. Marple: IEEE Trans. ASSP-**28**, 441–454 (1980)
7.4 H.J. Nussbaumer: *Fast Fourier Transform and Convolution Algorithms*, 2nd. ed., Springer Ser. Information Sci., Vol. 2 (Springer, Berlin, Heidelberg, New York 1982)
7.5 S.B. Kesler, S. Haykin, R.S. Walker: "Maximum-Entropy Field-Mapping in the Presence of Correlated Multipath", IEEE Intern. Conf. Acoust., Speech, Signal Processing, Atlanta, GA (1981) pp. 157–161
7.6 S. Kesler, S. Haykin: "Conventional, MLM and MEM Mapping of Inhomogeneous Spatial Field", First IEEE ASSP Workshop on Spectral Estimation, Hamilton, Ont. (1981) pp. 7.3.1–7
7.7 J. Makhoul: IEEE Trans. ASSP-**25**, 423–428 (1977)
7.8 T.E. Barnard: IEEE Trans. ASSP-**30**, 175–189 (1982)
7.9 S.B. Kesler: "Generalized Burg Algorithm for Beamforming in Correlated Multipath Field", IEEE Intern. Conf. Acoust., Speech, Signal Processing, Paris, France (1982) pp. 1481–1484
7.10 S. Haykin, H.C. Chan: IEE Proc. **172**, Pt. F, 464–470 (1980)
7.11 R. Kumaresan, D.W. Tufts: "Singular Value Decomposition and Spectral Analysis", First IEEE ASSP Workshop on Spectral Estimation, Hamilton, Ont. (1981) pp. 6.4.1–6.4.2
D.W. Tufts, R. Kumaresan: IEEE Proc. **70**, 975–989 (1982)
7.12 J.D. Markel, A.H. Gray, Jr.: *Linear Prediction of Speech* (Springer, Berlin, Heidelberg, New York 1976)
7.13 F.B. Hildebrand: *Introduction to Numerical Analysis* (McGraw-Hill, New York 1956)
7.14 C.L. Lawson, R.J. Hanson: Solving Least Squares Problems (Prentice-Hall, Englewood Cliffs, NJ, 1974)
7.15 V.C. Klemma, A.J. Laub: IEEE Trans. AC-**25**, 164–176 (1980)
7.16 E.A. Robinson: *Multichannel Time Series Analysis with Digital Computer Programs* (Holden-Day, Oakland, CA 1967; 2nd ed., Goose Pond Press, Houston, TX 1983)
7.17 R.L. Cupo: Computationally Efficient Methods of ARMA System Identification and Spectral Estimation, Ph. D. Thesis, E.E. Department, Cornell University, Ithaca, NY (1982)
7.18 J.P. Burg, D.G. Luenberger, D.L. Wenger: Proc. IEEE **70**, 963–974 (1982)
7.19 J.E. Shore: IEEE Trans. ASSP-**29**, 230–237 (1981)
7.20 D.J. Thomson: IEEE Proc. **70**, 1055–1096 (1982)
7.21 S.M. Kay, S.L. Marple, Jr.: IEEE Proc. **69**, 1380–1408 (1981)
7.22 IEEE First ASSP Workshop on Spectral Estimation, McMaster University, Hamilton, Ontario, Canada (1981)
7.23 IEEE Proc. Special Issue on Spectral estimation, **70**, 883–1127 (1982)
7.24 M.B. Priestley: *Spectral Analysis and Time Series*, Vol. 1 and 2 (Academic, New York 1981)

Subject Index

Topics in Applied Physics

Founded by **H.K.V. Lotsch**

Volume 6

Picture Processing and Digital Filtering

Editor: **T.S. Huang**

1979. 2nd corrected and updated edition. 1979. 113 figures, 7 tables. XIII, 297 pages. ISBN 3-540-09339-7

Contents: *T.S. Huang:* Introduction. – *H.C. Andrews:* Two-Dimensional Transforms. – *J.G. Fiasconaro:* Two-Dimensional Nonrecursive Filters. – *R.R. Read, J.L. Shanks, S. Treitel:* Two-Dimensional Recursive Filtering. – *B.R. Frieden:* Image Enhancement and Restoration. – *F.C. Billingsley:* Noise Considerations in Digital Image Processing Hardware. – *T.S. Huang:* Recent Advances in Picture Processing and Digital Filtering. – Subject Index.

Volume 42

Two-Dimensional Digital Signal Processing I

Linear Filters

Editor: **T.S. Huang**

1981. 77 figures. X, 210 pages. ISBN 3-540-10348-1

Contents: *T.S. Huang:* Introduction. – *R.M. Mersereau:* Two-Dimensional Nonrecursive Filter Design. – *P.A. Ramaoorthy, L.T. Bruton:* Design of Two-Dimensional Recursive Filters. – *B.T. O'Connor, T.S. Huang:* Stability of General Two-Dimensional Recursive Filters. – *J.W. Woods:* Two-Dimensional Kalman Filtering.

Volume 43

Two-Dimensional Digital Signal Processing II

Transforms and Median Filters

Editor: **T.S. Huang**

1981. 49 figures. X, 222 pages. ISBN 3-540-10359-7

Contents: *T.S. Huang:* Introduction. – *J.-O. Eklundh:* Efficient Matrix Transposition. –*H.J. Nussbaumer:* Two-Dimensional Convolution and DFT Computation. – *S. Zohar:* Winograd's Discrete Fourier Transform Algorithm. – *B.I. Justusson:* Median Filtering: Statistical Properties. – *S.G. Tyan:* Median Filtering: Deterministic Properties.

Volume 32

Image Reconstruction from Projections

Implementation and Applications

Editor: **G.T. Herman**

1979. 120 figures, 10 tables. XII, 284 pages. ISBN 3-540-09417-2

Contents: *G.T. Herman, R.M. Lewitt:* Overview of Image Reconstruction from Projections. – *S.W. Rowland:* Computer Implementation of Image Reconstruction Formulas. – *R.N. Bracewell:* Image Reconstruction in Radio Astronomy. – *M.D. Altschuler:* Reconstruction of the Global-Scale Three-Dimensional Solar Corona. – *T.F. Budinger, G.T. Gullberg, R.H. Huesman:* Emission Computed Tomography. – *E.H. Wood, J.H. Kinsey, R.A. Robb, B.K. Gilbert, L.D. Harris, E.L. Ritman:* Applications of High Temporal Resolution Computerized Tomography to Physiology and Medicine.

Volume 41

The Computer in Optical Research

Methods and Applications

Editor: **B.R. Frieden**

1980. 92 figures, 13 tables. XIII, 371 pages. ISBN 3-540-10119-5

Contents: *B.R. Frieden:* Introduction. – *R. Barakat:* The Calculation of Integrals Encountered in Optical Diffraction Theory. – *B.R. Frieden:* Computational Methods of Probability and Statistics. – *A.K. Rigler, R.J. Pegis:* Optimization Methods in Optics. – *L. Mertz:* Computers and Optical Astronomy. – *W.J. Dallas:* Computer-Generated Holograms.